BACTERIA and CANCER

*Based on the Proceedings of an
International Colloquium on Bacteria and Cancer,
held in Cologne, Germany from 16-18 March, 1982*

BACTERIA and CANCER

Edited by

J. JELJASZEWICZ
National Institute of Hygiene
Warsaw, Poland

G. PULVERER
Hygiene Institute, University of
Cologne, FRG

W. ROSZKOWSKI
National Institute of Tuberculosis
Warsaw, Poland

1982

ACADEMIC PRESS

A Subsidiary of Harcourt Brace Jovanovich, Publishers

London   New York

Paris   San Diego   San Francisco   São Paulo

Sydney   Tokyo   Toronto

ACADEMIC PRESS INC. (LONDON) LTD.
24/28 Oval Road
London NW1

United States Edition published by
ACADEMIC PRESS INC.
111 Fifth Avenue
New York, New York 10003

British Library Cataloguing in Publication Data
Bacteria and Cancer.
1. Cancer—Treatment—Congresses
2. Bacteria—Therapeutic use—Congresses
I. Jeljaszewicz, J.
II. Pulverer, G.
III. Roszkowski, W.
616.99 4069   RC271.B/

ISBN 0-12-383820-7

Printed in Great Britain by
St. Edmundsbury Press, Bury St. Edmunds, Suffolk

LIST OF PARTICIPANTS

ADLAM, C. *Wellcome Research Laboratories, Langley Court, Beckenham, Kent BR3 3BS, England*

BALDWIN, R.W. *Cancer Research Campaign Laboratories, University of Nottingham, Nottingham NG7 2RD, England*

BAŠIĆ, I. *Department of Animal Physiology, Faculty of Natural Sciences and Mathematics, University of Zagreb, 41000 Zagreb, Yugoslavia*

BRANDIS, H. *Institute of Medical Microbiology and Immunology, University of Bonn, 5300 Bonn-Venusburg, Federal Republic of Germany*

BRAUER, H.P. *Bundesärztekammer, Haedenkampstrasse 1, 5000 Cologne 41, Federal Republic of Germany*

BREDE, H.D. *Paul Ehrlich Institute for Chemotherapy, Paul Ehrlich Strasse 42-44, 6000 Frankfurt 70, Federal Republic of Germany*

CHEDID, L. *Division of Experimental Immunotherapy, Institut Pasteur, 28 rue du Docteur Roux, 75724 Paris Cedex 15, France*

EGGERS, H.J. *Institute of Virology, University of Cologne, Fürst-Pückler-Strasse 56, 5000 Cologne 41, Federal Republic of Germany*

FAINSTEIN, V. *Department of Developmental Therapeutics, Texas Medical Centre, Houston, Texas 77030, USA*

GOLDIN, A. *Division of Medical Oncology, Vincent T. Lombardi Cancer Research Centre, Georgetown University School of Medicine, 3800 Reservoir Road, Washington, DC 20007, USA*

HILL, M.J. *Bacterial Metabolism Research Laboratory, Public Health Laboratory Service, Porton Down, Salisbury, Wiltshire SP4 0JG, England*

JELJASZEWICZ, J. *Department of Bacteriology, National Institute of Hygiene, 24 Chocimska Street, 00-791 Warsaw 36, Poland*

KAN-MITCHELL, J. *University of Southern California, Comprehensive Cancer Centre, 2025 Zonal Avenue, Los Angeles, CF 90033, USA*

KAUFMANN, W. *Department of Internal Medicine, University of Cologne, Ostmerheimer Strasse 200, 5000 Cologne 91, Federal Republic of Germany*

KIRCHNER, H. *German Cancer Research Centre, Institute of Virus Research, Im Neuenheimer Feld, 6900 Heidelberg 1, Federal Republic of Germany*

KIRN, A. *Laboratory of Virology, University of Strassbourg, 3 rue Koeberle, 67000 Strassbourg, France*

KO, H.L. *Institute of Hygiene, University of Cologne, Goldenfelsstrasse 19-21, 5000 Cologne 41, Federal Republic of Germany*

KODAMA, M. *Department of Surgery, The Shiga University of Medical Sciences, Seta Tsukiwa-cho, Otsu 520-21, Japan*

KONDO, M. *Department of Medicine, Kyoto Prefectural University of Medicine, Kamikyo-ku, Kyoto 602, Japan*

KOTANI, S. *Department of Microbiology, Osaka University Dental School, 4-3-48 Nakanoshima, Kita-ku, Osaka 530, Japan*

MITCHELL, M.S. *University of Southern California, Comprehensive Cancer Centre, 2025 Zonal Avenue, Los Angeles, CF 90033, USA*

NAUMANN, P. *Institute of Medical Microbiology and Virology, University of Düsseldorf, Moorenstrasse 5, 4000 Düsseldorf, Federal Republic of Germany*

NAUTS, H.C. *Cancer Research Institute Inc., 1225 Park Avenue, New York, NY 10028, USA*

PETERS, G. *Institute of Hygiene, University of Cologne, Golden Eelsstrasse 19-21, 5000 Cologne 41, Federal Republic of Germany*

PICHLMAIER, H. *Department of Surgery, University of Cologne, Joseph-Stelzmann-Strasse 9, 5000 Cologne 41, Federal Republic of Germany*

PRAGER, M.D. *Department of Surgery, The University of Texas, Health Science Centre, 5323 Harry Hines Boulevard, Dallas Texas 75235, USA*

PULVERER, G. *Institute of Hygiene, University of Cologne, Goldenfelsstrasse 19-21, 5000 Cologne 41, Federal Republic of Germany*

ROSZKOWSKI, W. *Department of Immunology, National Institute of Tuberculosis, Plocka 26, 01-138 Warsaw, Poland*

SCHNEIDER, C. *Dr Madaus and Co., PO Box 910555, 5000 Cologne 91, Federal Republic of Germany*

SCHWAB, J.H. *Department of Bacteriology and Immunology, University of North Carolina, School of Medicine, Chapel Hill, NC 27514, USA*

SEDLACEK, H.H. *Behringwerke AG, PO Box 1140, 3550 Marburg 1, Federal Republic of Germany*

SIECK, R. *Dr Madaus and Co., PO Box 910555, 5000 Cologne 91, Federal Republic of Germany*

SINKOVICS, J. *Department of Virology and Epidemiology, Baylor College of Medicine, Texas Medical Centre, TX 77030, USA*

PREFACE

This book contains contributions specially prepared for publication presenting material delivered at the International
Colloquium on Bacteria and Cancer, held in Cologne on March
16–19, 1982. Discussions have been only slightly edited and
inserted after each session dealing with thematically associated topics.

The Colloquium was dedicated to the memory of Professor
Wolfgang Denk. On 21 March 1982 his 100th birthday anniver-
sary was celebrated in Vienna where Professor Denk worked all
his life. He contributed to the fundamentals of cancer
research including some prophetic views of immunoprophylaxis
of cancer. Together with Dr Coley, whose famous studies on
the influence of bacterial products on the course and beneficial treatment of cancer, he is one of the pioneering physicians in this field.

Bacteria and cancer are interrelated in many ways, most of
which are yet unknown or not clear. It is certain, however,
that bacteria may play a role in carcinogenesis. Microorganisms and their products may also exert beneficial effects in
cancer therapy. This book aims at summarizing the basic
findings of immunomodulation and other effects of bacteria
and their products on the course and development of cancer.
From these studies, some new approaches in cancer therapy
have been developed, such as the preparation and application
of synthetic compounds based on naturally occurring precursors. Some associated phenomena, such as immunosuppression,
bacteria or infections in cancer patients, have also been
discussed to show multiple interrelationships developing
between microorganisms and cancer.

We hope that this book presents a complete summary of the
most important developments in the field, thereby fruitfully
contributing not only to the therapy of malignant diseases
but also to a better understanding of them.

The Colloquium has been completely sponsored by the Dr
Madaus Company of Cologne to whom we are most thankful for
this contribution.

Warsaw and Cologne   J. Jeljaszewicz,
October, 1982   G. Pulverer and
                                               W. Roszkowski

CONTENTS

BACTERIAL PRODUCTS IN THE TREATMENT OF CANCER: PAST, PRESENT AND FUTURE

Helen Coley Nauts

Cancer Research Institute Inc.,
1225 Park Avenue,
New York, New York 10028, USA

For over 200 years physicians have observed dramatic "spontaneous" regressions of various types of neoplastic disease during or following acute concurrent infections (Nauts, 1980a). A large number of authors recorded their observations of the beneficial effects of such infections, as well as those in which inflammation, fever or incomplete surgery was involved. Our infection monograph contains 1032 references. Many more such cases were seen all over Europe and the United States but were never published (CRI, 1953-1982).

Vautier (1813) discussed the question of whether cancer may be cured by the sole forces of nature. He had found "several cases in searching the writing of the most careful observers in which cancer terminated happily by the development of gangrene". We have abstracted 22 such cases in which "gangrene" developed spontaneously or by inoculation, of which the majority were breast cancers (Nauts, 1980a, Series E, p. 118). Recent research on the Tumour Necrosis Factor suggests that in these patients a combination of bacteria had induced the tumour necrosis factor which these early physicians had designated as gangrene (Carswell *et al.*, 1975).

Such cases inspired physicians in the 18th and 19th centuries to induce "laudable pus, setons or issues" in their inoperable cancer patients as the first form of immunotherapy.

Tanchou, a prominent physician in Paris reported on 300 cases of breast cancer treated medically up to 1844 and included the early cases of a great many physicians in Europe (Tanchou, 1844).

The first was that of Schwenke (prior to 1744), an inoperable mammary carcinoma whose disease had progressed despite all the most effective remedies then in use (Nauts, 1980a).

Having lost all hope of cure she ceased treatment. Shortly
thereafter an abscess formed on her leg. As suppuration
became more abundant, the cancer diminished, then disappeared.
Against advice, the patient allowed the ulcer to heal. The
cancer then recurred. A new "issue" was then opened at the
site of the former abscess. When suppuration was well established, the breast cancer again gradually disappeared (cited
by Dupré de Lisle, 1774 and Tanchou, 1844).

Another French physician, Dussosoy, applied gauze dress-
ings soaked in gangrenous discharges on ulcerated breast
cancers or inoculated "gangrene" in a small incision. Tanchou
noted that success was complete. The ulceration destroyed
the entire tumour which sloughed off on the 19th day.
Dussosoy then concentrated on controlling the progress of the
gangrene, succeeded in doing so and in a few days the ulcer
became bright red and covered with healthy granulations.
Tanchou added: "Here gangrene seems to have replaced live
cautery, caustics or the scalpel" (Tanchou, 1844, pp. 192–
193; also cases 49—54, pp. 71—78.

Until our study began over 40 years ago noone had attempted:

a) to determine which types of tumours were most frequently
   benefited;
b) which types of infections or bacterial products were most
   apt to be effective;
c) whether other infections (viral or protozoal) might also
   be beneficial or might actually have a deleterious effect;
d) mechanisms of action: what was the optimum technique of
   administration of bacterial vaccines, as regards site,
   dosage, frequency and duration of injections.

Such fundamental studies seemed essential to the ultimate
development of the best products properly administered for
the treatment of neoplastic diseases not only in inoperable
cases but as adjuvants to surgery in operable cases and
ultimately as a possible means of preventing the disease
(Nauts, 1980b).

It was found that in all types of cancer the majority of
spontaneous regressions occurred following streptococcal
infections (principally erysipelas), next in frequency being
suppuration (staphylococcal or mixed infections). Many other
types were responsible for some dramatic regressions. However, the largest number of *permanent* results occurred following the more prolonged infections, principally staphyloccocal.
A total of 449 cases were analysed in our infection monograph
as follows:

Series A, 163 Cases: Pyogenic infections occurring spontaneously in inoperable cancer patients.

Series B, 58 Cases: Non-pyogenic infections occurring spontaneously in cancer patients (49 inoperable, 9 operable).

Series C, 117 Cases: Pyogenic infections occurring spontaneously before or after surgery in operable cancer patients (18 of these were apparently untreated, i.e. only biopsy was performed).

Series D, 5 Cases: Operable cancer patients in whom pyogenic infections were "encouraged" or actually induced.

Series E, 35 Cases: Inoperable cancer patients in whom pyogenic infections, mostly erysipelas, were actually induced.

Series F, 31 Cases: Inoperable cancer patients in whom attempts to inoculate or otherwise induce erysipelas failed.

Series G, 22 Cases: Mostly inoperable cancer patients in whom gangrene developed spontaneously or was induced.

Series H, 12 Cases: Mostly inoperable cancer patients in whom syphilis or malaria developed spontaneously or was induced.

Series I, 5 Cases: Inoperable Hodgkin's Disease.

Of these 449 patients approximately 250 had histologic confirmation of diagnosis and 125 of these were apparent cures, traced 5 to 54 years later. Of the failures, 21 survived far longer than expected — 5 to 30 years (Nauts, 1980a).

Although in recent years oncologists have concentrated largely on the use of BCG or *C. parvum* for immunotherapy, the literature clearly shows that concurrent or prior tuberculosis did not produce dramatic effects on cancer patients, although at least 36 authors beginning with Rokitansky in 1842 believed there was an apparent antagonism between tuberculosis and cancer (cited by Nauts, 1980a; Rokitansky, 1842). A few others discounted such a possibility. If one reviews the evidence carefully, it becomes evident that cancer and tuberculosis may occur in the same person, and if the tuberculosis is active and concurrent, it may cause partial or complete regression of the cancer. A few authors cited evidence suggesting the possible counteracting influence of tuberculosis on the incidence of cancer in various countries or localities, i.e. patients who had tuberculosis might be less apt to contract cancer (Tromp, 1954). In this connection it is of interest to see that in the United States since 1930 the incidence of lung cancer has steadily risen while the incidence of tuberculosis and respiratory infections has declined (see Fig. 1).

The possible antagonism between malaria and cancer was also reviewed. The majority of these reports appeared

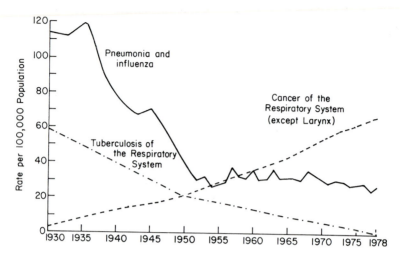

FIG. 1 *Death rates among white males USA 1930–1978.*

between 1900 and 1910. Trnka de Krzowitz (1783) reported a
case of scirrhus carcinoma of the breast who developed ter-
tian malaria, followed by complete regression of the cancer
in a few weeks (cited by Tanchou, 1844). Other such cases
were reported by Avramovici, Jovin and Portanova (cited by
Nauts, 1980a). All were fresh cases of malaria. On the
other hand, recent studies in Africa by Burkitt suggest that
chronic malaria which can be so debilitating, depresses the
reticuloendothelial system, and may predispose children to
Burkitt's lymphoma (Burkitt, 1969).

Amedée Latour was the first to suggest inoculating syphilis in inoperable cancer patients. The controversy as to the
possible antagonism between syphilis and cancer became quite
heated at the meetings of the Royal Academy of Medicine in
Brussels in 1851—52 (Didot, 1851—1852). Didot wished to *prevent* cancer by what he called syphilization (induced syphilis). He stated that since ancient times it has been known
that prostitutes never have uterine cancer, while most common
victims are good pious women given to celibacy (Didot, 1852).

Didot inoculated pus from chancres in the posterior cervical region in an inoperable breast cancer patient. Symptoms of constitutional syphilis developed successively for a
year while the breast cancer diminished in size and the pain
ceased. Mercurial treatment for the syphilis was then given
and the patient recovered. Auzias Turenne stated that in
his experience several cancer patients seemed to be immune
to syphilis. He emphasized that if one wished to induce
syphilis in cancer patients one should not wait too long and

one should not do so in terminal patients. In treating several inoperable cases he "had observed no unfavourable complications while a most notable amelioration occurred which was continued until constitutional syphilis was produced." These experiments were bitterly criticized by Ricord, a famous syphilologist (Didot, 1851).

To our knowledge no one else ever repeated these attempts. However, in 1883 Verneuil discussed the influence of spontaneously contracted syphilis on the course of the disease. He stated that the influence of syphilis in cancer seemed to be quite definite in certain cases and he cited several examples. He added that in many cases syphilis greatly ameliorated the pain of various neosplasms. Odier observed that sarcoma patients attacked by syphilis did not develop metastases (cited by Wolff, 1913—1914).

Epidemiological data assembled in the last 50 years indicate that in developed and developing countries the incidence of infection is decreasing at the same time that cancer is increasing, due to better control of infections and infectious diseases.

R.L. Smith *et al.* compared the incidence of cancer and infectious diseases among the White and Indian populations of the United States and Canada. They found the expected death rate for cancer was far lower in the Indians, while the rates for infective or parasitic diseases were over 6 times greater than the expected rate (Smith *et al.*, 1956, 1957).

M.G. Lewis, a pathologist who worked for many years in Kampala, Uganda, in a personal communication, mentioned the infrequency of metastases from malignant melanoma in Uganda even when the primary growth was extensive. He noted that in most of these patients, the lesions were infected and ulcerated when first seen, the majority being in the lower extremities (Nauts, 1980a).

A contributory factor to the increase in cancer might be the widespread use of antibiotics since 1940.

> The antibiotics may absolve the body of the need to bring the normal immunological mechanisms into use — a mechanism that has been acquired and perfected through millions of years of evolution (Meyer and Benjafield, 1955).

The beneficial effects of fever, heat and inflammation have also been studied not only on existing cancer, but on cancer incidence (Nauts, 1975a, 1975b, 1982a). In Japan the daily use of the hot full bath ($42°$—$48°C$) may destroy incipient cancers before they are clinically apparent. This practice may be one reason why the incidence of cancer of the breast, prostate, penis, testis and skin is much lower in

Japan than elsewhere (Nauts, 1975c). The intraductal, medul-
lary and colloid tumours are relatively more frequent in
Tokyo, and the tumours with circumscribed margins and a high
degree of cellular reaction are also more frequently seen
than in a similar representative series seen in Boston, Mas-
sachusetts where more invasive histological types occur more
frequently. Also the response to therapy for certain cancers,
such as carcinoma of the breast in Japan is considerably
better than in western countries (MacMahon *et al.*, 1973).

Huth, in Germany, was one of the first to report that
acute concurrent infections may increase the survival of
leukaemia patients (Huth, 1958). More recently it was found
that patients with acute leukaemia receiving *Pseudomonas
aeruginosa* vaccine to prevent infection maintained drug-
induced remissions considerably longer than controls treated
only by chemotherapy (Clarkson *et al.*, 1975).

Several thoracic surgeons, such as Sensenig, Cady, Takita,
Moore, Rucksdeschel and McKneally have reported a much higher
5-year survival in lung cancer patients who developed empyema
post-operatively (cited in Nauts, 1978 and Nauts, 1980a).

These data furnish considerable evidence that where bacterial infections or some infectious diseases are common,
cancer is apt to be rare, or if they do develop, they may
remain localized or they may respond more dramatically and
permanently to chemotherapy, or other forms of treatment.

William B. Coley, MD, a young New York Surgeon lost his
first case of cancer, a young woman of 19, in spite of early
and radical surgery for a sarcoma (Coley, 1891). Recognizing
that surgery alone was not the answer to the cancer problem,
and that he did not know enough about it, he studied about
100 histories of sarcoma patients treated in the New York
Hospital in the preceding decade. Among these was a case of
3 times recurrent round cell sarcoma of the neck who
developed erysipelas in the wound following a fourth incomplete removal in October 1884. The remaining tumour disappeared and there was no further recurrence. This led Coley
to try to induce erysipelas in inoperable cases. At the end
of a year in which he tried to do so on 10 terminal patients,
he realized that either one might hasten death or be unable
to induce an infection at all (Coley, 1893). He then decided
to use killed vaccine cultures but streptococci are very
thermolabile and the heat killed vaccine of streptococci was
totally inert; so was a filtered product. He then decided to
add the toxins of *Bacillus prodigiosus* (now known as *Serratia
marcescens*). The first case to receive this mixed filtered
vaccine was a bedridden young man of 19 with an inoperable
sarcoma of the abdominal wall and pelvis, involving the
bladder measuring 16 cm x 13 cm. He received injections

directly into the tumour mass for almost 4 months, with febrile reactions up to 40°C or more. No other treatment was
given. Complete regression occurred and the patient remained
well until sudden death from a heart attack over 26 years
later (Nauts *et al.*, 1953, Case 1, pp. 21—22).

In the last 41 years we have assembled detailed histories
of 896 cases of cancer treated by these mixed bacterial vaccines now known as MBV. These were microscopically proven
cases (Table 1).

The highest percentage of successes occurred in sarcoma of
soft tissues (Nauts, 1975c), lymphomas (Nauts and Fowler,
1969) including reticulum cell sarcoma of bone (Miller and
Nicholson, 1971). Complete or partial regressions also
occurred in inoperable or metastatic carcinomas of the breast
(Nauts, 1978, 1982b), colon (Fowler, 1969a, 1969b), head and
neck (Nauts, 1975), uterus, ovary (Nauts, 1977), kidney
(Nauts, 1973), neuroblastoma (Fowler and Nauts, 1970) and
various bone tumours (Nauts, 1974, 1975d, 1975e) (see Table
1). Other beneficial effects included marked decrease or
cessation of pain, improved appetite and weight gain (up to
50 pounds), reduction or disappearance of lymphedema, ascites or pleural effusion and remarkable regeneration of bone
(Nauts, 1975e, 1978). Significant palliation (symptom free
up to 5 years) and one apparently permanent result occurred
in multiple myeloma (Nauts, 1975f).

Most of the 126 osteogenic sarcoma cases occurred in the
long bones and received injections following surgery to prevent metastases. It is significant that, if injections were
given for at least 4 months, 85% remained well 4 to 60 years
later as compared to the 10—15% survival then achieved by
surgery alone. Those who died survived 4 to 13 years as compared to the average survival of 10 to 12 months following
amputation alone in the period in which these cases were
treated (Nauts, 1975d).

By using MBV therapy as an adjunct to *conservative* surgery,
in order to prevent recurrence or metastases, Coley was the
first to avoid amputation in patients with sarcoma or giant
cell tumour of the extremities. A few other surgeons followed
his lead. Of the 128 cases in which amputation was avoided,
57% remained well, as compared to 32% of the 166 cases in
which amputation *was* performed.

The highest percentage of successes with conservative surgery occurred in giant cell tumour of bone (96%) (Nauts,
1975e) and in the operable sarcomas of soft tissues (80%)
(Nauts, 1975c). The poorest occurred in Ewing's sarcoma as
most of these children had received excessive radiation prior
to MBV (Nauts, 1974).

TABLE 1

Five-year survival of 896 patients
with various types of tumours treated with
Coley toxins (mixed bacterial vaccines)

Type of tumour	Total No. of Cases	5-year survival Inoperable No.	%	Operable No.	%
Bone Tumours					
Ewing's Sarcoma	114	11/52	21	18/62	29
Osteogenic Sarcoma	162	3/23	13	43/139	31
Retic. Cell Sarcoma	72	9/49	18	13/23	57
Multiple Myeloma	12	4/8	50	2/4	50
Giant Cell Tumour	57	15/19	79	33/38	87
Soft Tissue Sarcomas					
Lymphosarcoma	86	42/86	49	—	—
Hodgkin's Disease	15	10/15	67	—	—
Other Soft Tissue Sarcomas	188	78/138	57	36/50	73
Gynaecological Tumours					
Breast Cancer	33	13/20	65	13/13	100
Ovarian Cancer	16	10/15	67	1/1	(100)
Cervical Carcinoma	3	2/3	67	—	—
Uterine Sarcoma	11	8/11	73	—	—
Other Tumours					
Testicular Cancer*	64	14/43*	34	15/21	71
Malignant Melanoma	31	10/17	60	10/14	71
Colorectal Cancer	13	5/11	46	2/2	(100)
Renal Cancer (adult)	8	3/7	43	1/1	(100)
Renal Cancer (Wilms' Tumour)	3	—	—	1/3	33
Neuroblastoma	9	1/6	17	2/3	67
TOTAL	896	238/523	46	190/374	51

*including 16 terminal cases

If MBV was begun prior to radiation, the response of the tumour was enhanced while protection of normal tissues was observed. Recent studies indicate that the response to chemotherapy is also enhanced by bacterial vaccines and hyperthermia.

Gershon reported that immunologic defence against metastases in tumour bearing hamsters is impaired if the tumour is removed before full development of immunity. In such cases metastases occur far more frequently than if the tumour is not removed too soon (Gershon and Carter, 1969). The clinical findings seem to bear this out, especially in osteogenic sarcoma and reticulum cell sarcoma of bone. Patients treated very soon after onset did less well than those amputated and given MBV 6 to 9 months after onset (Nauts, 1975d).

The studies of a great many investigators both in the laboratory and the clinic have stressed the vital importance of the immunological defences against cancer (Fowler, 1968, 1969a, 1969b; Fowler and Nauts, 1970; Miller and Nicholson, 1971; Nauts, 1975b,c,d,e,f, 1977, 1980a). Thus any agent which suppresses these defences can lead to a higher incidence of cancer, while those which stimulate, such as bacterial infections and microbial products can lessen the incidence and be beneficial as therapeutic modalities once cancer develops.

The concept of *induction and elicitation*, propounded by Mudd and Shayegani (1974) avoids the dangers of injecting the immunologically impaired cancer patient without abrogating the effectiveness of the antigenic stimulus provided by a former infection with living organisms. Mudd's concept utilizes the ubiquitous and naturally established delayed hypersensitivity of *Staphylococcus aureus* as the primary induction step for the more effective use of bacterial vaccines such as Staphage Lysate (SPL), a potent immunizing agent, in the subsequent elicitation of a nonspecific cell-mediated response.

Several oncologists have shown that cancer patients respond best to microbial products to which they have previously been exposed. *Staphylococcus* and *Pseudomonas* are now the most frequently isolated organisms. Using these and other bacterial toxins Waisbren (1982) has been treating cancer patients with encouraging results. He has also recently prepared the Coley formula (Tracy XI).

The first randomized trial of MBV was begun over 5 years ago at Memorial Sloan-Kettering Cancer Centre in New York City on advanced non-Hodgkins lymphoma: 40 patients had nodular poorly differentiated lymphoma, 5 had nodular mixed lymphoma and 3 had nodular histiocytic lymphoma: (Stage II in 4, Stage III in 23, and Stage IV in 21 patients). Half the patients received chemotherapy alone, and half received an injection of MBV before each cycle of chemotherapy. The MBV was a heat-killed preparation of *Streptococcus pyogenes* and *Serratia marcescens* (made according to the most effective of Coley's original formulae and known as Tracy XI). Radiation was given to initial areas of bulky disease or to nodal and extranodal sites responding partly to chemotherapy. Patients treated with MBV had a higher rate of complete remission (73% vs. 43%), longer duration of remission and longer survival (p=0.005) and no deaths, whereas 7 deaths occurred in those on chemotherapy alone (Kempin *et al.*, 1981 and personal communication from H. Oettgen, 1982).

The most dramatic of the so-called spontaneous regressions of cancer reported in the past 200 years occurred following acute *febrile* bacterial infections (Nauts, 1975a,b, 1980a). Patients receiving Coley's MBV had a higher percentage of permanent results if marked febrile reactions were elicited, especially the inoperable cases. Intratumoural or intravenous injections or larger doses given intradermally or intramuscularly elicited these reactions. (The subcutaneous route seemed least effective.) (Fowler, 1968, 1969a; Fowler and Nauts, 1970; Miller and Nicholson, 1971; Nauts, 1975b,c, d,e,f, 1977, 1978, 1980a).

Duration of such infections played a significant role in survival. A brief attack of erysipelas with marked fever produced the largest number of initial complete or partial regressions, but the largest number of *permanent* results occurred after the more prolonged suppurative types of infections such as staphylococcal or mixed pyogenic infections (Nauts, 1980a).

In this connection it was found that in patients receiving Coley's MBV duration was of *critical importance*: 14 of the 17 osteogenic sarcoma cases treated for at least 4 months following amputation were permanent successes (traced 4 to 55 years) and the 3 who died survived 4-13 years after onset (Nauts, 1975d).

SIGNIFICANT FACTORS AFFECTING SUCCESS OR FAILURE WITH MBV

Analysis of nearly 900 detailed histories of microscopically proven cancers so treated indicate that the essential factors were:

1. stage of disease and/or magnitude of tumour burden;
2. immune competence of the patient;
3. potency of vaccine preparations;
4. dosage, frequency and especially duration of injections;

5. close contact with tumour cells, i.e. site of injection;
6. timing in relation to surgery, radiation and/or chemotherapy.

Unfortunately, few who used the method recognized the importance of technique of administration, especially the need to produce adequate febrile reactions *consistently*. For example, in inoperable sarcoma of soft tissues, 60% complete regressions and permanent results were obtained if reactions averaged 38.5°–40.5°C as compared to 20% in those who had little or no febrile reactions (Nauts, 1975c).

The significance of the febrile reaction was not fully appreciated. Many patients, therefore, received inadequate doses of MBV given intramuscularly or subcutaneously producing slowly rising temperatures. We know that "slow heating" seems to induce thermal tolerance in some neoplasms. When given aggressively the rapid change in temperature suppresses the ability of neoplastic cells to adapt to higher temperatures.

MECHANISMS OF ACTION

No one using MBV empirically during Coley's lifetime or until recently knew that, in addition to producing fever, bacterial vaccines stimulate the reticuloendothelial system, activate macrophages, increase hematopoiesis and increase production of prostacyclin, endogenous interferon and endorphins. We now recognize that these far reaching effects were responsible for causing regression of extensive tumours and preventing recurrence and/or metastasis as well as immediate pain relief, improved blood picture, appetite and weight gain and rapid wound healing and regeneration of bone in cases in which bone had been destroyed by the tumour or removed at surgery.

Recent studies on prostacyclin are of special interest since this potent inhibitor of platelet clumping also is believed to prevent or decrease the formation of tumour thrombi (Honn, 1981). In this connection it is of interest to note that Coley's MBV was spectacularly successful in curing many patients with advanced thromboangiitis obliterans: severe pain ceased at once, gangrenous toes healed, and the patients remained well, with no recurrence working in war plants through World War II (Gray, 1935, and personal reports).

THE ROLE OF IRON

Iron is one of the most common single elements, making up one third of the earth. Its involvement in the early history of life is fascinating because the earliest living forms, the bacteria, have developed extremely efficient ways to utilize

it and clever biological systems to reduce it or solubilize
it. In many cases the capacity of bacteria, such as those
causing tuberculosis to multiply seems to be dependent on
their being in an environment with plentiful iron. The fact
is that many deadly bacteria need iron to be infectious.
Some bacteria apparently have the capacity to break down our
red cells during the process of infection. In this way they
release iron from the red cells and use it themselves. Bacterial systems like these which are iron-dependent have been
very carefully studied and beautifully analysed in the field
of microbiology.

Iron and the Immune System

Counter systems must clearly exist in higher animals to combat the very clever mechanisms whereby bacteria infect and
poison us. Some viruses have taken advantage of the iron
requirement of bacteria by using the iron receptors on the
bacterial cell surface to gain entry and infect it. We now
have reason to believe that our own immune system also
attacks some bacteria because of their need for iron. The
theory being developed suggests that certain cells of the
blood — the lymphocytes, the polymorphs and the macrophages —
are geared to detect the presence of iron. This is because
evolution may have taught these cells that iron is associated
with bacteria. Not only that, but immune cells are appearing
more and more to be the workhorses of an iron transport and
reutilization system absolutely essential for the control of
blood formation and possible other basic processes. The
complexity and ubiquity of these iron-based systems is just
beginning to be realized (de Sousa, 1981, 1982).
    For this iron commerce system to function, a series of
iron-binding and transporting proteins have been found to be
involved. These proteins, ferritin, transferrin and lacto-
ferrin, have long been known to be involved in iron storage
and transport, but now they are identified as the actual
molecular means used by the various white cells to deal with
iron.
    It has been estimated that one adult human male processes
approximately 2 billion old red blood cells every day. This
represents 38 mg of iron which has to be recycled by the
immune system. Ninety percent of this goes to new red cell
production, the other 10% remain in storage in the macrophages of the immune system and the liver, ready for reutilization as needed.
    Before the development of antisepsis, vaccines and anti-
biotics, we were exposed to all types of bacterial infections
and infectious diseases, and we were not given iron supple-

ments. At present, infections occur rarely and when they do, they are rapidly controlled by antibiotics. Iron supplementation of bread, milk and in multi-vitamin preparations ensures that the majority of the population in countries like the USA and Sweden is taking not only adequate amounts of iron, but in some cases (men and postmenopausal women) too much.

Our bodies' mechanisms for dealing with excess iron may break down or get overloaded. Control of iron levels is normally exercised at the point of entry in the small intestine. Normal people stop absorbing iron the moment they have sufficient iron in circulation and in storage. But a person with an abnormality of this controlling mechanism will be at risk if exposed to too much iron, i.e. in drinking water, eating too much meat, or taking too many iron supplements.

Iron and Cancer

Recent studies of the nutrient requirements of cancer cells *in vitro* and of growing tumours in animals (*in vivo*) have shown that cancer cells utilize iron in a fashion possibly similar to that of bacteria. It is suspected that the requirement of iron by cancer cells is related to its utilization for multiplication (DNA synthesis). Since iron is processed by the lymphoid system, particularly by the spleen where old red blood cells are broken down to have their iron reutilized, this is the system which gets overloaded first. A number of things happen as a result, the most obvious being that lymphocytes, which have been trained by evolution to go to iron-rich sites, "get stuck" in the spleen and proliferate abnormally. This in turn depletes their numbers in the rest of the body so it becomes vulnerable to infection. In addition, iron may directly produce a malignant change in the splenic macrophages producing the classic Reed-Sternberg cell which is diagnostic of Hodgkin's disease.

In summary, it is already known that iron inhibits some immunologic functions. Iron can be envisaged as critical to several steps of tumour growth and progression. In principle, intervention at each of the steps influenced by iron could arrest the growth of cancer effectively. In practice, some evidence that this is the case is beginning to emerge from cancer treatment with monoclonal antibodies against iron-associated proteins (ferritin and transferrin receptors) (de Sousa, 1981, 1982; Harman, 1981; Order *et al.*, 1981; Trowbridge and Domingo, 1981).

Such studies are vital now to our understanding of the role of iron in immunity and cancer and the work begun by de Sousa at Sloan-Kettering Institute in New York is fundamental. It will have wide ramifications, scientific and

clinical, in the understanding and treatment not only of cancer but of other diseases such as rheumatoid arthritis and
heart disease. It will also have important consequences for
nutrition and help our understanding of bacterial infections
and the way we combat them.

Dietary Considerations

A substantial amount of work in recent years has elucidated
the deleterious biological effects of free radical reactions
(Harman, 1981). Generation of some of the most toxic end
products of these reactions involve oxygen consumption and
are drastically enhanced by the presence of iron.

In communities where iron intake from diet is adequate,
and where, for lack of serious bacterial infections, all
available iron is either used to make new red cells or stays
in storage in the immune system, eventual accumulation with
age is inevitable, particularly in men and postmenopausal
women. It has recently been hypothesized that high iron
stores are responsible for the higher incidence of heart
disease which occurs in these two groups (Sullivan, 1981).
Thus dietary regimens reducing iron intakes and increasing
the intake of antioxidants such as Vitamins C and E can be
predicted to increase lifespan (Harman, 1981).

Experimental evidence for dietary antioxidants increasing
life span is well documented already. The possibility that
reduced iron intake would strengthen the beneficial action of
antioxidants seems at present to be a likely hypothesis and
from all that is known of the basic biochemistry of free
radical generation, well worth pursuing.

THE ROLE OF FEVER

What benefits do bacterial vaccines administered so as to
induce adequate febrile reactions of 39°—40.5°C offer which
cannot be obtained by using whole body hyperthermia produced
by hot air, or immersion in heated water or oil?

The available evidence suggests that modern whole body
hyperthermia usually requires immunosuppressive anaesthesia
or tranquillizing agents. Some of these heroic procedures
may in themselves be immunosuppressive (Yerushalmi, 1976).
On the other hand, local hyperthermia by radiofrequency appa-
ratus seems to offer greater promise and is better tolerated
by the patient, without any immunosuppressive effects.
Neither type appears to provide sufficient duration of therapy
to insure permanent results.

We believe that even better immediate and final results
may be possible by a judicious use of local hyperthermia

given weekly combined with injections of MBV, not only Coley's formula, but other types such as that used by Waisbren, given at least twice a week at first. These injections must then be continued after the local hyperthermia treatments are completed in order to stimulate the immune system to cope with absorption of necrotic tumour tissue, to destroy any residual neoplastic cells and to prevent recurrence and metastases by prolonged stimulation of the reticuloendothelial system, prostacyclin production and other factors in the host response to cancer. Vaccine therapy can be administered on an ambulatory basis after the initial week or two, and the family physician can be trained to carry on maintenance therapy tapering off to once a month if given for over 4 months.

STAPHYLOCOCCAL PROTEIN A

In addition to using bacterial vaccines such as MBV as discussed above we believe that more clinicians should consider using extracorporeal plasma perfusion over *Staphylococcal* protein A. This is a constituent of the cell wall of *Staphylococcus aureus* Cowans I, which reacts with the Fc region of immunoglobulins from many mammalian species (Forsgren and Sjöquist, 1966) and thus also combines with immune complexes in the serum. This procedure appears to nonspecifically remove blocking factors from the patient's plasma.

The first group to report their work with this technique was Bansal *et al*. (1978) who observed improvement in the general condition of one colon carcinoma patient as well as a decrease in tumour size and histological changes consistent with tumour destruction.

Ray and his colleagues in Philadelphia have further modified the immunoadsorption technique and extended their studies on 3 tumour models simultaneously, i.e. rat mammary adenocarcinoma (Ray *et al*., 1979; Ray, 1981; Ray *et al*., 1981b,c), canine venereal tumours (Ray, 1981a,b; Ray *et al*., 1981c; 1982a), and metastatic human colon carcinomas (Ray, 1981; Ray *et al*., 1981c; 1982a) and reported tumour regressions in all these tumour models.

Ray *et al*. have observed significant potentiation of antitumour reactivity in the tumour bearing hosts (Ray *et al*., 1981a; 1982). This treatment is associated with febrile reactions and shaking chills (Ray *et al*., 1982a,b,c). Histologic evaluation of biopsied tumours showed fibrotic and necrotic reactions and lymphocyte infiltrations (Ray *et al*., 1981d; 1982; 1982c). Complement activation, decreased blocking activity, increased skin reactivity to recall antigens (delayed type hypersensitivity reactions), elevation of

rosette forming cells, plasma-mediated potentiation of lymphocyte cytotoxicity, and increased antitumour antibody activity
were also reported by Ray *et al.* as the consistent findings
in their responding tumour bearing animals (Ray *et al.*, 1982
a,b,c,d,e).

Ray and Bandyopadhyay (1982d) have recently observed that
cancer patients' plasma can elute several types of biomolecules from *Staphylococcus aureus* during the *in vitro* adsorption procedure. Also they have observed that the eluted bacterial components can cause tumour necrosis in animals (Ray
et al., 1981d; 1982b,c).

Thus Ray believes that removal of blocking components from
the plasma on the one hand and stimulation of immune reactivity of the host by the leached bacterial components on the
other hand, may have contributed to the observed tumour
regressions in rat, dog and human tumours. Perhaps Ray and
others should now compare the effects of perfusion over Staph
protein A with injections of Staphage Lysate to see if the
perfusion procedure is essential to the beneficial effect.

Terman *et al.* used this procedure in dogs with spontaneous
mammary adenocarcinoma. A single nontoxic infusion of cytosine arabinoside was given after extracorporeal perfusion of
plasma over protein A bearing *S. aureus*. This resulted in a
necrotizing response rapid in onset and specific for tumour
tissue. Gross tumouricidal reactions 12 hours after this
combined treatment exceed the algebraic sum of responses to
either agent given alone in the same dogs, implying a synergistic effect (Terman, 1980).

He then extended this approach to patients with adenocarcinoma of the breast, with the initial objective of determining whether an antitumour effect could be demonstrated in
large easily measurable lesions. Five consecutive such cases
who had relapsed while under conventional therapy were given
autologous plasma or plasma from similar patients which had
been perfused over purified protein A immobilized in a collo-
dion-charcoal matrix (PACC) (Terman, 1981). This was sometimes followed by intravenous cytarabine, 2 to 20 mg/kg of
body weight. These procedures were repeated and objective
partial remission or improvement was observed in 4 of the 5
patients. They concluded that these preliminary results
warranted further study of the mechanism involved in the
antitumour effect as well as the potential value of the technique.

In Sweden at least 2 groups are working on this approach.
(Forsgren and Sjöquist, 1966, and personal communication from
S. Jonnson, 1982).

Another study at Memorial Sloan-Kettering Cancer Center
by Jones *et al.* (1980) concerns the treatment of feline

lymphosarcoma, a rapidly fatal disease. Preliminary results show disease regression in 11 of 12 cats lasting up to 3 years. This is the first report of actual cure in this tumour as well as the first report of therapeutic clearance of a persistent viraemia by this procedure (personal communication, 1982).

WHAT OF THE FUTURE?

We believe that never before has there been such tremendous opportunities to utilize the knowledge we have reviewed to help control and prevent neoplastic diseases with bacteria and bacterial products.

Clinicians must be encouraged to advise their patients as to proper nutrition and the most effective vitamin supplements, especially A, C and E and *avoidance of iron*. In planning overall treatment, localized hyperthermia should be used as well as bacterial vaccines administered so as to invoke systemic fever. These should be given before and after debulking surgery, irradiation or chemotherapy, as required. In patients at risk of developing infections, antibiotics should be avoided whenever possible, relying instead on the use of mixed bacterial vaccines. *Antipyretics should be avoided since fever is beneficial*. Ambulant patients should be given hot baths once or twice a day (to 45°C) and returned to a warm bed preferably with electric blankets or hot water bottles to maintain heat. This is especially helpful during the febrile reaction occurring with bacterial vaccine therapy, or with extracorporeal perfusion over *Staphylococcus* protein A.

In using these microbial products we must remember that they stimulate the production of endogenous interferon, some more than others (Matsubana *et al.*, 1977; Singer *et al.*, 1971; Stinebring and Youngner, 1974; Youngner and Stinebring, 1965).

By combining endotoxin and BCG in the treatment we may get better results due to production of tumour necrosis factor (TNF).

TNF is now being studied intensively in Tokyo headed by Haranaka who worked with Old at Sloan-Kettering Institute. TNF is now being used on patients in Japan.

Finally, the use of Staph A as outlined above should be included in order to remove blocking factors and to benefit from its host stimulating effects.

Bacterial vaccines, to be most effective, must not merely be used as a last resort when the immune status of cancer patients is seriously compromised, but should be initiated *prior* to surgery, radiation, chemotherapy or hyperthermia and continued during and after these modalities. Let us evaluate many different bacterial vaccines, including Staphage Lysate, MBV, Pseudogen (Parke Davis), *Escherichia coli* (Difco 0127B8), Hollister-Stier's Mixed Respiratory Vaccines and possibly *Lactobacillus bulgaricus* which is being evaluated in Bulgaria and the USSR (Cancer Research Institute Records: personal communications). Let us also consider wider use of TNF, and of yeast extracts such as glucan which is now being evaluated clinically by Krementz at Tulane University (personal communication).

Some patients may respond better to one type rather than another due to prior sensitization. One may expect better responses in populations which are still exposed to more infections. Cooperative studies on an international scale should speed progress towards more effective combined therapies for control of cancer and ultimately perhaps for prevention of the disease. Persons at high risk of cancer might be given periodic courses of MBV to stimulate the various host resistance mechanisms which we have discussed.

The fact that thousands more cancer patients were not given the benefit of such therapy since it was first initiated by Coley 90 years ago, may be one of the greatest tragedies in medical history. Many factors contributed to this, but the most serious recent one was the incredible barriers created by the FDA (Food and Drug Administration) since 1963. To rule Coley's MBV as a "new drug" when it had been used for 70 years in nearly 900 patients was incredible. The red tape involved in securing Investigational New Drug licences (INDs) for new vaccines now is preventing many clinical investigators from undertaking such studies.

Fortunately, the Government of West Germany recognized that the new laws they had put into effect after the thalidomide disaster were too stringent and they were repealed.

Progress will now have to come from clinical trials in China, Japan, Poland, West Germany and other European countries which do not have such paralysing restrictions. We appeal to you all to exert every effort to speed progress.

ACKNOWLEDGEMENTS

The author wishes to express her profound appreciation to Lloyd J. Old, MD and other members of the Scientific Advisory Council of Cancer Research Institute for their invaluable encouragement and guidance over the last 30 years, and to Rosemary Leeder for typing the manuscript.

REFERENCES

Bansal, S.C., Bansal, B.R., Thomas, H.L. *et al.* (1978). *Ex vivo* removal of serum IgG in a patient with colon carcinoma: some biochemical, immunological and histological observations. *Cancer* **42**, 1-18.

Burkitt, D. (1969). Aetiology of Burkitt's lymphoma – an alternative hypothesis to a vectored virus. *J. Nat. Cancer Inst.* **42**, 19-28.

Cancer Research Institute Records (1953-1982). Personal communications.

Carswell, E.A., Old, L.J., Kassel, R.L. *et al.* (1975). An endotoxin induced serum factor that causes necrosis of tumours. *Proc. Natl. Acad. Sci. USA* **72**, 3666-3670.

Clarkson, B., Dowling, M.D., Gee, T.S. *et al.* (1975). Treatment of acute leukaemia in adults. *Cancer* **36**, 775-795.

Coley, W.B. (1891). Contributions to the knowledge of sarcoma. *Ann. Surg.* **14**, 199-220.

Coley, W.B. (1893). The treatment of malignant tumours by repeated inoculations of erysipelas; with a report of 10 original cases. *Amer. J. Med. Soc.* **105**, 487-511.

Coley, W.B. (1894). The treatment of inoperable malignant tumours with toxins of erysipelas and the Bacillus prodigiosus. *Trans. Amer. Surg. Assn.* **12**, 183-212.

Coley, W.B. (1909-1910). The treatment of inoperable sarcoma by bacterial toxins (the mixed toxins of Streptococcus of erysipelas and the Bacillus prodigiosus). *Proc. Roy. Soc. Med., Surg. Sect.* **3**, 1-48. (Also in *Practitioner, London* (1909) **83**, 589).

de Sousa, M. (1981). "Lymphocyte Circulation. Experimental and Clinical Aspects". John Wiley and Sons, New York. pp. 197-297.

de Sousa, M., da Silva, B., Donner, M. *et al.* (1982). Iron and the lymphomyeloid system: rationale for considering iron a target for immunosurveillance. *In* "Proteins of Iron Storage and Transport" (Ed. P. Saltman), Elsevier/North-Holland.

Didot, A. (1851-1852). Essai sur la prophylaxie du cancer par la syphilization artificielle. *Bull. Acad. Roy. Belge, Brussels* **11**, 100-172.

Dupré de Lisle (1774). "Traité du Vice cancéreux". Paris, Couturier fils.

Forsgren, A. and Sjöquist, J. (1966). "Protein A" from *S. aureus*. I. Pseudo-immune reaction with human gamma-globulin. *J. Immunol.* **97**, 822-827.

Fowler, G.A. (1968). "Testicular Cancer Treated by Bacterial Toxin Therapy as a Means of Enhancing Host Resistance.

End Results in 63 Determinate Cases". Monograph No. 7, New York Cancer Research Institute*. New York.

Fowler, G.A. (1969a). "Beneficial Effects of Acute Bacterial Infections or Bacterial Toxin Therapy on Cancer of the Colon or Rectum". New York Cancer Research Institute*. Monograph No. 10, New York.

Fowler, G.A. (1969b). "Enhancement of Natural Resistance to Malignant Melanoma with Special Reference to the Beneficial Effects of Concurrent Infections and Bacterial Toxin Therapy". New York Cancer Research Institute*. Monograph No. 9, New York.

Fowler, G.A. and Nauts, H.C. (1970). "The Apparently Beneficial Effects of Concurrent Infections, Inflammation or Fever and of Bacterial Toxin Therapy on Neuroblastoma". New York Cancer Research Institute *. Monograph No. 11, New York.

Gershon, R.K. and Carter, R.L. (1969). Factors controlling concomitant immunity in tumour bearing hamsters: effects of prior splenectomy and tumour removal. *J. Nat. Cancer Inst.* **43**, 533–543.

Gray, H.J. (1935). Thromboangiitis: Significant findings and theory of aetiology (nonspecific protein therapy with Coley's erysipelas and prodigiosus toxins). *Med. Bull. Vet. Admin.* **11**, 16–23, and personal communications (1946).

Harman, D. (1981). The aging process. *Proc. Natl. Acad. Sci. USA* **78**, 7124–7128.

Hoffman, M.K., Oettgen, H.F. and Old, L.J. *et al.* (1978). Induction and immunological properties of tumour necrosis factor. *J. Reticul. Soc.* **23**, 307–319.

Honn, K.V., Cicone, B. and Skoff, A. (1981). Prostacyclin: A potent anti-metastatic agent. *Science* **212**, 1270–1273.

Huth, E.F. (1958). Zum Antagonismus zwischen bakteriellen Infektionen und malignen Erkrankungen. *Med. Klinik* **53**, 2173.

Jones, F.R., Yoshida, L.H., Ladiges, W.C. *et al.* (1980). Treatment of feline leukaemia and reversal of FeLV by *ex vivo* removal of IgG: a preliminary report. *Cancer* **46**, 675–684.

Kempin, S., Cirrincione, C., Straus, D.L. *et al.* (1981). Improved remission rate and duration in nodular non-Hodgkin's lymphoma (NHL) with the use of mixed bacterial vaccines (MBV). *Proc. Amer. Soc. Clin. Oncol.* **22**, 514.

Klein, G. (1975). Immunological Surveillance Against Neoplasia. The Harvey Lectures, Series 69, 71–102.

---

*
Name changed to Cancer Research Institute Inc. in 1973.

Matsubana, S., Suzuki, F. and Ishida, N. (1977). Induction
    of interferon in mice by a streptococcal preparation,
    Picibanil. Igaku-no-Ayumi "Advances in Medicine" **102**, 536-
    537.
MacMahon, B., Morrison, A.S., Ackerman, L.V. *et al.* (1973).
    Histologic characteristics of breast cancer in Boston and
    Tokyo. *Int. J. Cancer* **11**, 338-344.
Matthews, N. (1979). Tumour necrosis from the rabbit. III.
    Relationship to interferons. *Brit. J. Cancer* **40**, 534-539.
Meyer, B.A. and Benjafield, J.D. (1955). Carcinoma and anti-
    biotics. *Med. Press* **234**, 206-208.
Miller, T.N. and Nicholson, J.T. (1971). End results in
    reticulum cell sarcoma of bone treated by toxin therapy
    alone or combined with surgery and/or radiation (47 cases)
    or with concurrent infection (5 cases). *Cancer* **27**, 524-
    548.
Mudd, S. and Shayegani, M. (1974). Delayed type hypersensi-
    tivity to *Staph. aureus* and its uses. *Ann. New York Acad.
    Sci.* **236**, 244-251.
Nauts, H.C., Fowler, G.A. and Bogatko, F.H. (April, 1953).
    A review of the influence of bacterial infection and of
    bacterial products (Coley's toxins) on malignant tumours
    in man. *Acta Medica Scandinavica* **145**, Supplement 276,
    Stockholm.
Nauts, H.C. and Fowler, G.A. (1969). "End Results in Lympho-
    sarcoma Treated by Toxin Therapy Alone or Combined with
    Surgery and/or Radiation or with Concurrent Bacterial
    Infection". New York Cancer Research Institute*. Mono-
    graph No. 6, New York.
Nauts, H.C. (1973). "Enhancement of Natural Resistance to
    Renal Cancer: Beneficial Effects of Concurrent Infections
    and Immunotherapy with Bacterial Vaccines". New York
    Cancer Research Institute*. Monograph No. 12, New York.
Nauts, H.C. (1974). "Ewing's Sarcoma of Bone: End Results
    Following Immunotherapy (Bacterial Toxins) Combined with
    Surgery and/or Radiation". Cancer Research Institute Inc.
    Monograph No. 14, New York.
Nauts, H.C. (1975a). Pyrogen Therapy of Cancer: An Historical
    Overview and Current Activities. Trans. International
    Symposium on Cancer Therapy by Hyperthermia and Radiation,
    Washington, D.C., April.
Nauts, H.C. (1975b). Immunotherapy of cancer by microbial
    products. *In* "Host Defence Against Cancer and its Poten-
    tiation". (Eds D. Mizuno *et al.*), University of Tokyo
    Press, Tokyo and University Park Press, Baltimore, 337-351.

---

*Name changed to Cancer Research Institute Inc. in 1973.

Nauts, H.C. (1975c). "Beneficial Effects of Immunotherapy (Bacterial Toxins) on Sarcoma of Soft Tissues. End Results in 186 Determinate Cases with Microscopic Confirmation of Diagnosis — 49 Operable, 137 Inoperable". Cancer Research Institute Inc. Monograph No. 16, New York.

Nauts, H.C. (1975d). "Osteogenic Sarcoma: End Results Following Immunotherapy (Bacterial Vaccines) 165 Cases, or Concurrent Infections, Inflammation or Fever, 41 Cases". Cancer Research Institute Inc. Monograph No. 15, New York.

Nauts, H.C. (1975e). "Giant Cell Tumour of Bone: End Results Following Immunotherapy (Coley Toxins) Alone or Combined with Surgery and/or Radiation (66 Cases) or with Concurrent Infection (4 Cases)". Cancer Research Institute Inc. Monograph No. 4, 2nd Ed., New York.

Nauts, H.C. (1975f). "Multiple Myeloma: Beneficial Effects of Acute Infections or Immunotherapy (Bacterial Vaccines)". Cancer Research Institute Inc. Monograph No. 13, New York.

Nauts, H.C. (1977). "Beneficial Effects of Acute Concurrent Infection, Inflammation, Fever or Immunotherapy (Bacterial Toxins) on Ovarian and Uterine Cancer". Cancer Research Institute Inc. Monograph No. 17, New York.

Nauts, H.C. (1978). Bacterial vaccine therapy of cancer. *In* "Proc. Symposium on Biological Preparations in the Treatment of Cancer, London, April 13–15, 1977". Developments in Biological Standardization **38**, 487–494, S. Karger, Basel.

Nauts, H.C. (1980a). "The Beneficial Effects of Bacterial Infections on Host Resistance to Cancer. End Results in 449 Cases. A Study and Abstracts of Reports in the World Medical Literature (1775–1980) and Personal Communications". Cancer Research Institute Inc. Monograph No. 8, 2nd Ed. (1,032 references), New York.

Nauts, H.C. (1980b). "Bibliography of Reports Concerning Experimental or Clinical Use of Coley's Toxins (*Streptococcus pyogenes* and *Serratia marcescens*) 1893–1980". Cancer Research Institute Inc. (388 references including 124 by W.B. Coley), New York.

Nauts, H.C. (1982a). Bacterial pyrogens: Beneficial effects on cancer patients. International Symposium on Biomedical Thermology, June 30 — July 4, 1981, Strasbourg, France (in press).

Nauts, H.C. (1982b). "Immunological Factors Affecting Incidence, Prognosis and Survival in Breast Cancer". Cancer Research Institute Inc. Monograph No. 18, New York. (To be published).

Old, L.J. and Boyse, E.A. (1973). Current Enigmas in Cancer Research. The Harvey Lectures, Series 67, 273–315.

Order, S.F., Klein, S.L. and Leicher, D.K. (1981). Antiferri-
  tin antibody for isotopic cancer therapy. *Oncology* **38**,
  154-160.
Ray, P.K., Cooper, D.R., Bassett, J.G. *et al.* (1979). Anti-
  tumour effect of *Staphylococcus aureus* organisms. *Fed.*
  Proc. **38**, 4558.
Ray, P.K. (1981a). *In* "Proc. Workshop on Therapeutic Plasma-
  pheresis and Cytapheresis, April 25, 26, 1979". (Eds
  A.J. Nemo and H. Taswell), pp. 334-339. US Dept. of Health
  and Human Services, NIH, October.
Ray, P.K., Idiculla, A., Rhoads, J.E. Jr. *et al.* (1981b).
  Extracorporeal immunadsorption using protein A — containing
  Staphylococcus aureus column. A method for the quick
  removal of abnormal IgGs or its complexes from the plasma.
  In "Plasma Exchange Therapy" International Symposium,
  Weisbaden 1980, 150-154. (Eds H. Borberg and P. Reuther).
  Georg Thieme Verlag, Stuttgart/New York.
Ray, P.K., McLaughlin, D., Mohammed, J. *et al.* (1981c). *Ex*
  vivo immunoadsorption of IgG or its complexes — a new
  modality of cancer treatment. *In* "Immune Complexes and
  Plasma Exchanges in Cancer Patients". (Eds B. Serrou and
  C. Rosenfeld), I: 197-207. Elsevier/North Holland Biomed.
  Press.
Ray, P.K., Idiculla, A., Clarke, L. *et al.* (1981d). Immuno-
  adsorption of IgG and/or its complexes from colon carcinoma
  patients — adjunct therapy for cancer. *In* "Proc. Intern
  International Conference on the Adjuvant Therapy of Cancer"
  March 18-21, p.29, Tucson, Arizona.
Ray, P.K., Clarke, L., McLaughlin, D. *et al.* (1982a). Immuno-
  therapy of cancer: extracorporeal adsorption of plasma-
  blocking factors using non-viable *Staphylococcus aureus*
  Cowan I. *In* "Plasma Exchange Symposium". (Ed. S. Nagel),
  Gottingen, S. Karger, Munich.
Ray, P.K. (1982b). Suppressor control as a modality of can-
  cer treatment. *Plasma Therapy* **3**,
Ray, P.K. and Raychaudhuri, S. (1982c). "Immunotherapy of
  Cancer — Present Status and Future Trends". Pergamon
  Press, New York. (in press).
Ray, P.K. and Bandyopadhyay, S. (1982d). Plasma elution of
  biomolecules from nonviable *Staphylococcus aureus* Cowan I.
  Fed. Amer. Soc. Exp. Biol. **41**,
Ray, P.K., Idiculla, A., Mark, R. *et al.* (1982e). Extra-
  corporeal immunoadsorption of plasma from a metastatic
  colon carcinoma patient by protein A nonviable *Staphylo-*
  coccus aureus. Clinical, biochemical, serological and
  histological evaluation of the patient's response. *Cancer*
  49,

Rokitansky, C.F. (1842-1846). Handbuch der pathologischen
    Anatomie. Wien, Braumiller u. Seidel.
Singer, S.H., Hardegree, C., Duffin, N. *et al.* (1971). Induc-
    tion of interferon by bacterial vaccines and allergenic
    extracts. *J. Allergy* **47**, 332-340.
Smith, R.L., Salsbury, C.G. and Gilliam, A.G. (1956).
    Recorded and expected mortality among the Navejo, with
    special reference to cancer. *J. Nat. Cancer Inst.* **17**, 77-
    89.
Smith, R.L. (1957). Recorded and expected mortality among the
    Indians of the United States with special reference to
    cancer. *J. Nat. Cancer Inst.* **18**, 385-396.
Stinebring, W.R. and Youngner, J.S. (1964). Patterns of
    interferon appearance in mice injected with bacteria or
    bacterial endotoxin. *Nature* **204**, 712.
Sullivan, J.L. (1981). Iron and sex difference in heart
    disease risk. *Lancet* **i**, 1293-1294.
Tanchou, S.L. (1844). "Recherches sur le Traitement médical
    des Tumeurs cancéreuses du Sein. Ourvrage pratique basé
    sur trois cents Observations. (Extraits d'un grand Nombre
    d'Auteurs)". Germer Baillière, Paris.
Terman, D.S., Tamamoto, T., Mattioli, M. *et al.* (1980).
    Extensive necrosis of spontaneous canine mammary adeno-
    carcinoma after extracorporeal perfusion over *Staphylo-
    coccus aureus* Cowans I. I. Description of acute tumouri-
    cidal response, morphologic, histologic, immunohistochemi-
    cal, immunologic, and serologic findings. *J. Immunol.*
    124, 795-805.
Terman, D.S. (1981). Preliminary observations of the effects
    on breast adenocarcinoma of plasma perfused over immobi-
    lized protein A. *New Engl. J. Med.* **305**, 1195-1199.
Trnka de Krzowitz, W. (1783). Historia febris hecticaeomnis
    aevi observata medica continens. (History of remittent
    fevers). Vindobonae: apud R. Graefferum. (Cited by
    Tanchou, 1844).
Tromp, S.W. (1954). Possible counteracting influence of
    tuberculosis on development of cancer in the Netherlands.
    Amer. J. Clin. Path. **34** (Suppl. No. 8), 35.
Trowbridge, I.S. and Domingo, D.L. (1981). Anti-transferrin
    receptor monoclonal antibody and toxin-antibody conjugates
    affect growth of human tumour cells. *Nature* **194**, 171-173.
Vautier, A.H. (1813). Vue générale sur la maladie cancéreuse.
    Thèse de Paris 43, 11.
Verneuil (1883). Influence de la syphilis sur la marche du
    cancer. *J. de Med. et de Chir. prat. Paris* **54**, 398-400.
Waisbren, B.A. Monograph to be published 1982 and personal
    communications, 1977-1982.

Wolff, J. (1911-1914). Die Lehre von der Krebskrankheit von den altesten Zeiten bis zur Gegenwart. *Jena* **2**, 104 and **3**, 482-405.

Yerushalmi, A. (1976). Influence on metastatic spread of whole-body or local tumour hyperthermia. *Eur. J. Cancer* **12**, 455-463.

Youngner, J.S. and Stinebring, W.R. (1965). Interferon appearance stimulated by endotoxin, bacteria or viruses in mice pretreated with *Escherichia coli* endotoxin or infected with mycobacterium tuberculosis. *Nature* **208**, 456-458.

BACTERIA AND HUMAN CARCINOGENESIS

M.J. Hill

*PHLS Centre for Applied Microbiology and Research,
Bacterial Metabolism Research Laboratory,
Porton Down, Salisbury SP4 0JG*

I. THE NORMAL HUMAN BACTERIAL FLORA

All body surfaces are colonized by bacteria, the composition of the flora being dependent on the conditions at that surface. The major factors determining the composition of the flora are the supply of nutrients, the time available for growth, the pH, the oxygen tension and E_h, the secretion of antibacterial substances (such as lysozyme, antibody, bile, pancreatic enzymes etc.), production of antibacterial substances by other bacteria (such as bacteriophage, bacteriocines), ability to produce induced enzymes, and the ability to attach to and colonize surfaces. Table 1 describes the flora of various sites on the body surface and the factors controlling that flora; the subject has been reviewed extensively elsewhere (Skinner and Carr, 1974). By far the most important sites are the digestive tract and the skin but other sites may be heavily colonized under certain circumstances.

A. The Stomach

In western persons the normal gastric mucosa secretes large amounts of hydrochloric acid, maintaining the pH of the gastric contents close to 2. As the mucosa ages it atrophies and the rate of secretion of acid progressively decreases. When the acid secretion is insufficient to achieve bactericidal levels a resident flora is able to establish itself; this flora is acid-tolerant (streptococci, micrococci, lactobacilli and veillonellae) at lower pH values but the more acid sensitive putrifactive organisms increase in importance as acid secretion decreases. Empirical observation on the gastric flora indicate that pH values below 4 are bactericidal,

TABLE 1

*The bacterial flora of various body surfaces
and the factors controlling it.*

Site	Dominant flora	Major controls
Saliva	cocci (eg streptococci, micrococci, veillonellae, neisseria), principally Gram +ve	lysozyme, immune controls
Stomach	Normally no bacterial flora	Gastric acid
Small intestine	Few organisms and no "normal flora"	Gastric acid, bile, pancreatic juice
Large intestine	Very rich flora, dominated by anaerobic rod-shaped bacteria	Redox potential, nutrients, stasis
Skin	Diptheroids, micrococci and staphylococci	Fatty acids, desiccation
Vagina	Lactobacilli and streptococci	pH

TABLE 2

The relationship between gastric pH and the composition of the gastric juice bacterial flora.

pH	Composition of the flora
Less than 4	Very few live organisms
Between 4 and 5	Acid-tolerant organisms (lactobacilli and streptococci and micrococci)
Between 6 and 8	Largely enteric flora containing faecal streptococci, bacteroides etc., as well as the lactobacilli and oral streptococci

those between 4 and 5 permit the growth only of acid tolerant
organisms whilst above pH 5 faecal-type organisms flourish
(Table 2). In the achlorhydric stomach the oxygen-tension is
sufficient to prevent the growth of the more fastidious
anaerobic bacteria (clostridia are very rarely isolated from
gastric juice for example) and the aerobic organisms are normally more numerous than other anaerobes. In western countries the proportion of persons with a resting gastric juice
pH greater than 4 is very small in young adults but has been
estimated at about 10% at 50 years old and about 50% of 70
year old persons. In contrast, in parts of Colombia and Chile
in South America, or in Vellore, South India the majority of
young adults are achlorhydric.

B. Small Intestine

Since the gastric acid effectively sterilizes the gastric
contents the material entering the small intestine contains
very few bacteria. Further, the mucosa is continually
flushed by the large volumes of fluid secreted into the small
intestine as part of the digestive process; many of these
(e.g. bile) are powerful antibacterial agents. Consequently
the small intestine is normally only very sparsely populated
except at the beginning of a meal (Drasar *et al.*, 1969) or in
persons who are achlorhydric or in conditions where there is
an intestinal stasis.

C. Large Intestine

Whereas the time taken for the intestinal contents to travel
the length of the small intestine is only 2—4 hours, the
large bowel transit time is normally 20—80 hours in western
persons and so there is ample time for a rich bacterial flora
to develop. Although digestion is very effective and most of
the nutrients in the diet are absorbed from the small intestine, there is a constant supply of nutrients to the large
intestine in the form of indigestible dietary components
(e.g. "fibre") and desquamated gut mucosal cells. The
absence of glucose allows the production of inducible enzymes
and so full utilization of these nutrients is possible. The
large intestine is highly anoxic and so the dominant organisms are the non-sporing strictly anaerobic rod-shaped bacteria (Table 3). Approximately 50% of the dry weight of
faeces is accounted for by the gut bacteria (Stephen and
Cummings, 1980) and so the number of bacteria present in the
normal colon is approximately 10^{14}.

TABLE 3

*The normal faecal flora of humans eating a western diet
compared with those eating a vegetarian diet
based on Matoke (a banana preparation).
Data from Aries* et al. *(1969).*

	Western diet[*]	Matoke diet
Anaerobic organisms		
Bacteroides spp	10^{10}	10^8
Bifidobacterium spp	10^{10}	10^9
Clostridium spp	10^4	10^4
Veillonella spp	10^4	10^5
Facultative organisms		
Escherichia coli	10^7	10^8
Streptococcus virdans	10^7	10^8
Streptococcus faecalis	10^6	10^7
Microaerophilic organisms		
Lactobacillus spp	10^6	10^7

[*] $\log_{10}$ (number of organisms/g wet weight).

D. Urinary Tract

Although the urinary tract is normally free from bacteria, in
this case the abnormal is not rare. In a rural general practice in England the proportion of persons who had a urinary
tract infection at some time during the course of a single
year was 20% (Sinclair and Tuxford, 1971); most of these
would have been asymptomatic and due to a single infecting
organism; 80% of urinary tract infections are due to *Escherichia coli*, with *Proteus* spp, *Staph. albus* and *Strep. faecalis*
accounting for most of the remainder. A proportion will,
however, be chronic and due to a mixed population of organisms; such infections are usually secondary to anatomical
abnormalities in the bladder, bladder tumours, urethral
stricture or bladder bilharzia etc. Under the circumstances
a profuse mixed flora of 10^8 organisms per ml urine is not
uncommon.

E. Genital Tract

The normal vagina is acidic and has in consequence an acid-tolerant flora of lactobacilli and streptococci (Hurley *et al.*, 1974). In women with *Trichomonas vaginalis* infection, however, the vaginal pH is close to 7 and the flora is richer and more putrifactive (Harington *et al.*, 1973).

F. Biliary Tract

The normal biliary tract is sterile, but it can be infected with bile-resistant organisms (e.g. *Salmonella typhi*) and carriage of typhoid organisms in the bile is a common sequel to typhoid infection. The bile is often infected secondary to gallstones.

G. Summary

Bacteria are widely distributed around the human body either as normal flora or as the result of chronic infection (as in the urinary bladder and the gall bladder) or as the result of the failure of a bactericidal defence mechanism (as in the stomach).

II. THE ABILITY OF BACTERIA TO PRODUCE CARCINO-GENESIS *IN VITRO*

Bacteria are able to produce a wide range of carcinogens, mutagens or tumour promoters from a wide range of substrates. In this section I will only discuss those that might be produced by the normal flora from substrates which would be available to them. They have been reviewed more extensively by Hill (1975).

A. Tryptophan Metabolites

Tryptophan is metabolized by bacteria to a very wide range of metabolites, most of which are also produced by the liver. Of the range of tryptophan metabolites excreted in normal human urine some are produced only by the bacteria (e.g. indole, quinaldic acid) whilst all of the remainder could be produced by either the liver or bacteria; there are no data concerning the relative contribution of bacteria and the liver to the urinary tryptophan metabolite pool. A range of tryptophan metabolites, including indole, 3-hydroxykynurenine, 3-hydroxyanthranilic acid, kynurenine, quinaldic acid, 8-hydroxyquinaldic acid and xanthrenuric acid, have been claimed to be carcinogenic or mutagenic or tumour promoters (Table 4).

TABLE 4

*Urinary tryptophan metabolites shown to be carcinogenic,
mutagenic or tumour promoting.
(for details and references see Bryan, 1971).*

Metabolite	Test system	Indication
Indole Indoleacetic acid	A.A.F. treated rats	cocarcinogenic?
3-hydroxykynurenine 3-hydroxyanthranilic acid 8-hydroxyquinaldic acid xantrenuric acid quinaldic acid	bladder implantation	carcinogenic? cocarcinogenic?
3-hydroxykynurenine 3-hydroxyanthranilic acid	cultured mammalian tissue cells	mutagenicity

B. Volatile Phenols

Urine contains 2 major volatile phenols (*p*-cresol and phenol)
and these are produced in the colon by bacterial action on
the phenolic amino acids tyrosine and phenylalanine. The
daily output of urinary volatile phenols (UVPs) is 50—100 mg
(Schmidt, 1949; Bone *et al.*, 1976) 90% of which is *p*-cresol;
germ-free animals do not excrete UVPs and the amount in
human urine is greatly decreased by preoperative bowel preparation or in persons with an ileostomy (Bone *et al.*, 1976)
and increased by increasing the amount of dietary protein
(Cummings *et al.*, 1979).

C. Ethionine

Fisher and Malleter (1961) have shown that many species of
bacteria, including *E. coli*, when grown in mineral salts
medium containing glucose and methionine are able to synthesize the s-ethyl analogue, ethionine. Farber *et al.* (1967)
showed that ethionine is carcinogenic in rats.

D. N-nitroso Compounds

Secondary amines react with nitrous acid at pH 2, to yield
N-nitroso compounds, but Sander (1968) showed that the reaction could take place at pH 7 in the presence of bacteria,
and this has been confirmed by many groups since. Secondary

amines are synthesized in the human colon by bacterial action; lysine is decarboxylated then deaminated and cyclized to yield piperidine whilst arginine and ornithine are similarly cyclized to pyrrolidine. Lecithin is hydrolysed by bacterial lecithinase to yield choline which is then dealkylated to yield dimethylamine. These secondary amines are absorbed from the colon and excreted in the urine but are also secreted in a number of body fluids including gastric juice. Nitrate is present in large amounts in the diet and is readily reduced to nitrite by bacterial action. A wide range of species of bacteria present in the human bacterial flora are able to catalyse the N-nitrosation reaction (Hawksworth and Hill, 1971).

E. Hydrolysis of Conjugated Carcinogens

Cycasin is a naturally occuring β-glucoside present in cycad nuts; it is hydrolysed by bacterial p-glucosidase to yield methylazoxymethanol which is hepatotoxic in all animals tested (including humans) and is carcinogenic in rodents. When cycasin is fed to rodents the carcinomas are formed in the intestine; when administered intravenously cycasin is nontoxic and noncarcinogenic and it is similarly benign when given orally or by any other route to germ-free rodents. Thus the toxicity is entirely due to the metabolic activity of the gut flora in releasing the toxic aglycone. The mammalian small intestine produces a β-glucosidase but it is substrate-specific and is inactive against cycasin. The toxicology of cycasin has been reviewed by Laqueur (1968). It is not known whether cycasin is a member of a whole class of plant β-glucosides with carcinogenic aglycones or whether it is unique in this respect.

Polycyclic aromatic hydrocarbons present in the diet are absorbed by passive diffusion from the small intestine and transported to the liver by the portal blood system. There they are hydroxylated (if necessary) and then conjugated as glucuronides or glutathione derivatives before being secreted in the bile and returned to the intestine. On reaching the bacterially colonized part of the gut the glucuronides are hydrolysed by bacterial β-glucuronidase and the aglycone is then absorbed and returned to the liver to undergo another enterohepatic cycle. The enterohepatic circulation of polycyclic aromatic hydrocarbons (PAHs) has been reviewed by Smith (1966), but there have been 2 recent observations of interest in the context of this presentation. Renwick and Drasar (1976) reported that not only were the conjugated PAHs hydrolysed by bacterial β-glucuronidase but the released hydroxylated PAH could undergo further dehydroxylation to

release the parent carcinogen. Kinoshita and Gelboin (1978)
noted that whereas the aglycone released by hydrolysis of
benz(a)pyrene-3-glucuronide is not mutagenic or carcinogenic
and does not bind to DNA, an active intermediate produced
during the hydrolysis binds covalently to DNA added to the
incubation mixture.

F. Bile Acid Metabolites

Bile acids are important in the digestion of fat and are synthesized by the liver and secreted in bile. The human liver
synthesizes 2 bile acids — cholic acid and chenodeoxycholic
acid — which are conjugated with the amino acids glycine and
taurine before secretion in bile. The bile acids undergo
enterohepatic circulation, the bile acid conjugates being
recovered from the terminal ileum by an active transport
mechanism which is more than 95% efficient. The residue of
bile conjugates entering the large intestine are subjected to
a range of metabolic reactions by the gut bacteria including
deconjugation, hydroxyl dehydrogenation and inversion, dehydroxylation and nuclear dehydrogenation (Table 5). The
enterohepatic circulation of the bile acids has been reviewed
by Hofmann (1977) whilst the metabolism of the bile acids by
gut bacteria has been reviewed by Hill (1975).

TABLE 5

*The major bile salt degrading enzymes
produced by gut bacteria*

Enzyme	Reaction	Species producing the enzyme
Cholylglycine hydrolase	Deconjugates bile salts	Most of the major genera of anaerobes + *Strep. faecalis*
Hydroxysteroid oxidoreductase	Produces keto bile acids and inverted hydro- xyl groups	Most genera except lactobacilli and oral streptococci
7α-dehydroxylase	Removes the 7α hydroxyl group	Most of the major genera of anaerobes
Δ^4dehydrogenases	Produces unsatu- rated bile acids from 3-oxo sub- strates	A few clostridial species

The first interest in the possible role of bile acids in
carcinogenesis was aroused by the demonstration by Wieland
and Dane (1933) that deoxycholic acid could be converted
chemically into the extremely potent carcinogen 20-methyl-
cholanthrene. In 1940 Badger *et al.* showed that deoxycholic
acid was carcinogenic when painted on the skin of rodents in
an oily vehicle and since then the evidence has steadily
accumulated that bile acids, particularly deoxycholic and
lithocholic acids (the 2 principle bile acids excreted by
western persons) are tumour promoters and are mutagenic
(Table 6). The unsaturated bile acids have not been studied
extensively but preliminary studies (McKillop, 1981) suggest
that some may be direct acting mutagens, as might be suspected from their chemical structure.

TABLE 6

*Evidence that bile acids are mutagenic,
carcinogenic or tumour promoters.*

Type of study	Bile acid tested	Observation
Skin-painting on rats	Deoxycholic acid	cocarcinogenic
	Range of other bile acids	inactive
Rectal instillation in rats treated with dimethylhydrazine	Deoxycholic acid Lithocholi acid	both cocarcinogenic
	Other bile acids	inactive
Salmonella mutagenesis	Deoxycholic acid Lithocholic acid	both comutagenic
Drosophila muta-genesis	Deoxycholic acid	mutagenic

III. BACTERIAL METABOLITES AND HUMAN CANCER

Although a role for bacterial metabolites in the causation of
cancer of the large bowel and of the stomach has received
considerable attention, the part played by bacterial metabolites in the causation of cancer at other sites has not been
widely studied. Nevertheless they may be implicated in the
causation of cancer at many sites.

A. The Large Bowel

Large bowel cancer is one of the "diseases of western civili-
zation" and is thought to be associated with a western diet
which is rich in meat and fat and therefore is relatively
deficient in carbohydrate. The epidemiology of the disease
has been reviewed recently by Correa and Haenszel (1978). A
role for bacteria and bile acids was first postulated by
Aries *et al.* (1969) and the first comparison of populations
was reported by Hill *et al.* (1971a). Since then a considerable body of evidence in favour of a role for bile acid metabolites has accumulated (Table 7) from comparisons of populations, from case-control studies and from studies of patients
with diseases which put them at high risk of developing colorectal cancer. The bacterial enzymes implicated are the 7α-
dehydroxylase and the Δ^4-dehydrogenase (Table 8). The role
of bacterial metabolites of the bile acids in the causation
of large bowel cancer has been reviewed recently by Hill
(1980, 1981a, 1982).

TABLE 7

*Evidence that bile acids are implicated in the
causation of colorectal cancer*

1. Comparison of populations with different incidences of colorectal cancer	8 studies from 4 different research groups show LBC incidence related to faecal bile acid concentration.
2. Case-control studies	2 studies showed a difference between cases and controls, 3 did not.
3. Studies of adenoma patients	Study of 143 patients with adenomas showed correlation between faecal bile acid concentration and adenoma size and therefore malignant potential).

All of the bacterial metabolites described in the previous
section may be produced in the large bowel and so may be
implicated in the causation of colorectal cancer; the evidence has been reviewed by Hill (1977, 1981a) but is weak for
all of the metabolites in comparison with that for the bile
acids.

TABLE 8

Evidence implicating 7α-dehydroxylase and
Δ⁴-dehydrogenase in the causation
of large bowel cancer

---

1. Activity of 7α-dehydroxylase   Mastromarino *et al.* (1976)
   greater in cancer patients
   than in controls

2. The concentration of the pro-   Hill *et al.* (1971a)
   duct of 7α-dehydroxylation of
   cholic acid (deoxycholic acid)
   correlates with bowel cancer
   incidence.

3. The products of 7α-dehydroxy-   see Table 6
   lation of the primary bile
   acids are both promoters of
   carcinogenesis

4. The carriage rate of   Hill *et al.* (1975)
   clostridia with Δ⁴-dehydro-   Blackwood *et al.* (1978)
   genase is higher in colo-
   rectal cancer cases than in
   controls

B. The Stomach

The incidence of stomach cancer is highest in countries in
eastern Asia, in the Andean countries of south and central
America, in Eastern Europe and in Iceland; in all of these
areas the incidence of large bowel cancer is low and in those
countries where large bowel cancer is common the incidence of
gastric cancer is relatively low. The epidemiology of gastric cancer has been reviewed recently by Correa (1982).

Cancer of the stomach undoubtedly has many causes and has
at least 2 histological types (the intestinal type and the
diffuse type on the classification of Lauren, 1965). Correa
et al. (1975) postulated the aetiology of gastric cancer of
the intestinal type, in which the first stage was the development of gastric atrophy, followed by atrophic gastritis,
intestinal metaplasia, dysplasia of increasing severity and
finally carcinoma. In addition, there have been a number of
studies implicating dietary nitrate in the causation of gastric cancer, reviewed by Fraser *et al.* (1980). Correa *et al.*
postulated that nitrate may in some way be responsible for

the initial lesion, the gastric atrophy which results in a profuse resident flora being established in the stomach. This resident flora would then reduce dietary nitrate to nitrite and then catalyse the N-nitrosation of secondary amines in the gastric juice. The N-nitroso compounds would be responsible for the development of intestinal metaplasia, then dysplasia and finally neoplasia in the gastric mucosa.

If this hypothesis is correct 2 conclusions can be drawn immediately. The first is that gastric atrophy (and eventually pernicious anaemia) should be more prevalent in populations exposed to high intakes of nitrate than in the general population; the second is that persons with gastric achlorhydria for any reason should be at high risk of developing gastric cancer. The second of these conclusions is undoubtedly true. It is well established that persons with pernicious anaemia have a risk of gastric cancer 4–6 times that in the general population (Blackburn *et al.*, 1968; Mosbech and Videback, 1950); similarly it is now established that persons who have had a Polya partial gastrectomy have an increased risk of gastric cancer after a latent period of 15 –20 years (Taksdal and Stalsberg, 1973). Ruddell *et al.* (1978) have shown that the gastric juice analyses are compatible with the cause of the increased malignancy in pernicious anaemia being due to the formation of N-nitroso compounds (Table 9) and similarly evidence has been obtained

TABLE 9

The evidence for N-nitroso compounds as a cause of the high risk of gastric cancer associated with pernicious anaemia (P.A.)
(*Ruddell* et al., *1978*)

---

1. Patients with P.A. have a profuse gastric flora rich in nitrate reductase. Control persons have a very sparse bacterial flora.

2. The concentration of nitrite is high in the gastric juice of P.A. patients but very low in controls. The nitrite concentration in P.A. patients increases with the degree of epithelial dysplasia.

3. Gastric juice is rich in N-nitrosatable amines.

4. Gastric juice of P.A. patients contains high N-nitroso compound concentrations.

---

from partial gastrectomy patients. The aetiology of the precursor lesion is less clear. There is no evidence yet that,
for example, persons industrially exposed to nitrate have an
increased risk of pernicious anaemia and it is more likely
that the very high prevalence of gastric achlorhydria
observed in young adults in the Andean countries is due to
malnutrition (Thomasen *et al.*, 1981). There have been no
reports on the effect of high nitrate intake on the gastric
mucosa of experimental animals. The role of bacteria and
nitrate in human gastric cancer has been reviewed recently by
Hill (1981b).

C. The Biliary Tree and Pancreas

Lowenfels (1978) postulated that cancer of the biliary tree
was caused by bacterial metabolites of the bile acids produced *in situ* in the gall bladder or biliary tree. In support
of this he noted that a high proportion of persons with cancer of the biliary tree have associated infections of the
gall bladder. These infections could have been secondary to
the tumour and so it is of interest that Welton *et al.* (1979)
showed that carriers of *Salmonella typhi* (where the organisms
are resident in the biliary tract) have a subsequent risk of
cancer of the biliary tract 6 times that in the normal population. Gall stones predispose to gall bladder cancer and
they are also associated with gall bladder infection.
   Although the basic hypothesis is attractive, the proposed
mechanism is less so; the bile acids are tumour promoters
rather than initiators and they are relatively slowly degraded
by bacteria even under intestinal conditions. However,
Kinoshita and Gelboin (1978) showed that the deconjugation of
benz(a)pyrene-3-glucuronide by bacteria led to the formation
of a short-lived intermediate able to bind covalently to DNA
(and therefore potentially carcinogenic). Many environmental
carcinogens are enterohepatically circulated and are secreted
in bile as glucuronide conjugates. In bile they are at
relatively high concentration and the organisms commonly
associated with biliary infection are good producers of β-
glucuronidase (although it has still to be demonstrated that
the enzyme is active in whole bile). These appear, therefore, to be better candidates for the bacterial release of
carcinogens in the biliary tract.

D. The Breast and Endometrium

The case for suspecting that bacterial metabolites produced
in the colon might be implicated in the causation of cancer
of the breast and of the endometrium is largely epidemio-

TABLE 10

The association between cancers of the
colon, breast and endometrium.

Geographical ·	All 3 cancers have the same distribution internationally, within Europe and within the British Isles.
Relation to social class	All 3 cancers increase in risk with increasing socio-economic status.
Urban-rural differences	All 3 are more common in urban than in rural areas.
Migrant studies	When migrants move from a low incidence to a high incidence area their risk of all 3 cancers increases within their lifetime.
Association within persons	A woman who has had a primary cancer successfully treated in one site is at above average risk of developing a primary in the other 2 sites.

logical. Cancers of the colon, the breast and the endometrium are associated geographically and demographically
(Table 10). In families with the "cancer family syndrome"
women who develop colon cancer have an extremely high risk of
developing a subsequent endometrical cancer (Lynch *et al.*,
1979). Intestinal bacteria are able to produce oestrogen
analogues from the biliary steroids (Goddard and Hill, 1972)
and it has been postulated that these might be implicated in
the promotion of oestrogen-dependent breast cancers (Hill *et
al.*, 1971b) and presumably also of oestrogen-dependent endo-
metrial cancers. In support of this, Blackwood *et al.* (1978)
have shown that the carriage rate of the organism to produce
oestrogen analogues was very much higher in breast cancer
patients than in controls.

E. The Bladder

Bryan (1971) has reviewed the data implicating urinary tryptophan metabolites (UTM) in the causation of bladder cancers
not associated with industrial exposures and there was

considerable interest in this hypothesis at that time. Many
of the UTMs were shown to be bladder carcinogens using the
bladder implantation technique which implies that chronic
irritation of the bladder may also be necessary. There is no
evidence of this and so interest in the role of UTMs has
waned in recent years.

In contrast, the role of bacteria in the bladder carcinogenesis associated with bilharzia is arousing considerable
current attention following the report by Hicks *et al.* (1977)
that patients with bilharzial infection of the bladder have
an associated bacterial cystitis and high levels of urinary
N-nitroso compounds. Bacterial infection of the bladder has
been associated with the demonstration of the presence of
urinary N-nitroso compounds (Table 11); this would be expected

TABLE 11

Demonstration of N-nitroso compounds of urine
of patients with urinary tract infection.

Reference	Observation
Brooks *et al.* (1972)	N-nitrosodimethylamine in urine of patients with Proteus infection of the urine.
Hicks *et al.* (1977)	N-nitroso compounds in the urine of persons with bacterial urinary tract infection secondary to bilharzia.
Radomski *et al.* (1978)	N-nitroso compounds in urine of persons with bladder infection.
Kazikoe *et al.* (1979)	Volatile N-nitroso compounds in urine of patients with bladder cancer.

since the urine is the major route of excretion of nitrate
and of secondary amines and other N-nitrosatable compounds.
In western populations most urinary tract infections are with
Escherichia coli (80% of cases) or *Proteus* spp (10% of cases),
which are producers of a highly active nitrate reductase and
which catalyse the N-nitrosation of secondary amines. However, in these simple infections the N-nitroso compounds are
not activated and this is in accord with the epidemiological
evidence which shows no association between urinary tract
infection and subsequent bladder cancer. In contrast, the
bladder infections associated with bilharzia are mixed ones

involving many bacterial species and so there is ample oppor-
tunity not only for the production of N-nitroso compounds but
also for their subsequent activation.

If the bladder cancer associated with bilharzia infection
is due to N-nitroso compounds, then it might be preventable
by the administration of nitrite scavengers such as ascorbic
acid or α-tocopherol.

F. The Colon in Patients with Ureterocolic Anastomosis

In patients who have had urine diversion to the sigmoid colon
the situation is an extreme form of that in bilharzial
patients. Thus the urine, rich in nitrate and nitrosatable
amino compounds, is conducted into the sigmoid colon with its
rich and diverse bacterial flora. These are ideal conditions
for the bacterial production of N-nitroso compounds. In addition, the urine is the final route of excretion of the glucuronide and sulphate conjugates of the polycyclic aromatic
hydrocarbons. If bacteria are implicated in human carcinogenesis then these patients surely are the most likely to
provide clear cut evidence. It was not surprising, therefore, that these patients have a very high risk of colon
carcinogenesis at the anastomotic site (Stewart *et al.*, 1981)
with a latency of 15—20 years (note the similarity to the
latency in patients with Polya partial gastrectomy where N-
nitroso compounds were also implicated).

The role of bacteria and N-nitroso compounds in the carcinogenesis in these patients is currently being investigated;
the ureterocolic anastomosis has been superseded as the operation of choice by the ileal loop and, more recently, the
colon loop. If bacteria are implicated, then preliminary
results on the bacteriology of loop urine indicates that the
colon loop should be by far the safest operation from the
carcinogenesis point of view.

G. The Uterine Cervix

The incidence of cancer of the cervix is very high in the
Bantu populations of Southern Africa, where there is also a
high prevalence of infected cervical discharges, principally
due to the flagellate *Trichomonas vaginalis*. This apparent
association was investigated by Allsobrook *et al.* (1975) who
demonstrated the presence of N-nitroso compounds on the
vaginal discharge. Since nitrate is secreted in many other
body secretions (e.g. sweat, tears, saliva, gastric juice) it
might be expected to present in vaginal secretion and similarly secondary amines might also be expected to be present.
The normal vaginal flora contains very few organisms able to

reduce nitrate but in *T. vaginalis* infection the bacterial
flora is modified to contain more of the nitrate-reducing
species (e.g. micrococci, *E. coli*). Thus, the conditions are
suitable for the production of N-nitroso compounds and if
these are responsible for the cervical cancers then a latency
of 15—20 years following *T. vaginalis* infection would be
expected. The relevant epidemiological studies have yet to
be done. However, if future research establishes a causal
relationship between *T. vaginalis* infection and cervical cancer mediated by the bacterial production of N-nitroso compounds, it clearly implies that more vigorous early treatment
of vaginal infection is necessary in these high risk populations.

IV. CONCLUSIONS

In this brief review of the possible role of bacteria in
human cancer I have summarized the bacterial flora at various
sites in the body, then described the data on the production
of carcinogens, promoters or mutagens by bacteria and then
considered the possible role of bacteria in the causation of
cancer at various sites. This is a very young subject and we
have still to demonstrate that bacteria are important in carcinogenesis at any site. However, it is important that the
possibility be investigated fully, because if such a role for
bacteria can be shown, then such cancers, like any other bacterial disease, are potentially preventable.

V. ACKNOWLEDGEMENTS

This work has been financially supported by the Cancer
Research Campaign to whom I express my gratitude.

VI. REFERENCES

Allsobrook, A.J.R., Du Plessis, L.S., Harington, J.S., Nunn,
    A.J. and Nunn, J.R. (1975). *In* "N-nitroso Compounds in
    the Environment" (Eds P. Bogovski and E.A. Walker), p. 197.
    IARC, Lyon.
Aries, V.C., Crowther, J.S., Drasar, B.S., Hill, M.J. and
    Williams, R.E.O. (1969). Bacteria and the aetiology of
    cancer of the large bowel. *Gut* 10, 334-335.
Badger, G.M., Cook, J.W., Hewett, C.L., Kennaway, E.L.,
    Kennaway, N.M., Martin, R.H. and Robinson, A.M. (1940).
    Proc. Roy. Soc. B. **129**, 439-467.

Blackburn, E.L., Callender, S.T., Dacie, J.V. *et al.* (1968).
Possible association between pernicious anaemia and leukaemia: a prospective study of 1625 patients with a note on
the very high incidence of gastric cancer. *Int. J. Cancer*
3, 163-170.

Blackwood, A., Murray, W.R., Mackay, C. and Calman, K. (1978).
Faecal bile acids and clostridia in the tiology of colorectal cancer and breast cancer. *Br. J. Cancer* **38**, 175.

Bone, E., Tamm, A. and Hill, M. (1976). The production of
urinary phenols by gut bacteria and their possible role in
the causation of large bowel cancer. *Am. J. Clin. Nutr.*
29, 1448-1554.

Brooks, J.B., Cherry, W.B., Thacker, L. and Alley, C.C.
(1972). Analysis by gas chromatography of amines and
nitrosamines produced *in vivo* and *in vitro* by *Proteus
mirabilis*. *J. Inf. Dis.* **126**, 143.

Bryan, G.T. (1971). The role of urinary tryptophan metabolites in the aetiology of bladder cancer. *Am. J. Clin.
Nutr.* **24**, 841-847.

Correa, P. (1982). Epidemiology of gastric cancer and its
precursor lesions. *In* "Gastrointestinal Cancer" (Eds.
J. De Coss and P. Sherlock), Martinus Nijhoff, Amsterdam,
pp. 119-130.

Correa, P. and Haenszel, W. (1978). The epidemiology of
large-bowel cancer. *Adv. Cancer Res.* **26**, 1-141.

Correa, P., Haenszel, W., Cuello, C., Tannenbaum, S. and
Archer, M. (1975). A model for gastric cancer epidimi-
ology. *Lancet* **ii**, 58-60.

Cummings, J.H., Hill, M.J., Bone, E.S., Branch, W.J. and
Jenkins, D.J.A. (1979). The effect of meat protein with
and without dietary fibre on colonic function and metabolism. II. Bacterial metabolites in faeces and urine. *Am.
J. Clin. Nutr.* **32**, 2094-2101.

Drasar, B.S., Shiner, M. and McLeod, G.M. (1969). The bacterial flora of the gastrointestinal tract in healthy and
achlorhydric persons. *Gastroenterology* **56**, 71-77.

Farber, E., McConomy, J., Franzen, B., Marroquin, F., Stewart,
G.A. and Magee, P.N. (1967). Interaction between ethionine and rat liver RNA and protein *in vivo*. *Cancer Res.*
27, 1761-1772.

Fisher, J.F. and Malleter, M.F. (1961). The natural occurrence of ethionine in bacteria. *J. Gen. Physiol.* **45**, 1-13.

Fraser, P., Chilvers, C., Beral, V. and Hill, M.J. (1980).
Nitrate and human cancer: a review of the evidence. *Int.
J. Epidim.* **9**, 3-11.

Goddard, P. and Hill, M.J. (1972). Degradation of steroids
by intestinal bacteria. IV. The Aromatization of ring A.
Biochim. Biophys. Acta **280**, 336-342.

Harington, J.S., Nunn, J.R. and Irwig, L. (1973). Dimethyl-
    nitrosamine in the human vaginal vault. *Nature* (London),
    241, 49-50.

Hawksworth, G.M. and Hill, M.J. (1971). Bacteria and N-nitro-
    sation of secondary amines. *Br. J. Cancer* **25**, 520-526.

Hicks, R.M., Walters, C.L., Elsebai, I., El Aasser, A.-B.,
    El Merzebani, M. and Gough, T.A. (1977). Demonstration of
    nitrosamines in human urine. Preliminary observation on a
    possible aetiology for bladder cancer in association with
    chronic urinary tract infections. *Proc. Roy. Soc. Med.*
    70, 413-416.

Hill, M.J. (1975). The aetiology of colon cancer. *Crit. Rev.
    Toxicol.* **4**, 31-82.

Hill, M.J. (1977). The role of unsaturated bile acids in the
    tiology of large bowel cancer. *In* "Origins of Human Can-
    cer" (Eds H. Hiatt, J. Watson and J. Winsten), Cold Spring
    Harbor Lab. Press, New York. pp. 1627-1640.

Hill, M.J. (1980). Conservation of bile acids. *In* "Drugs
    Affecting Lipid Metabolism" (Eds F. Fumagalli, D. Kritch-
    evsky and R. Paoletti), Elsevier, Amsterdam. pp.89-96.

Hill, M.J. (1981a). Metabolic epidemiology of large bowel
    cancer. *In* "Gastrointestinal Cancer" (Eds J. De Cosse and
    P. Sherlock), Martinus Nijhoff, The Hague, pp. 187-226.

Hill, M.J. (1981b). Nitrates and bacteriology. Are these
    important aetiological factors in gastric carcinogenesis?
    In "Gastric Cancer" (Eds J. Fielding, C. Newman, C. Ford
    and B. Jones), Pergamon, Oxford, pp.35-46.

Hill, M.J. (1982). Lipids, intestinal flora and large bowel
    cancer. *In* "Dietary Fats and Health" (Eds E.G. Perkins,
    W.J. Visek and J.C. Lyon), A.O.C.S. (In press).

Hill, M.J. (1982a). Genetic and environmental factors in
    human colorectal cancer. *In* "Colon Carcinogenesis" (Eds
    R. Malt and R. Williamson), MTP, Lancaster, pp. 73-81.

Hill, M.J. (1982b). Influence of nutrition on intestinal
    flora. *In* "Colon and Nutrition" (Eds H. Goebell and H.
    Kaspar), MTP, Lancaster. pp. 37-46.

Hill, M.J., Drasar, B.S., Aries, V.C., Crowther, J.S.,
    Hawksworth, G.M. and Williams, R.E.O. (1971a). Bacteria
    and aetiology of cancer of large bowel. *Lancet* i, 95-100.

Hill, M.J., Goddard, P. and Williams, R.E.O. (1971b). Gut
    bacteria and aetiology of cancer of the breast. *Lancet*
    ii, 472-473.

Hill, M.J., Drasar, B.S., Williams, R.E.O., Meade, T.W., Cox,
    A.G., Simpson, E.P. and McKorson, B.C. (1975). Faecal
    bile acids and clostridia in patients with cancer of the
    large bowel. *Lancet* i, 535-538.

Hofmann, A.G. (1977). The enterohepatic circulation of bile
    acids in man. *Clin. Gastroenterol.* **6**, 3-24.

Hurley, R., Stanley, V., Leask, B. and DeLouvois, J. (1974).
    Microflora of the vagina during pregnancy. *In* "The Normal
    Microbial Flora of Man" (Eds E. Skinner and L. Carr),
    Academic Press, London, pp. 155–186.
Kazikoe, T., Wang, T.T., Eng, V.W., Furrer, R., Dion, P. and
    Bruce, W.R. (1979). Volatile N-nitrosamines in the urine
    of normal donors and of bladder cancer patients. *Cancer
    Res*. **39**, 829–832.
Kinoshita, N. and Gelboin, H. (1978). β-glucuronidase cata-
    lysed hydrolysis of benzo(a)pyrene-3-glucuronide and bind-
    ing to DNA. *Science* **199**, 307–309.
Laqueur, G.L. and Spatz, M. (1968). Toxicology of cycasin.
    Cancer Res. **28**, 2262–2267.
Lauren, P. (1965). The two histological main types of gastric
    carcinoma: diffuse and so-called intestinal type. *Acta
    Path. Microbiol. Scand*. **64**, 31–49.
Lowenfels, A. (1978). Does bile promote extra-colonic cancer?
    Lancet **ii**, 239–241.
Lynch, H.T., Giurgis, H.A., Harris, R.E., Lynch, P.M., Lynch,
    J.F., Elston, R.C., Go, R.C.P. and Kaplan, E. (1979).
    Clinical, genetic and biostatistical progress in the cancer
    family syndrome. *Front. Gastro. Res*. **4**, 142–150.
McKillop, C. (1981). Ph.D. Thesis, Liverpool Polytechnic.
Mastromarino, A., Reddy, B.S. and Wynder, E.L. (1976). Meta-
    bolic epidemiology of colon cancer: enzymic activities of
    the faecal flora. *Am. J. Clin. Nutr*. **29**, 1455–1460.
Mosbech, J. and Vedebaek, A. (1950). Mortality from and risk
    of gastric carcinoma among patients with pernicious
    anaemia. *Br. Med. J*. **ii**, 390–394.
Radomski, J.L., Greenwald, D., Hearn, W.L., Block, N.L. and
    Woods, F.M. (1978). Nitrosamine formation in bladder
    infections and its role in the aetiology of bladder cancer.
    J. Urol. **120**, 48.
Renwick, A.G. and Drasar, B.S. (1976). Environmental carcino-
    gens and large bowel cancer. *Nature* **263**, 234–235.
Ruddell, W.S.J., Bone, E.S., Hill, M.J. and Walters, S.C.L.
    (1978). Pathogenesis of gastric cancer in pernicious
    anaemia. *Lancet* **i**, 521–523.
Sander, J. (1968). Nitrosaminsynthese durch Bakterien.
    Hippe Seylers *Z. Physiol. Chem*. **349**, 429–432.
Schmidt, E.G. (1949). Urinary phenols. IV. The simultaneous
    determination of phenol and p-cresol in urine. *J. Biol.
    Chem*. **179**, 211.
Sinclair, T. and Tuxford, A.F. (1971). Bladder infection at
    rate of 184/1000 per annum in a rural general practice.
    The Practitioner **207**, 81–90.
Skinner, F.A. and Carr, J.G. (1974). The Normal Microbial
    Flora of Man. Academic Press, London.

Smith, R.L. (1966). The biliary secretion and enterohepatic circulation of drugs and other organic compounds. *Progr. Drug Res.* **9**, 300-360.

Stephen, A.M. and Cummings, J. (1980). Mechanism of action of dietary fibre in the human colon. *Nature* **284**, 283-284.

Stewart, M., Pugh, R. and Hill, M.J. (1981). *Br. J. Urol.* **53**, 115-118.

Taksdal, S. and Stalsberg, T. (1973). Histology of gastric carcinoma occurring after gastric surgery for benign conditions. *Cancer* **32**, 162-166.

Thomasen, H., Burke, V., Gracey, M. (1981). Impaired gastric function in experimental malnutrition. *Am. J. Clin. Nutr.* **34**, 1278-1280.

Welton, J.C., Marr, J.S. and Friedman, S.M. (1979). An association between hepatobiliary cancer and the typhoid carrier state. *Lancet* **i**, 791-794.

Wieland, H. and Dane, E. (1933). The constitution of bile acids. III. The place of attachment of the side chain. *Z. Physiol. Chem.* **219**, 240-244.

POTENTIAL USE OF MURAMYL PEPTIDES IN CANCER THERAPY AND PREVENTION

Louis Chedid, Andre Morin and Nigel Phillips

Immunotherapie Experimentale, Institut Pasteur, 28, rue du Dr Roux, 75724 Paris Cedex 15, France

I. INTRODUCTION

It is well established that the use of whole microorganisms endowed with immunomodulating properties such as BCG or *C. parvum* can be beneficial in a number of cancer patients (for reviews see Holmes *et al.*, 1977; Baldwin and Pimm, 1978; Baldwin and Byers, 1979). However, these agents can elicit a variety of side effects such as granuloma formation, lymphoid hyperplasia, adjuvant arthritis, sensitization to mycobacterial antigens and potentiation of the toxic effects of some endotoxins (Chedid *et al.*, 1978). Isolating and characterizing the minimal bacterial cell wall structure necessary for biological activity but which is devoid of most of the untoward side-effects has represented a major advance in the field of biological adjuvants deriving from bacteria (Johnson *et al.*, 1978). The smallest biological-active cell wall moiety has been shown to be a simple glycopeptide containing a bacterial-specific carbohydrate, muramic acid, and two amino acids, L-alanine and D-isoglutamine (Lederer, 1980). This glycopeptide (muramyl dipeptide or MDP) has been synthesized, and to date, more than 300 analogues have been prepared (Dukor *et al.*, 1979; Lefrancier and Lederer, 1981; Adam *et al.*, 1981) from which a large number has been tested for activity in our laboratory (Chedid *et al.*, 1978).

Since the original molecule has pyrogenic activity, it has been of interest that certain immunologically active derivatives are devoid of pyrogenicity (Chedid *et al.*, 1982). One of these, the butyl ester derivative of MDP, MDP (Gln)-O*n*Bu, has already been administered to humans without producing any toxic side-effects or alteration of clinical or biological parameters (Oberling *et al.*, 1981). This paper will focus on the potential utilization of the muramyl peptides both as

therapeutic and as preventative agents in the treatment of
human malignancies.

II. ANTITUMOUR ACTIVITIES OR MURAMYL PEPTIDES

The ability of muramyl peptides to nonspecifically inhibit
tumour growth in a variety of experimental models has been
recently reviewed (Lederer and Chedid, 1982; Chedid and Morin,
1982). For the induction of specific or nonspecific cell-
mediated immune response — a mechanism primarily involved in
tumour rejection — the use of muramyl peptides in water-in
mineral oil-emulsions, e.g. Freund's complete adjuvant, or
more metabolizable oils such as squalane has usually been
required. Thus, when injected in combination with trehalose-
dimycolate (TDM), a glycolipid also derived from mycobacterial
cell wall, MDP and a range of analogues have been shown to
inhibit the growth of a hepatocarcinoma intradermally transplanted in guinea pigs (McLaughlin *et al.*, 1980; Yarkoni *et
al.*, 1981). However, neither of the two agents alone could
exert such an antitumour effect.

   Inclusion of MDP within liposomes results in nonspecific
tumour activity. In a metastatic melanoma model in mice,
Fidler *et al.*(1981) have demonstrated that repeated i.v.
injections of liposome-encapsulated MDP, following primary
tumour excision, increased both the number of animals free of
lung metastasis and the survival time. Since the same
authors have found that the tumouricidal activity of alveolar
macrophages could be increased by treatment with MDP in liposomes to a higher degree than with free MDP (Sone and Fidler,
1981), the activation of these cells is likely to be involved,
at least in part, in the beneficial antitumour effects
observed with liposome-encapsulated MDP. In addition, Sone
and Fidler (1980) have reported that a combination of sub-
threshold amounts of MDP and macrophages activating factor
(MAF) could act synergistically in inducing rat alveolar
macrophages to become tumouricidal for syngeneic, allogeneic
and xenogeneic target cells. Encapsulation of both agents
within liposomes resulted in a far higher degree of macrophage activation than that obtained with the free agents.
Although a larger clinical use of liposomes is still a matter
of controversy, such carriers have been satisfactorily employed in humans. Since it has been shown that the butyl
ester derivative of MDP can be safely used in humans, it is
possible to consider the evaluation of Fidler's observations
in human cancers relevant to the murine melanoma model studied.
   Specific immunization against tumour cells using lipophilic
muramyl peptide derivatives such as 6-*0*-acetyl esters and
quinonyl esters of MDP has also been studied. Both types of

derivative were capable of inhibiting in mice the growth of
transplanted hepatoma or methylcholanthrene-induced fibrosarcoma tumours upon simultaneous injection with the tumour
cells. These derivatives have also been shown to augment
cell-mediated cytotoxicity to allogeneic target cells in mice
(Yamamura et al., 1977; Azuma et al., 1979; Uemiya et al.,
1979; Saiki et al., 1981). These observations could be of
interest where specific immunization against certain neoplasms
is considered as a therapeutic procedure. However, such an
approach may be dependent on the purification of neoantigens
associated with a particular malignancy.

An inhibitory activity of muramyl dipeptides on the growth
of murine lymphoid tumours has recently been observed
(Phillips et al., 1982). This study evaluated the ability of
two muramyl dipeptides, MDP and its adjuvant-inactive stereo-
isomer MDP(D-D) (N-acetylmuramyl-D-alanyl-D-isoglutamine), to
inhibit the in vivo growth of a number of murine ascitic cell
lines. The lymphomas studied, a range of thymoma cell lines
possessing phenotypic antigens characteristic of helper
(LYT-1^+2^-), cytotoxic/suppressor (LYT-1^-2^+) or precursor
(LYT-1^+2^+) T-cells (Mathieson et al., 1978), could be of use
not only in characterizing the tumour cells for muramyl
dipeptide activity, but also the efficacy by which such
adjuvants are able to modulate their growth. The occurrence
within the same mouse strain of B-lymphocyte and plasma cell
lymphomas has enabled a comparative study of the effects of
muramyl dipeptides on tumours of both T- and B-lymphocyte
lineage to be carried out. The ability of the two muramyl
dipeptides to affect the growth of the lymphomas was compared
with that of the known cytotoxic agent, cyclophosphamide.
The doses required to effect 50% inhibition (ID_{50}) of ascitic
tumour growth are shown in Table 1. The ascitic growth of
the cell line expressing the helper phenotype was not inhibited by MDP or MDP(D-D) in the dose range studied (0.06 to
5 mg/kg/day via the i.p. route). The growth of 2 out of 3
cell lines expressing the cytotoxic/suppressor phenotype and
both cell lines expressing the precursor phenotype was inhibited by MDP. MDP(D-D) had no effect on the growth of these
cell lines. Cyclophosphamide treatment in the dose range
1.25 to 20 mg/kg/day resulted in a significant inhibition of
growth of all the thymoma cell lines, with almost identical
ID_{50} values being observed. In contrast to the results
obtained with the thymoma cell lines, MDP had no significant
effect on the growth of the B-cell or plasmacytoma cell lines.
However, significant inhibition of the growth of these cell
lines was observed with MDP(D-D).

A number of interesting observations have arisen from this
study. Firstly, where a tumour cell line is responsive to

TABLE 1

Inhibition of murine growth in vivo by muramyl dipeptides

Tumour	Phenotype[1]	ID_{50}, mg/kg/day i.p.[2]			ID_{50} cyclophosphamide / ID_{50} muramyldipeptide	
		MDP	MDP(D-D)	Cyclo-phosphamide	MDP	MDP(D-D)
BALENTL 13	Helper (LYT-1$^+$2$^-$)	Inactive[3]	Inactive	3.0	-	-
BALENTL 3	Cytotoxic suppressor (LYT-1$^-$2$^+$)	Inactive	Inactive	4.5	-	-
BALENTL 5	Cytotoxic suppressor (LYT-1$^-$2$^+$)	0.35	Inactive	2.0	7.0	-
BALENTL 7	Cytotoxic suppressor (LYT-1$^-$2$^+$)	0.10	Inactive	2.0	20	-
BALENTL 9	Precursor (LYT-1$^+$2$^+$)	0.15	Inactive	3.0	20	-
P1798	Precursor (LYT-1$^+$2$^+$)	0.20	Inactive	4.0	20	-
ABPL 2	IgM - cytoplasmic	Inactive	0.60	8.0	-	13
MOPC 173	IgM - cytoplasmic + secreted	Inactive	0.20	20.0	-	20

[1] LYT-1$^+$2$^-$ is a phenotypic marker of T-helper cells; LYT-1$^+$2$^+$ is a phenotypic marker of T-cyto-toxic/suppressor cells; LYT-1$^-$2$^+$ is a phenotypic marker of T-precursor cells. ABPL 2 is a B-cell lymphoma, MOPC 173 is a plasmacytoma.

[2] ID_{50} = dose necessary to obtain 50% inhibition of ascitic tumour growth for a dose period of 4-8 days.

[3] No significant inhibition of growth found at up to 5.0 mg/kg/day.

treatment with MDP or MDP(D-D), the muramyl dipeptide is some
7 to 10-fold more effective than cyclophosphamide in inhibiting its growth. Secondly, the ability of MDP to inhibit the
growth of the thymoma cell lines does not appear to correlate
with the LYT-phenotype expressed by these cells. The thymoma
cell lines studied have been shown to have a wide variation
in the expression of the enzyme terminal deoxy-nucleotide
transferase (TdT) (Jin Kim *et al.*, 1978). The 2 thymoma cell
lines where MDP was inactive have little or reduced TdT activity in comparison with the high levels found for those cell
lines where MDP was active. Such levels are characteristic
of immature T-lymphocytes (Kung *et al.*, 1975), and the ability
of MDP to inhibit thymoma growth may correlate with the degree
of maturity of the cell, rather than the observed LYT-phenotype expressed by such cells. Thirdly, the results would
suggest that specific inhibition of lymphoma growth relative
to the lineage of the tumour cell (T- or B-lymphocyte) may be
possible using different muramyl dipeptides, offering the
potential of selective chemotherapy. Whilst it is difficult
to extrapolate to other tumour systems, such selectivity and
efficacy may have advantages over the use of established
chemotherapeutic agents in the treatment of lymphoid tumours.
Whilst the mechanisms by which the muramyl dipeptides exert
their antitumour effect is not known, preliminary studies
have indicated that the resident macrophage population within
the ascites plays a central role in the expression of activity.

III. MECHANISM OF ACTION OF MURAMYL PEPTIDES

A number of macrophage functions or activities can be stimulated by muramyl peptides including: (1) the production of
monokines such as the interleukin 1 (IL-1) lymphocyte activating factor (LAF) or endogenous pyrogens (EP); (2) increased
metabolic activities such as incorporation of glucosamine and
glucose oxidation; and (3) stimulation of prostaglandin,
collagenase and cyclic-AMP production. Stimulation of effector functions such as tumouricidal, phagocytic and bactericidal activities have also been described (for an extensive
review, see Leclerc and Chedid, 1981). Macrophage activation
does account, at least in part, for the nonspecific antitumour properties, and, as discussed below, the nonspecific
anti-infectious and adjuvant effects of muramyl peptides. It
seems that muramyl peptides can stimulate the cytotoxic
effects observed in systems which are thought to be involved
in host antitumour defences, such as antibody-dependent cell-
mediated cytotoxicity (ADCC) (Leclerc *et al.*, 1979) and T-
cell dependent cytotoxicity (Matter, 1979). Recently, Sharma
et al. (1981) have reported that MDP was also capable of

enhancing in mice the natural killer (NK) activity of spleen
cells. This was observed in CBA/J mice given high doses of
MDP either by the i.p. or i.v. route 3 days prior to testing
of splenic NK activity. Another interesting finding was that
a combined treatment with MDP prior to LPS resulted in a
higher splenic NK activity than that obtained by either of
the 2 agents alone.

In our own studies with another mouse strain (Morin and Le
Garrec, to be published) we have found that MDP could either
enhance or inhibit splenic NK activity, depending on the time
of administration. MDP(D-D) and MDP(Gln)-O*n*-Bu were able to
augment NK activity when injected 1 or 3 days before the test
respectively, whereas MDP conjugated to a synthetic carrier
(poly(DL-alanyl) -- poly(L-lysine, MDP-A-- L) was inhibitory
when injected 24 h previously. Experiments performed in
order to elucidate the mechanisms by which MDP exerts such a
dual modulating effect, indicated the following: (1) Both the
in vitro removal of adherent spleen cells by nylon wool filtration and the *in vivo* administration of indomethacin at
various times prior to MDP or MDP-A--L treatment were unable
to reverse the inhibitory effects of those agents, suggesting
that neither suppressor macrophages nor their prostaglandin
production are predominantly involved in the inhibition of NK
activity; (2) No circulating interferon could be detected in
the serum of mice treated with MDP 3 days before testing NK
activity. However, injection of a sheep antimouse-interferon
antibody simultaneously with MDP resulted in the abolition of
MDP-induced stimulation. These preliminary results suggest
that a local production of interferon might be involved in
the stimulation of NK activity by MDP.

IV. ENHANCEMENT OF RESISTANCE TO INFECTION BY
MURAMYL PEPTIDES

At the present time, intercurrent fatal infections are one of
the major threats that clinicians dealing with cancer patients
have to face. This is caused by the immunosuppression arising from antineoplastic therapy and perhaps by the neoplasia
itself. Muramyl peptides could be useful in the treatment
of patients with cancer for reasons other than the antitumour
activities described above. Particularly, such agents could
improve the anti-infectious defences in immunocompromised
patients. This could be achieved both through the enhancement of nonspecific resistance and through an augmentation of
specific immunization against a number of pathogens.

A. Enhancement of Nonspecific Anti-infectious Resistance

Patients treated by immunosuppressive drugs face a high risk of infection by a number of nosocomial agents including Gram-negative rods such as *Klebsiella pneumoniae, Pseudomonas aeruginosa, Escherichia coli, Staphylococcus aureus, Aspergillus* and *Candida* (Singer *et al.*, 1977; Young *et al.*, 1977; Edwards *et al.*, 1978). *Streptococcus pneumoniae* represents also a major cause of overwhelming postsplenectomy infections (O.P.S.I.) (Sullivan *et al.*, 1978). In addition, viral infections due to herpes-zoster virus and especially varicella in childhood leukaemia and other malignancies (Feldman *et al.*, 1975) are most common.

Since another facet of the biological activities of muramyl peptides is their ability to nonspecifically increase resistance against a number of experimental infections, this property could very well be taken of advantage in immunocompromised cancer patients. The protective capacity of MDP was first demonstrated in normal adult mice infected with *Klebsiella pneumoniae,* provided the adjuvant was administered either before or at the same time as the bacterial challenge (Chedid *et al.*, 1977). Interestingly, the nonspecific enhancement of resistance could be observed even when MDP was administered orally. Moreover, in contrast to LPS, MDP and some of its derivatives have been shown to exert a protective effect in neonate mice challenged with the same organism, and even with a strain rendered antibiotic-resistant by plasmid transfer (Parant *et al.*, 1978; Parant, 1979). MDP was also found capable of augmenting nonspecific resistance to a variety of organisms commonly found in immunocompromised hosts. These pathogens included *Salmonella typhimurium* (Dietrich *et al.*, 1980), *Streptococcus pneumoniae* (Humphres *et al.*, 1980), *Pseudomonas aeruginosa, Candida albicans* and *Listeria monocytogenes* (Fraser-Smith and Matthews, 1981). A protective effect of MDP has also been reported against parasitic infections by *Trypanosoma cruzii* (Kierszenbaum and Ferraresi, 1979) and *Toxoplasma gondii* (Krahenbuhl *et al.*, 1981). Such nonspecific anti-infectious effects of MDP have also been observed in animal models of immunosuppression, which are of special relevance to certain clinical situations. It is of interest to note that the administration of MDP or certain derivatives can reverse the increased susceptibility of immunosuppressed mice to a number of organisms (Parant, 1979; Parant and Chedid, 1980; Sackmann and Dietrich, 1980). In addition, both in normal and in immunocompromised infected mice, a synergistic effect between muramyl peptides and antibiotics has been demonstrated

(Dietrich *et al.*, 1980; Sackmann and Dietrich, 1980; Fraser-
Smith and Matthews, 1981).

B. Enhancement of Specific Immunization Against Pathogens

The adjuvant properties of muramyl peptides have been demonstrated not only with a range of antigens but also with
several vaccines (Chedid *et al.*, 1978; Gisler *et al.*, 1979;
Parant, 1979). The incidence of *Pseudomonas aeruginosa*
infection in patients with haematologic malignancies accounts
for about 50% of all bacterial septicemias, and greater than
50% of all bacterial pneumonias. Forty per cent of all
patients with Pseudomonas sepsis still die in spite of
improved anti-infectious management, and with *P. aeruginosa*
pneumonia, fatality may be as high as 80% (Reynolds *et al.*,
1975). For these reasons, the search for vaccines against
this pathogen has been actively pursued and a number are now
under evaluation. However, the vaccine antigens generally
contain lipopolysaccharide (endotoxin) and their injection is
associated with side-effects such as local pain and fever.

A nontoxic high-molecular weight polysaccharide has been
recently isolated from the slime of *P. aeruginosa*. Used as
an immunizing agent in mice, it induced a protective immunity
against a lethal challenge with the live organism, and in
human volunteers it evoked a significant increase in binding
and opsonic antibody titres (Pier *et al.*, 1978a, 1978b; Pier,
1982). The use of such an immunogen administered with a nontoxic adjuvant will be advantageous. In mice and rabbits,
MDP has been shown to be a suitable adjuvant when associated
with another type of vaccine preparation — a *P. aeruginosa*
toxoid obtained by formalin treatment of toxin A — which is
weakly immunogenic when injected alone (Cryz *et al.*, 1981).
Studies in humans of the potential of this type of adjuvant
associated with an antipseudomonas vaccine would appear to be
particularly justified.

Muramyl peptides could also be of interest in the specific
immunization against pneumococcal infections. Since streptococcal pneumonia represents the major cause of O.P.S.I.,
patients with anatomical or functional asplenia should receive
pneumococcal polysaccharide vaccines. Splenectomized patients
with Hodgkin's disease who have been treated with combined
radiation and chemotherapy have especially poor responses to
this vaccine (Sullivan *et al.*, 1978). They therefore represent a population which is likely to benefit from the co-ad-
ministration of such an adjuvant as MDP, since this type of
agent meets the requirements for safety and effectiveness
recently reviewed by Edelman *et al.*, 1980).

V. OTHER IMMUNOLOGICAL APPROACHES FOR THE PREVENTION AND THERAPY OF CANCER

A. Immunization against Hepatitis B Virus (HBV) Infection as a Means of Preventing Primary Hepatocellular Carcinoma (PHC)

A number of epidemiological studies have demonstrated a striking correlation between the occurrence of PHC and the presence of hyperendemic HBV. The bulk of evidence recently presented in a workshop held in Dakar, Senegal (April, 1980) clearly supports HBV as being an aetiological agent of PHC.

This cancer is uncommon in Europe and the USA (1.3/100,000) but has a high incidence in Tropical Africa and the Far East (up to 150/100,000). The incidence of HBV infection in European and American populations is low (0.1-1% HBsAg positive in blood donors) with an incidence of 5.3% HBsAg in PHC patients. In contrast, 10-25% of African and Far Eastern populations have HBs antigenemia (Maupas and Melnick, 1981). It is therefore reasonable to suppose that active immunization in early life against HBV infection would allow some control of PHC in such high risk populations. Because of the cost and scarcity of HBs antigen, a reduction in the amount of material required for immunization would be attractive. Such a goal could be achieved by association of the antigen with a potent adjuvant. Another alternative would be to obtain synthetic antigens. The first example of a successful immunization against diphtheria using an entirely synthetic vaccine in conjunction with muramyl dipeptide has recently been reported by Audibert *et al.* (1981). Other synthetic bacterial or viral antigens have since been described (Beachey *et al.*, 1981; Arnon *et al.*, 1980; Müller *et al.*, 1982). HBsAg peptides have also been synthesized, which not unexpectedly are weakly immunogenic (Lerner *et al.*, 1981; Dreesman *et al.*, 1982; Prince *et al.*, 1982). Therefore, the association with an adjuvant offers a number of potential advantages. In the study performed by Dreesman *et al.* (1982), mice were indeed immunized with uncoupled synthetic HbsAg peptides in the presence of different types of adjuvants, including MDP in liposomes.

B. Immunization against Hormones

Synthetic hormones such as the terminal fragment of human β-chorio-gonadotropin (β-HCG) or the hypothalamic factor LH-RH have been used as immunogens in view of either controlling fertility or increasing body weight of cattle by immuno-

logical castration. Such vaccines may, however, have a potential use in cancer therapy, chorio-eptitheliomas for instance.
Recently, studies in mice have shown that the association of
MDP with the hypothalamic factor LH-RH resulted in the production of anti-LH-RH antibodies. These antibodies were
capable of binding to LH-RH, and effected a castration of
male mice, as judged by vesicular weight, histological changes
occurring in testicular interstitial cells and seminiferous
tubules, and the disappearance of germ cells (Carelli *et al.*,
1982). Coupling of LH-RH with a muramyl dipeptide derivative,
N-acetyl-muramyl-L-alanine-D-isoglutamine-L-lysine (MTP)
resulted in still higher antibody titres. Interestingly,
coupling of the muramyl dipeptide derivative resulted in a
loss of pyrogenic activity. Moreover, no anti-MDP antibodies
were detected. Since the elevated levels of anti-LH-RH antibodies correlated reciprocally with a decrease in the level
of circulating LH, the immunological castration observed is
likely to be due, at least in part, to an inhibition of
pituitary LH-release. In those malignancies where an excess
of hormonal production, dependent on an hypersecretion of
hypothalamic LH-RH, is thought to be involved, the use of
synthetic hormone vaccines could thus be of use in performing
a selective immunological hypophysectomy rather than a global
one through surgery or irradiation.

VI. SUMMARY AND CONCLUSIONS

Due to their large spectrum of activity as demonstrated in a
variety of animal models, muramyl peptides may be particularly
applicable for the treatment of certain types of human malignancies. Several approaches for antitumourigenic control
can be envisaged. Nonspecific activation of macrophage
tumouricidal activity in metastatic sites, as demonstrated by
the use of liposome-encapsulated MDP, may be of use where
there is involvement of the lung and where adjuvant targeting
would be of considerable benefit. The marked synergism
between MDP and MAF demonstrated by Fidler and coworkers
should also be exploited. The specific immunization against
tumour cells demonstrated by the use of lipophilic derivatives
such as 6-*0*-acyl esters of MDP or quinonyl-MDP is also of
interest, but may well depend on the eventual identification
and purification of tumour-specific antigens. Such antigens
are likely to prove to be weakly immunogenic, and their
association with a safe and potent adjuvant will be of primary
importance. The possibility of utilizing muramyl peptides as
immunomodulating agents able to activate a number of host
defence mechanisms thought to be involved in the control of
neoplasia (NK activity, cytotoxic T-cell and killer cells) is

also of considerable interest. The demonstration of selectivity of muramyl dipeptides for lymphomas of T- or B-cell
lineage may offer the potential for the selective use of such
adjuvants against specific types of tumours.

Another major rationale for exploiting muramyl peptides in
the treatment of patients with cancer is that of stimulating
resistance to infection. The immunodepression observed in
cancer patients, resulting from tumour-induced mechanisms and
the effects of treatment with chemotherapeutic agents, may be
reversed by the administration of muramyl peptides. The
stimulation of anti-infectious resistance observed in immunocompromised host vis-à-vis a number of pathogens, and the
reduction in the amount of antibiotics necessary to protect
against infection gives the muramyl peptides an activity profile which may be of considerable benefit for cancer patients.

Muramyl peptides may be realistically considered as a
useful preventative agent of primary hepatocarcinoma, through
their potential for improving immunization against HBV infection. Such a preventive approach may be important not only
in terms of the cost and availability of purified antigenic
material, but also as a more general approach for other human
cancers associated with specific viral causative agents.

The demonstration of an effective immunological castration
resulting from the injection of muramyl dipeptide coupled to
LH-RH has illustrated the possibility of a more selective
approach. The immunological hypophysectomy thus obtained
could be advantageous as an alternative approach to more
standard procedures. Antibodies to human β-HCG have also been
obtained with MDP associated to a synthetic β-HCG fragment.
Such antibodies could be of use not only as a diagnostic tool,
but also as a therapeutic agent in cancers associated with an
excess of β-HCG production such as chorio-epithelioma.

The retention of the biological properties without associated pyrogenic activity in some muramyl dipeptide derivatives,
notably MDP(Gln)-OnBu, has opened up the possibility of their
therapeutic use in man. Initial toxicological studies have
shown that in rats and cynomolgus monkeys, this derivative
is without adverse side-effects at doses of up to 1 mg/kg
daily for 3 months (unpublished data). This tolerance has
now been confirmed in a phase I study performed initially in
23 cancer patients receiving up to 100 μg/kg of MDP(Gln)-OnBu
as a single subcutaneous injection. Since this study, additional patients have been administered the agent in the presence of an antigen in order to evaluate its adjuvant potency
(Oberling et al., to be published).

In conclusion, the multifaceted activity of structurally
well-defined agents such as muramyl peptides, coupled with
the absence of toxic side-effects observed with the use of

whole microorganisms, offers the potential for their employment in a controlled, safe and specific manner. Their antitumour activity, both specific and nonspecific anti-infectious
and immunomodulatory properties, present a number of possibilities not only for the rational treatment of cancer, but
also its control in those populations identified as being at
risk from aetiological agents.

VIII. REFERENCES

Adam, A., Petit, J.F., Lefrancier, P. and Lederer, E. (1981).
    Muramyl peptides: chemical structure, biological activity
    and mechanism of action. *Mol. Cell. Biochem.* **41**, 27-47.
Arnon, R., Sela, M., Parant, M. and Chedid, L. (1980). Anti-
    viral response elicited by a completely synthetic antigen
    with built-in adjuvanticity. *Proc. Natl. Acad. Sci. USA*
    77, 6769-6772.
Audibert, F., Jolivet, M., Chedid, L., Alouf, J.E., Boquet,
    P., Rivaille, P. and Siffert, O. (1981). *Nature* **289**, 593-
    594.
Azuma, I., Yamawaki, M., Uemiya, M., Saiki, I., Tanio, Y.,
    Kobayashi, S., Fukuda, T., Imada, I. and Yamamura, Y.
    (1979). Adjuvant and antitumour activities of quinonyl-
    N-acetylmuramyl dipeptides. *Gann* **70**, 847-848.
Baldwin, R.W. and Pimm, M.V. (1978). BCG in tumour immuno-
    therapy. *Adv. Cancer Res.* **28**, 91-147.
Baldwin, R.W. and Byers, V.S. (1979). Immunoregulation by
    bacterial organisms and their role in the immunotherapy of
    cancer. *Springer Semin. Immunopathol.* **2**, 79-100.
Beachey, E.H., Seyer, J.M., Dale, J.B., Simpson, W.A. and
    Kang, A.H. (1981). Type-specific protective immunity
    evoked by synthetic peptide of *Streptococcus pyogenes* M-
    protein. *Nature* **292**, 457-459.
Carelli, C., Audibert, F., Chedid, L. and Gaillard, J. (1982).
    Immunological castration of male mice by a totally syn-
    thetic vaccine (LH-RH conjugated to MDP) administered in
    saline. (Submitted).
Chedid, L., Parant, M., Parant, F., Lefrancier, P., Choay, J.
    and Lederer, E. (1977). Enhancement of nonspecific
    immunity to *Klebsiella pneumoniae* infection by a synthetic
    immunoadjuvant (N-acetyl-muramyl-L-alanyl-D-isoglutamine)
    and several analogues. *Proc. Natl. Acad. Sci. USA* **74**,
    2089-2093.
Chedid, L., Audibert, F. and Johnson, A.G. (1978). Biological
    activities of muramyl dipeptide, a synthetic glycopeptide
    analogous to bacterial immunoregulating agents. *Progr.
    Allergy* **25**, 63-105.

Chedid, L. and Morin, A. (1982). Current status of muramyl peptides. *In* "International Symposium on Current Concepts in Human Immunology and Cancer Immunomodulation" (Eds B. Serrou, C. Rosenfeld and J. Daniels), Elsevier/North Holland Biomedical Press, Amsterdam, (In press).

Chedid, L.A., Parant, M.A., Audibert, F.M., Riveau, G.J., Parant, F.J., Lederer, E., Choay, J.P. and Lefrancier, P.L. (1982). Biological activity of a new synthetic muramyl peptide adjuvant devoid of pyrogenicity. *Infect. Immun.* **35**, 417–424.

Cryz, S.J., Friedman, R.L., Pavlovskis, O.R. and Iglewiski, B.H. (1981). Effect of formalin toxoiding on *Pseudomonas aeruginosa* Toxin A: biological, chemical and immunochemical studies. *Infect. Immun.* **32**, 759–768.

Dietrich, F.M., Sackmann, W., Zak, O. and Dukor, P. (1980). Synthetic muramyl dipeptide immunostimulants: protective effects and increased efficacy of antibiotics in experimental bacterial and fungal infections in mice. *In* "Current Chemotherapy and Infectious Diseases" Proc. 11th ICC and 19th ICAAC (Eds J.D. Nelson and C. Grassi), American Society for Microbiology, Washington, **2**, 1730–1732.

Dreesman, G.R., Sanchez, Y., Ionescu-Matiu, I., Sparrow, J.T., Six, H.R., Peterson, D.L., Hollinger, F.B. and Melnick, J.L. (1982). Antibody to hepatitis B surface antigen after a single inoculation of uncoupled synthetic HBsAg peptides. *Nature* **295**, 158–160.

Dukor, P., Tarcsay, L. and Baschang, G. (1979). Immunostimulants. *Annu. Rep. Med. Chem.* **14**, 146–167.

Edelman, R., Hardegree, M.C. and Chedid, L. (1980). News: summary of an international symposium on potentiation of the immune response to vaccines. *J. Infect. Dis.* **141**, 103–112.

Edwards Jr, J.E., Lehrer, R.I., Stiehm, E.R., Fischer, T.J., and Young, L.S. (1978). Severe candidal infections: clinical perspective, immune defence mechanisms and current concepts of therapy. *Ann. Intern. Med.* **89**, 91–106.

Feldman, S., Hughes, W.T. and Daniel, C.B. (1975). Varicella in children with cancer: seventy seven cases. *Pediatrics* **56**, 388–397.

Fidler, I.J., Sone, S., Fogler, W.E. and Barnes, Z.L. (1981). Eradication of spontaneous metastases and activation of alveolar macrophages by intravenous injection of liposomes containing muramyl dipeptide. *Proc. Natl. Acad. Sci. USA* **78**, 1680–1684.

Fraser-Smith, E.B. and Matthews, T.R. (1981). Protective effects of muramyl dipeptides analogues against infections of *Pseudomonas aeruginosa* or *Candida albicans* in mice. *Infect. Immun.* **34**, 676–683.

Gisler, R.H., Dietrich, F.M., Baschang, G., Brownbill, A., Schumann, G., Staber, F.G., Tarcsay, L., Wachsmuth, E.D. and Dukor, P. (1979). New developments in drugs enhancing the immune response: activation of lymphocytes and accessory cells by muramyl dipeptide. *In* "Drugs and Immune Responsiveness" (Eds J.L. Turk and D. Parker), Macmillan Press Ltd., London, 133–160.

Holmes, E.C., Morton, D.L. and Eilber, F.R. (1977). Immunotherapy of cancer. *West. J. Med.* **126**, 102–109.

Humphres, R.C., Henika, P.R., Ferraresi, R.W. and Krahenbuhl, J.L. (1980). Effects of treatment with muramyl dipeptide and certain of its analogues on resistance to *Listeria monocytogenes* in mice. *Infect. Immun.* **30**, 462–466.

Jin Kim, K., Neinbaum, F.I., Mathieson, B.J., McKeever, P.E. and Asofsky, R. (1978). Characteristics of Balb/c T cell lymphomas grown as continuous *in vitro* lines. *J. Immunol.* **121**, 339–344.

Johnson, A.G., Audibert, F. and Chedid, L. (1978). Synthetic immunoregulating molecules: a potential bridge between cytostatic chemotherapy and immunotherapy of cancer. *Cancer Immunol. Immunother.* **3**, 219–227.

Kierszenbaum, F. and Ferraresi, R.W. (1979). Enhancement of host resistance against *Trypanosoma cruzi* infection by the immunoregulatory agent muramyl dipeptide. *Infect. Immun.* **25**, 273–278.

Krahenbuhl, J.L., Sharma, S.D., Ferraresi, R.W. and Remington, J.S. (1981). Effects of muramyl treatment on resistance to infection with *Toxoplasma gondii* in mice. *Infect. Immun.* **31**, 716–722.

Kung, P.C., Silverstone, A.E., McCaffrey, R.P. and Baltimore, D. (1975). Murine terminal deoxynucleotidyl transferase: cellular distribution and response to cortisone. *J. Exp. Med.* **141**, 855–865.

LeClerc, C. and Chedid, L. (1981). Macrophage activation by synthetic muramyl peptides. *Lymphokine Rep.* **3**, 1–21.

LeClerc, C., Juy, D., Bourgeois, E. and Chedid, L. (1979). *In vivo* regulation of humoral and cellular immune responses of mice by a synthetic adjuvant, N-acetyl-muramyl-L-alanyl-D-isoglutamine, Muramyl dipeptide for MDP. *Cell. Immunol.* **45**, 199–206.

Lederer, E. (1980). Synthetic immunostimulants derived from the bacterial cell wall. *J. Med. Chem.* **23**, 819–825.

Lederer, E. and Chedid, L. (1982). Immunomodulation by synthetic muramyl peptides and trehalose diesters. *In* "Immunological Approaches to Cancer Therapeutics". (Ed. E. Mihich), Wiley Interscience, New York, Chap. 4, 107–135.

Lefrancier, P. and Lederer, E. (1981). Chemistry of synthetic immunomodulant muramyl peptides. *Prog. Chem. Org. Natur. Prod.* **40**, 1-47.

Lerner, R.A., Green, N., Alexander, H., Liu, F.T., Sutcliffe, J.G. and Schinnick, T.M. (1981). Chemically synthesized peptides predicted from the nucleotide sequence of the hepatitis B virus genome elicit antibodies reactive with the native envelope protein of Dane particles. *Proc. Natl. Acad. Sci. USA* **78**, 3403-3407.

Mathieson, B.J., Campbell, P.S., Potter, M. and Asofsky, R. (1978). Expression of Ly 1, Ly 2, Thy 1 and TL differentiation antigens on mouse T cell tumours. *J. Exp. Med.* **147**, 1267-1279.

Matter, A. (1979). The effects of muramyl dipeptide (MDP) in cell-mediated immunity. A comparison between *in vitro* and *in vivo* systems. *Cancer Immunol. Immunother.* **6**, 201-210.

Maupas, P. and Melnick, J.L. (1981). Hepatitis B infection and primary liver cancer. *Prog. Med. Virol.* **27**, 1-5.

McLaughlin, C., Schwartzman, S.M., Horner, B.L., Jones, G.H., Moffatt, J.G., Nestor Jr, J.J. and Tegg, D. (1980). Regression of tumours in guinea pigs after treatment with synthetic muramyl dipeptides and trehalose dimycolate. *Science* **208**, 415-416.

Müller, G.M., Shapira, M. and Arnon, R. (1982). Anti-influenza response achieved by immunization with a synthetic conjugate. *Proc. Natl. Acad. Sci. USA* **79**, 569-573.

Oberling, F., Bernard, C., Chedid, L., Choay, J., Giron, C. and Lang, J.M. (1981). Phase I study of MDP in man. *In* "International Symposium on Immunomodulation by Microbial Products and Related Synthetic Compounds". Osaka, Japan. Abst. IX.6, p.41.

Parant, M. (1979). Biologic properties of a new synthetic adjuvant, muramyl dipeptide (MDP). *Springer Semin. Immunopathol.* **2**, 101-118.

Parant, M. and Chedid, L. (1980). Enhancement of nonspecific resistance to infections by synthetic adjuvants in newborn or in immunodeficient adult mice. *In* "International Symposium on Infections in the Immunocompromised Host", Veldhoven, The Netherlands. Abst. 28, p.44.

Parant, M., Parant, F. and Chedid, L. (1978). Enhancement of the neonate's nonspecific immunity to Klebsiella infection by muramyl dipeptide, a synthetic immunoadjuvant. *Proc. Natl. Acad. Sci. USA* **75**, 3395-3399.

Pier, G.B. (1982). Safety and immunogenicity of high-molecular weight polysaccharide vaccine from immunotype 1 *Pseudomonas aeruginosa*. *J. Clin. Invest.* (In press).

Pier, G.B., Sidberry, H.F., Zolyomi, S. and Sadoff, J.C. (1978a). Isolation and characterization of a high-molecular weight polysaccharide from the sline of *Pseudomonas aeruginosa*. *Infect. Immun*. **22**, 908-918.

Pier, G.B., Sidberry, H.F. and Sadoff, J.C. (1978b). Protective immunity induced in mice by immunization with high-molecular weight polysaccharide from *Pseudomonas aeruginosa*. *Infect. Immun*. **22**, 919-925.

Phillips, N., Paraf, A., Bahr, G., Modabber, F. and Chedid, L. (1982). Modulation of murine lymphoma growth *in vivo* by MDP, MDP(D-D) and cyclophosphamide. (Submitted).

Prince, A.M., Ikram, H. and Hopp, T.P. (1982). Hepatitis B virus vaccine: identification of HBsAg/a and HBsAg/d but not HBsAg/y subtype antigenic determinants on a synthetic immunogenic peptide. *Proc. Natl. Acad. Sci. USA* **79**, 579-582.

Reynolds, H.Y., Levine, A.S., Wood, R.E., Zierdt, C.H., Dale, D.C. and Pennington, J.E. (1975). *Pseudomonas aeruginosa* infections: persisting problems and current research to find new therapies. *Ann. Intern. Med*. **82**, 819-831.

Sackmann, W. and Dietrich, F.M. (1980). The effect of nor-muramyl dipeptide, a synthetic immunostimulant, on host defence mechanisms and efficacy of antibiotics in immuno-suppressed mice. *In* "International Symposium on Infections in the Immunocompromised Host", Veldhoven, The Netherlands, Abst. 27, p.44.

Saiki, I., Tanio, Y., Yamawaki, M., Uemiya, M., Kobayashi, S., Fukuda, T., Yukimasa, H., Yamamura, Y. and Azuma, I. (1981). Adjuvant activities of quinonyl-N-acetyl muramyl peptides in mice and guinea pigs. *Infect. Immun*. **31**, 114-121.

Sharma, S.D., Van Tsai, Krahenbuhl, J.L. and Remington, J.S. (1981). Augmentation of mouse natural killer cell activity by muramyl dipeptide and its analogues. *Cell. Immunol*. **62**, 101-109.

Singer, C., Kaplan, M.H. and Armstrong, D. (1977). Bacteremia and fungemia complicating neoplastic disease. A study of 364 cases. *Am. J. Med*. **62**, 731-742.

Sone, S. and Fidler, I.J. (1980). Synergistic activation by lymphokines and muramyl dipeptide of tumouricidal properties in rat alveolar macrophages. *J. Immunol*. **125**, 2454-2460.

Sone, S. and Fidler, I.J. (1981). *In vitro* activation of tumouricidal properties in rat alveolar macrophages by by synthetic muramyl dipeptide encapsulated in liposomes. *Cell. Immunol*. **57**, 42-50.

Sullivan, J.L., Ochs, H.D., Schiffman, G., Hammerschlag, M.R., Miser, J., Vichinsky, E. and Wedgwood, R.J. (1978). Immune response after splenectomy. *Lancet* **1**, 178-181.

Uemiya, M., Sugimura, K., Kusama, T., Saiki, I., Yamawaki, M., Azuma, I. and Yamamura, Y. (1979). Adjuvant activity of 6-*O*-mycoloyl derivatives of N-acetylmuramyl-L-seryl-D-isoglutamine and related compounds in mice and guinea pigs. *Infect. Immun.* **24**, 83-89.

Yamamura, Y., Azuma, I., Sugimura, K., Yamawaki, M., Uemiya, M., Kusumoto, S., Okada, S. and Shiba, T. (1977). Immunological and anti-tumour activities of synthetic 6-*O*-mycoloyl-N-acetylmuramyl dipeptides. *Proc. Japan Acad.* **53**, 63-66.

Yarkoni, E., Lederer, E. and Rapp, H.J. (1981). Immunotherapy of experimental cancer with a mixture of synthetic muramyl dipeptide and trehalose dimycolate. *Infect. Immun.* **32**, 273-276.

Young, L.S., Martin, W.J., Meyer, R.D., Weinstein, R.J. and Anderson, E.T. (1977). Gram-negative rod bacteremia: microbiologic, immunologic and therapeutic considerations. *Ann. Intern. Med.* **86**, 456-471.

NONSPECIFIC AND ANTIGEN-SPECIFIC STIMULATION OF HOST DEFENCE MECHANISMS BY LIPOPHILIC DERIVATIVES OF MURAMYL DIPEPTIDES

Shozo Kotani[1], Haruhiko Takada[1], Masachika Tsujimoto[1],
Takao Kubo[1], Tomohiko Ogawa[1], Ichiro Azuma[2],
Hidemasa Ogawa[3], Kensuke Matsumoto[3], Wasim A. Siddiqui[4],
Atshushi Tanaka[5], Shigeki Nagao[5], Osamu Kohashi[6],
Seizaburo Kanoh[7], Tetsuo Shiba[8] and Shoichi Kusumoto[8]

[1]*Department of Microbiology, Osaka University Dental School,
Kita-ku, Osaka 530, Japan;*

[2]*Institute of Immunological Sciences,
Hokkaido University, Kita-ku, Sapporo 060, Japan;*

[3]*Research Institute of Daiichi Seiyaku,
Edogawa-ku, Tokyo 132, Japan;*

[4]*Department of Tropical Medicine and Medical Microbiology,
School of Medicine, University of Hawaii at Manoa,
Honolulu, Hawaii 96916, USA;*

[5]*Department of Biochemistry, Shimane Medical University,
Izumo, Shimane 693, Japan;*

[6]*Research Institute for Disease of the Chest, Faculty of
Medicine, Kyushu University, Higashi-ku, Fukuoka 812, Japan;*

[7]*Department of Pharmacology, National Institute of Hygienic
Sciences, Osaka Branch, Higashi-ku, Osaka 540, Japan;*

[8]*Faculty of Science, Osaka University,
Toyonaka, Osaka 560, Japan*

I. INTRODUCTION

A number of studies over the past 10 years have revealed that bacterial cell wall peptidoglycans and compounds synthesized after the model of a part of the peptidoglycan hold manifold biological activities, mainly those to stimulate host defence mechanisms in various ways. Listed in Tables 1 and 2 are activities detected in *in vivo* and *in vitro* assays, respectively. Some of them may be beneficial to the host, while others are more detrimental than beneficial, although it is not easy to make a clear distinction between beneficial and detrimental effects.

TABLE 1

Biological activities of cell walls
and muramyl peptides (in vivo).

---

1. Modulation of immune responses (Chedid *et al.*, 1978a; Kotani *et al.*, 1981; Lederer, 1980a,b; Parant, 1979).

    a) Potentiation of antibody-mediated immune responses (Azuma *et al.*, 1976a; Ellouz *et al.*, 1974; Kotani *et al.*, 1975; Morisaki, 1980).

        Sup.1: Increasing serum IgG_2 level of guinea pigs (Kotani *et al.*, 1975; Stewart-Tull *et al.*, 1975).
        Sup.2: Increasing serum IgG_1 level of mice (Heymer *et al.*, 1978).
        Sup.3: Increasing serum IgE level of mice (Ohkuni *et al.*, 1977, 1979).
        Sup.4: Activation of helper T cells (Löwy *et al.*, 1977; Sugimoto *et al.*, 1978).

    b) Potentiation of cell-mediated immune responses (Azuma *et al.*, 1976a; Ellouz *et al.*, 1974; Kotani *et al.*, 1975; Morisaki *et al.*, 1980).

        Sup. : Induction of cytotoxic effector T cells (Azuma *et al.*, 1978; Taniyama *et al.*, 1975).

    c) Inhibition of antibody-mediated immune responses (Chedid *et al.*, 1976; Dziarski, 1978; Leclerc *et al.*, 1979b).

        Sup. : Suppression of IgE response of mice (Kishimoto *et al.*, 1979).

2. Stimulation of reticuloendothelial system (Tanaka *et al.*, 1979; Waters and Ferraresi, 1980).

3. Induction of monocytosis (Kato *et al.*, 1982).

4. Increase of interferon production (Barot-Ciorbaru *et al.*, 1978).

5. Increase of natural resistance to microbial infections and tumour development.

    a) Microbial infections (Chedid *et al.*, 1977; Franser-Smith and Mattews, 1981; Kierszenbaum and Ferraresi, 1979; Krahenbuhl *et al.*, 1981; Matsumoto *et al.*, 1981; Misaki *et al.*, 1966).

(TABLE 1 Contd.)

---

b) Tumour development (Azuma *et al.*, 1979a,b,c; Johnson *et al.*, 1978; Saiki *et al.*, 1981; Yamamura *et al.*, 1981).

    Sup.1: Induction of activated macrophages (Fidler *et al.*, 1981; Juy and Chedid, 1975; Namba *et al.*, 1978).

    Sup.2: Induction of natural killer cells (Sharma *et al.*, 1981).

6. Induction of experimental autoimmune diseases

a) Experimental allergic encephalomyelitis (Nagai *et al.*, 1978a,b).

b) Adjuvant polyarthritis (Koga *et al.*, 1976; Kohashi *et al.*, 1976, 1980; Nagao and Tanaka, 1980).

7. Pyrogenicity (Dinarello *et al.*, 1978; Kotani *et al.*, 1976; Mašek *et al.*, 1978; Riveau *et al.*, 1980).

8. Induction of acute inflammatory reaction and chronic granulomatous lesions (Emori and Tanaka, 1978; Maeda *et al.*, 1980; Schwab, 1979; Yamamota *et al.*, 1980).

9. Provocation of necrotic inflammation at the site prepared with tubercle bacilli (Nagao *et al.*, 1982).

10. Increase in vascular permeability (Ohkuni and Kimura, 1976).

11. Immunogenicity (Abdulla and Schwab, 1965; Nguyen-Huy *et al.*, 1976; Reichert *et al.*, 1980).

Among the activities shown in Table 1, the following 2 are attracting the attention of research workers from a practical viewpoint, with the object of making use of peptidoglycans and related synthetic compounds for the following purposes (1) stimulation of antigen-specific immune responses, either humoral or cell-mediated, and (2) enhancement of nonspecific resistance to endogenous as well as exogenous harmful agents. The former activity can lead us to improve the potency of vaccines including possible antitumour vaccines, and to develop synthetic vaccines, a model of which has recently been reported by Audibert and Jolivet (1982). The latter activity may serve to cope with microbial infections which stubbornly resist chemotherapy alone and frequently encountered in cancer patients of the terminal stage, and also to regress or suppress tumour development.

TABLE 2

*Biological activities of cell walls
and muramyl peptides* (in vitro)

1. Modulation of cells involved in natural and acquired
immunity:

   a) Monocytes and macrophages:

   Chemotactic activity (Ogawa *et al.*, 1982c); Enhancement
   of adherence and spreading on surfaces (Pabst and Johnston, 1980; Tanaka *et al.*, 1980); Inhibition of migration (Adam *et al.*, 1978; Yamamoto *et al.*, 1978);
   Stimulation of differentiation (Akagawa and Tokunaga,
   1980); Inhibition of DNA synthesis (Tanaka *et al.*,
   1982); Increase of glucosamine uptake (Imai *et al.*,
   1980; Takada *et al.*, 1979a); Stimulation of lysosomal
   enzyme release (Imai and Tanaka, 1981); Increase in
   superoxide and glucose oxidase levels (Imai *et al.*,
   1980; Pabst *et al.*, 1980; Pabst and Johnston, 1980);
   Stimulation (Hadden, 1978; Nozawa *et al.*, 1980) or
   inhibition (Smialowicz and Schwab, 1978) of phagocytosis or killing of microbes; Induction of cytotoxicity
   or cytostasis on tumour cells (Juy and Chedid, 1975;
   Sone and Fidler, 1980; Taniyama and Holden, 1979);
   Production of collagenase (Wahl *et al.*, 1979); Stimulation of monokine production (Iribe *et al.*, 1981;
   Oppenheim *et al.*, 1980; Rook and Stewart-Tull, 1976;
   Tenu *et al.*, 1980).

   b) B lymphocytes:

   Mitogenesis (Azuma *et al.*, 1976b; Damais *et al.*, 1975,
   1977; Takada *et al.*, 1977); Polyclonal activation
   (Saito-Taki *et al.*, 1980; Specter *et al.*, 1977) and
   inhibition of mitogen-induced polyclonal activation
   (Löwy *et al.*, 1980); Enhancement (Leclerc *et al.*,
   1979a; Sugimura *et al.*, 1979) or inhibition (Leclerc
   et al., 1979c) of antigen-specific antibody formation;
   [replacement of the helper function of T cells (Watson
   and Whitlock, 1978)].

   c) T lymphocytes:

   Mitogenesis (Azuma *et al.*, 1976b; Takada *et al.*, 1979b);
   Enhancement of helper function (Sugimura *et al.*, 1979);
   Stimulation of differentiation (Prunet *et al.*, 1978);
   Induction of cytotoxic effector cells (Igarashi *et al.*,
   1977); Stimulation of mixed lymphocyte reactions
   (Sharma *et al.*, 1975).

(TABLE 2 Contd.)

---

d) Neutrophil leucocytes:

Stimulation (Ishihara *et al.*, 1982) or inhibition (Jones and Schwab, 1970) of phagocytosis.

e) Basophil leucocytes:

Increase of histidine uptake (Ogawa *et al.*, 1982b).

f) Mast cells:

Stimulation of histamine release (Kimura *et al.*, 1982).

g) Thrombocytes:

Liberation of serotonin (Harada *et al.*, 1982; Kotani *et al.*, 1981).

h) Enhancement of cell viability (Leclerc *et al.*, 1978).

2. Activation of complement system through alternative and classical pathways (Barker *et al.*, 1977; Greenblatt *et al.*, 1978; Kawasaki, 1982; Kotani *et al.*, 1981).

3. Stimulation of bone resorption (Dewhirst, 1982).

4. Gelation of amoebocyte lysate of horse-shoe crab (Kotani *et al.*, 1977b).

5. Contraction of ileal strips (Ogawa *et al.*, 1982a).

6. Binding with cytoplasmic membranes of mammalian cells and stabilization of artificial cell membranes (Stewart-Tull *et al.*, 1978).

The minimal effective structure responsible for most of the biological activities listed in Tables 1 and 2 is well established to be *N*-acetylmuramyl-L-alanyl-D-isoglutamine (MDP), namely the structural unit common to cell wall peptidoglycans of essentially all of bacterial species parasitic to mammals. However, the practical use of beneficial activities of MDP molecule in clinical and preventive medicine has been limited for several reasons: pyrogenicity; failure to manifest its full antigen-specific immunostimulating activity without help of irritative vehicles for administration; possibility of giving rise to immunological diseases; and so on. As a result of collaborative studies to pursue MDP analogues or derivatives with the maximum beneficial and the minimal deleterious biological activities, we have tentatively chosen the following 3 lipophilic derivatives of MDP (Fig. 1) as compounds which could be competent for use in

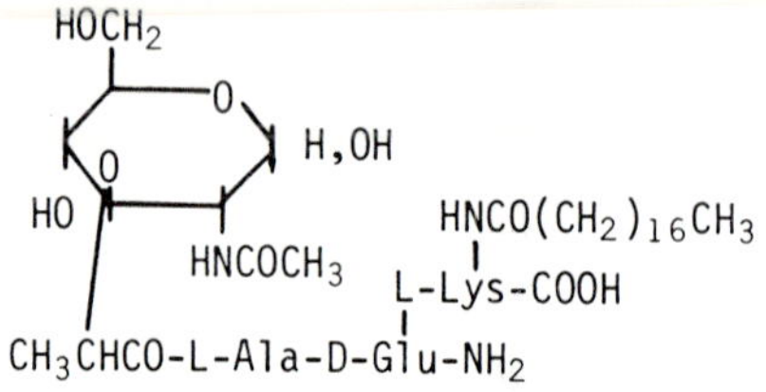

MDP-Lys(L18) N^{α}-(N-acetylmuramyl-L-alanyl-D-isoglutaminyl)-
                N^{ε}-stearoyl-L-Lys

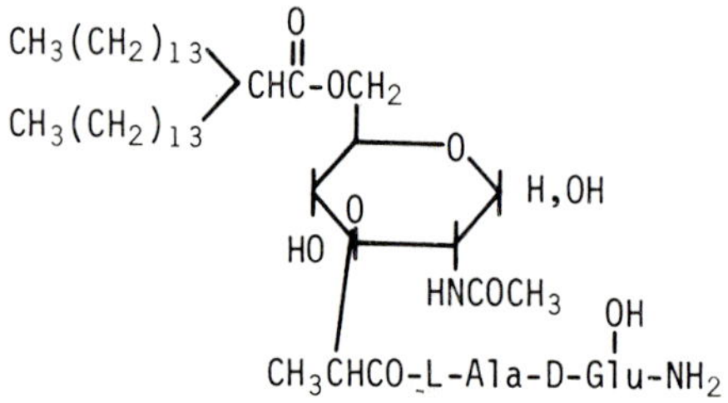

B30-MDP   6-O-(2-tetradecylhexadecanoyl)-N-acetylmuramyl-
           L-alanyl-D-isoglutamine

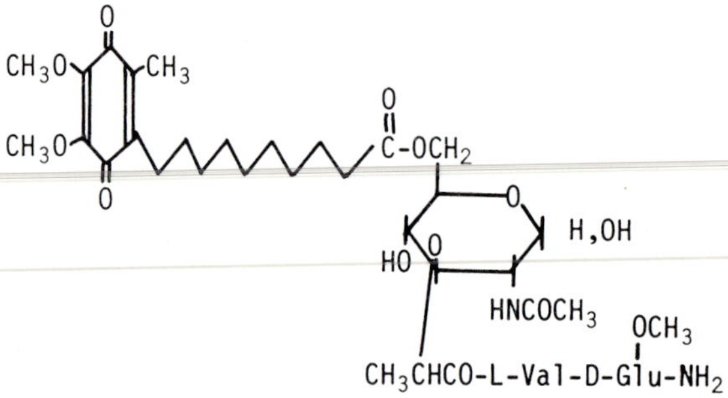

Quinonyl-MDP-66   6-O-QS10-N-acetylmuramyl-L-valyl-D-isoglutamine
                   methyl ester

FIG. 1 *Structure of lipophilic derivatives of muramyl dipeptides used in this study.*

controlled clinical trials: (1) N^{α}-(N-acetylmuramyl-L-alanyl-
D-isoglutamine)-N^{ε}-stearoyl-L-lysin for enhancement of nonspecific host resistance to harmful agents, (2) 6-O-(2-tetra-
decylhexadecanoyl)-MDP for stimulation of antigen-specific
immune responses, both humoral and cell-mediated, and (3)

benzoquinonyl derivative of N-acetylmuramyl-L-valyl-D-iso-
glutamine methyl ester for tumour immunotherapy.

II. ENHANCEMENT OF NONSPECIFIC HOST RESISTANCE TO MICROBIAL INFECTIONS BY N^{α}-(N-ACETYLMURAMYL-L-ALANYL-D-ISOGLUTAMINYL)-N^{ε}-STEAROYL-L-LYSIN (ABBREVIATED AS MDP-LYS(L18))

In 1966, Misaki *et al.* fractionated cellular constituents of
Mycobacterium bovis to locate an active principle capable of
enhancing nonspecific resistance of mice to staphylococcal
infection, and found that cell walls, particularly their peptidoglycan moiety, were responsible for the above biological
effect. The activity of cell walls and related compounds to
enhance host resistance to microbial infections has been the
subject of extensive and productive studies of Chedid and his
collaborators for the last several years, by use of water-
soluble adjuvants (WSA) derived from whole cells or cell
walls, synthetic MDP, β-D-aminophenyl glycoside of MDP,
1-O-(MDP-L-alanyl)-glycerol-3-mycolate and others (Chedid *et
al.*, 1977, 1978b, 1979; Elin *et al.*, 1976; Parant *et al.*,
1978, 1980). Along a similar line of approach, Ogawa and his
co-workers have investigated the structure-anti-infectious
activity relationships of a series of MDP analogues, especially 6-O-acyl derivatives. It was found that 6-O-stearoyl-
MDP, L18-MDP, was one of the most promising compounds in
enhancing the resistance of mice, either normal or immunosuppressive, to the sepsis type of *Escherichia coli* infection
(Matsumoto *et al.*, 1978; Osada *et al.*, 1982a,b). The study
has been extended by use of 3 types of synthetic MDP derivatives having substituted groups in the γ-carboxyl group of
D-isoglutamine residue: (1) γ-alkyl-amides, (2) γ-esters, and
(3) γ-(N^{α}-MDP-N^{ε}-acyllysyl) derivatives, totalling 21 compounds. Comparative studies on their resistance-enhancing
activity have led to a tentative conclusion that MDP-Lys(L18)
is the compound of choice (Matsumoto *et al.*, in prep.).
    Table 3 shows that both MDP-Lys(L18) and L18-MDP are more
effective than MDP in enhancement of the resistance of mice
to septic infections with various microbes. The greatest
protection was obtained against infections with *E. coli* and
Staphylococcus aureus. Appreciable protective effect was
seen with infections with 2 strains of *Pseudomonas aeruginosa*
and one strain (sk) of *Klebsiella pneumoniae*, whereas the
infection with a wild type strain (3167) of the latter was
quite unaffected. Against *Candida albicans* infection, MDP-
Lys(L18) alone produced significant protection. The influence of inoculum size, administration route and dosage on
the anti-infectious activity of MDP-Lys(L18), L18-MDP and

KOTANI *et al.*

TABLE 3

Effect of MDP analogues on susceptibility of mice to infections with various pathogenic microbes

Species (strain)	Inoculum size (cells/mouse)	% survival			
		MDP	L18–MDP	MDP–Lys(L18)	PBS
E. coli	1.2 x 10	70	80	90	10
(E77156)	6.0 x 10	n.d.	30	45	0
P. aeruginosa	3.0 x 10	45	55	60	0
(No. 15)	6.0 x 10	12	27.5	n.d.	0
P. aeruginosa	2.0 x 10	n.d.	40	60	0
(P-I-III)	4.0 x 10	n.d.	20	30	0
K. pneumoniae (sk)	1.0 x 10	12.6	37	n.d.	0
K. pneumoniae (3167)	3.8 x 10	0	0	10	0
S. aureus	1.6 x 10	n.d.	90	95	30
E46	3.2 x 10	n.d.	80	60	0
C. albicans D12	3.0 x 10	10	10	40	0

Groups of 20 outbred Slc:ddY mice (5 week old, male) were intraperitoneally injected with 0.1 mg of test compounds 24 h before infection. Infection was performed by subcutaneous inoculation with bacterial suspension in nutrient broth, except that *C. albicans* was injected by intravenous route.

MDP was studied on *E. coli* infection (Table 4). Good dose-dependency of the activity was noted between 0.01 and 0.0001 mg/mouse of MDP-Lys(L18), and between 0.1 and 0.01 mg of both L18-MDP and MDP. The minimal effective dose inducing apparent protection was around 0.001, 0.01 and 0.03 mg for MDP-Lys(L18), L18-MDP and MDP, respectively. The peak activity was obtained at a dose of 0.01 mg with MDP-Lys(L18) and 0.1 mg with L18-MDP and MDP. A noteworthy finding that the resistance of mice against *K. pneumoniae* infection was effectively enhanced by oral administration of MDP was first reported by Dr Chedid's group (1977). The present study showed that oral administration of L18-MDP and MDP-Lys(L18) increased more effectively the resistance of mice against *E. coli* septic infection than MDP, although at a dosage level of 0.1 mg no differences in the efficacy were observed among the 3 compounds by administration via other routes. With

TABLE 4

Effect of the inoculum size of challenge, the dose and route of administration of MDP analogues on their protective activity on E. coli *infection*

Inoculum size (cells/mouse)	Dose (mg/mouse)	Route	Difference of % survival[a]		
			MDP	L18-MDP	MDP-Lys(L18)
2.5×10^6			70.0	80.0	n.d.
5.0×10^6			65.0	80.0	90.0
7.5×10^6	0.1	i.p.	40.0	80.0	n.d.
1.0×10^7			30.0	50.0	60.0
2.0×10^7			0	15.0	n.d.
	0.5		n.d.	n.d.	95.0
	0.3		67.5	80.0	n.d.
	0.1		65.0	85.0	90.0
5.0×10^6	0.03	i.p.	35.0	50.0	n.d.
	0.01		10.0	30.0	90.0
	0.001		0	0	35.0
	0.0001		0	0	5.0
	1.0	p.o.	45.0	72.7	70.0
5.0×10^6	0.1	i.p.	65.0	83.1	70.0
	0.1	s.c.	62.5	76.0	85.0
	0.1	i.v.	55.0	78.0	65.0

Groups of Slc:ddY mice were treated with test compounds 24 h before *E. coli* infection.

[a]The difference between the percentages of survival in treated groups and those of the respective controls on day 7. Data represent cumulative result of 3 experiments, each of which was performed with 20 mice under essentially the same experimental conditions.

regard to time interval between treatment and infection, the peak activity was obtained by the treatment 24 h before infection with all 3 specimens (data are not shown). The activity decreased when the interval was more than 3 days, or less than 24 h before infection.

Table 5 summarizes synergistic effects of combined therapy of MDP-Lys(L18) and chemotherapeutic agents against various septic infections in mice. The ED_{50} value of cefazolin against *E. coli* infection was 27.4 mg/kg, whereas that of the combined therapy was 11.1 mg/kg. With *P. aeruginosa* infec-

TABLE 5

Synergistic, curative effect of MDP-Lys(L18) and chemotherapeutic agent on various microbial infections

Infection Species (strain)	Route	Inoculum size (MLD/mouse)	Treatment	ED_{50}	(95% confidence limit, mg/kg)
E. coli (E77156)	i.p.	8	Cefazolin alone	27.4	(16.1 – 33.6)
			Cefazolin + MDP-Lys(L18)	11.1	(7.4 – 16.5)
P. aeruginosa (P-I-III)	s.c.	4	Gentamicin alone	4.0	(2.8 – 5.9)
			Gentamicin + MDP-Lys(L18)	1.4	(0.8 – 2.7)
S. aureus (E46)	i.p.	8	AB-PC alone	5.63	(3.7 – 6.6)
			AB-PC + MDP-Lys(L18)	1.07	(0.67 – 1.7)
C. albicans (D12)	i.v.	1	AMPH alone	7.08	(3.79 – 13.24)
			AMPH + MDP-Lys(L18)	3.98	(2.05 – 7.73)

Groups of 10 Slc:ddY mice were treated subcutaneously with 0.1 mg of MDP-Lys(L18) 24 h before infection, and then treated subcutaneously with therapeutic agents (intravenously with AMPH B) 2 h after infection. ED_{50} values were calculated by the Probit method. AB-PC: aminobenzyl-penicillin; AMPH: amphotericin B.

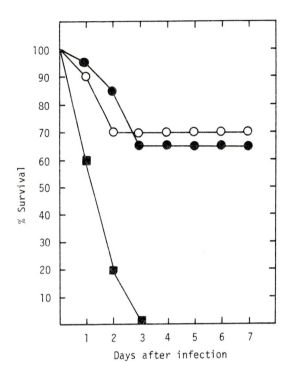

FIG. 2 *Effect of MDP-Lys(L18) on resistance of mice immuno-compromised by cyclophosphamide to infection with* E. coli *E77156. Two groups of 20 ddY male mice were treated intra-peritoneally with 100 mg/kg of cyclophosphamide. One group (●) was treated subcutaneously with 0.1 mg/mouse of MDP-Lys(L18) 2 h later, and then infected subcutaneously 24 h later with 1.5 x 10⁶ cells/mouse, which was 1/4 MLD for a normal untreated mouse. The other immunocompromised group (■) receiving no MDP-Lys(L18) treatment was similarly infected. A group of 20 mice not treated with either cyclophosphamide or MDP-Lys(L18) was submitted to the similar infection (○).*

tion, ED_{50} value of gentamicin was 4.0 mg/kg while that of combined use of gentamicin and MDP-Lys(L18) was 1.4 mg/kg. Likewise, the ED_{50} value of aminobenzylpenicillin was reduced from 5.63 mg to 1.07 mg/kg by combined use of MDP-Lys(L18) against *S. aureus* infection, and that of amphotericin B against *C. albicans* infection was decreased from 7.08 mg to 3.98 mg/kg. MDP-Lys(L18) was further shown to have the effect of restoring the normal resistance level against *E. coli* infection of mice, the defence mechanisms of which had

been compromised by prior administration of cyclophosphamide
(Fig. 2).

II. STIMULATION OF ANTIGEN-SPECIFIC IMMUNE RESPONSES
    BY 6-*O*-(2-TETRADECYLHEXADECANOYL)-MDP (ABBREVIATED
    AS B30-MDP)

MDP is known to require a set of administration conditions,
as exemplified by injection as water-in-mineral oil emulsion,
to exert its full immunostimulating activity, particularly
the adjuvancy to induce cell-mediated immune responses such
as delayed-type-hypersensitivity (DTH). Several years ago,
Kinoshita (1978) revealed that a water-in-squalane emulsion or
a double emulsion prepared using squalane served as an useful
vehicle for demonstration of the adjuvancy of MDP in both
induction of DTH and stimulation of antibody production
against ovalbumin in guinea pigs. Squalene, however, was
found inactive in his study. Usefulness of squalane as a
substitute for mineral oil of water-in-oil emulsion to support the adjuvancy of MDP, particuarly of its lypophilic
derivatives, was confirmed later (Carelli *et al.*, 1981).
      Further studies of our research group have demonstrated
that some of 6-*O*-acyl-MDP, that is, compounds possessing
linear, α-branched, or α-branched and β-hydroxylated higher
fatty acid at C-6 position of the muramic acid residue, exhibited strong immunostimulating activity to test protein antigens (ovalbumin, bovine serum albumin and keyhole limpet
haemocyanin) with various vehicles for administration, which
were far less irritative than water-in-mineral oil emulsion,
but not effective in supporting the adjuvancy of MDP (Kotani
et al., 1977a, 1978; Tsujimoto, 1981). Thus, taking together
the minimum effective dose, the extent of stimulation of cellular and humoral immunity, and accompanying detrimental
effects (local irritative action, pyrogenicity and others)
and the adaptability to a wide range of vehicles for administration, we have tentatively reached a conclusion that B30-
MDP is the most favourable MDP derivative for possible human
use to potentiate antigen-specific immune responses, especially cell mediated immunity.
      A few experimental results supporting the above conclusion
will be described in some detail. Figure 3 illustrates that
B30-MDP, at a dosage of several to a few tens of µg/animal
can effectively cause the induction of DTH to ovalbumin (in
terms of positive corneal reaction) when it was administered
to guinea pigs together with nonirritative vehicles such as
liposomes, a squalene-in-water emulsion, Intralipid®, and
plain phosphate buffered saline (PBS). Production of circulating antibody was similarly stimulated by B30-MDP administered with the above vehicles (Fig. 4). Figure 5 shows time

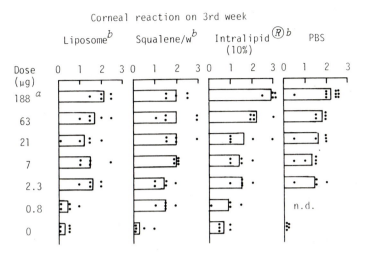

FIG. 3 *Immunostimulating activity of B30-MDP with nonirritative vehicles to induce a delayed-type-hypersensitivity to ovalbumin in guinea pigs. Groups of 5 outbred Hartley guinea pigs (female, 300 g) were immunized by intra-footpad injection of 1 mg (2 mg with PBS) of ovalbumin and an indicated dose of B30-MDP in various vehicles. Induction of DTH was evaluated by corneal test 3 weeks after the immunization. (●) Score of an individual animal, (▭) average score of each group. [a]Equimolar with 100 µg of MDP, [b]Liposomes consisted of lecithin and cholesterol, squalene/w was made of 5 vol of squalene and 95 vol of 0.2 Tween 80 in PBS, and Intralipid® (10%) is 10% intravenous fat emulsion for clinical use.*

courses of reactions at the injection site and regional lymph nodes of guinea pigs receiving the intra-footpad injection of 63 µg of B30-MDP with one or 2 mg of ovalbumin. The reaction (in terms mainly of swelling) of the site injected with B30-MDP and ovalbumin in vehicles other than water-in-mineral oil emulsion was nothing serious. Swelling of the regional lymph nodes was sometimes considerable for one to 2 weeks after the injection, but tended to subside rapidly during the following few weeks. Enlargement of the regional lymph nodes caused by B30-MDP seems to be due to an inherent activity of this lypophilic derivative of MDP, in view of the finding of Tanaka (1982) that B30-MDP added to PBS was highly active in induction of massive epitheloid granuloma in guinea pigs, unlike MDP which required water-in-mineral oil emulsion as a vehicle for granuloma formation.

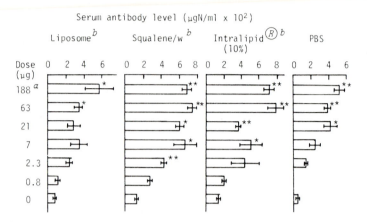

FIG. 4 *Immunostimulating activity of B30-MDP with nonirritative vehicles to stimulate circulating antibody formation to ovalbumin in guinea pigs. Legends are the same as those in Fig. 3 except the following: Serum specimens were taken from each animal 4 weeks after the immunization to estimate anti-ovalbumin precipitin level. (☐━┤) Average and ±1 S.E. of antibody content (μg N/ml) of each group. *, ** The difference between the test and respective control groups is statistically significant (**P<0.01, *P<0.05).*

Studies to make use of the adjuvant activity of B30-MDP in enhancing the potency of vaccines will be described below. One is concerned with a highly purified influenza vaccine (HA-NA vaccine) which was prepared from a virus pool harvested from chick embryos by a series of zonal centrifugation, Triton treatment and sucrose density gradient centrifugation, and consisted almost exclusively of hemagglutinin and neuraminidase. The addition of B30-MDP with either squalene-in-water emulsion, Intralipid® or PBS as adjuvant enhanced significantly the potency of B/Kanagawa/3/76 HA-NA vaccine (0.02 μg protein/animal x 2) to produce virus-neutralizing (Fig. 6) and haemagglutination-inhibiting antibodies (Fig. 7) in guinea pigs. Significant potentiation by B30-MDP was also obtained with A/Kumamoto/122/76 HA-NA vaccine (data are not shown).

The other study is a preliminary trial of Siddiqui *et al.* (1978, 1982) to enhance the protective activity (based on cell-mediated immunity) of *Plasmodium falciparum* vaccine in *Aotus* monkeys. The vaccine was prepared from the FVO strain which had been maintained by continuous *in vitro* cultivation for the past 5 years. Table 6 shows a design of the

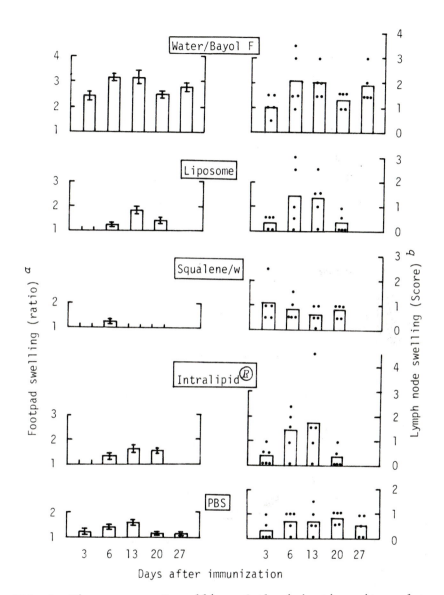

FIG. 5 *Time course of swelling of the injection site and regional lymph nodes of guinea pigs (5 per group) receiving 1 mg (2 mg with PBS) of ovalbumin and 63 μg of B30-MDP in various vehicles. [a]Ratio of the cross section of the injected foot and respective control foot. [b]Arbitrarily graded from 0 for no swelling to 3 − 3.5 for marked swelling. (●) Score of an individual animal and (▭) average score of each group.*

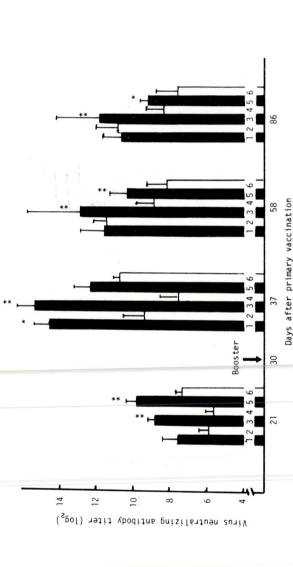

FIG. 6 *Enhancement of the potency of influenza HA-NA vaccine (B/Kanagawa/3/76, 0.02 µg protein) by B30-MDP (63 µg) in various vehicles to produce virus-neutralizing antibody. Groups of 5 out-bred Hartley guinea pigs were immunized by subcutaneous injection of the vaccine with or without B30-MDP and boosted with the vaccine alone on day 30. Closed column — Test groups receiving the vaccine and B30-MDP in squalene-in-water emulsion (1), Intralipid® (3) and PBS (5). Open column — Control groups immunized with the vaccine alone in squalene-in-water emulsion (2), Intralipid® (4) and PBS (6). (⊤) +1 S.E. **, * The difference between the test and respective control groups is statistically significant (**P<0.01, *P<0.05).*

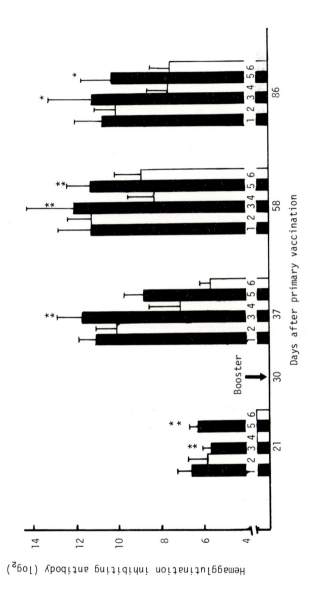

FIG. 7 Enhancement of the potency of influenza HA-NA vaccine (B/Kanagawa/3/76, 0.02 μg protein) by B30-MDP (63 μg) in various vehicles to produce haemagglutination-inhibitory antibody. Legends are the same as those in Fig. 6.

TABLE 6

Use of B30-MDP in the vaccination of Aotus *monkeys (Type VI) against* Plasmodium falciparum *(Vietnam-Oak Knoll strain).*

Monkey No.	Composition of the vaccine (i.m. injection/day 0 & 21)		Challenge (day 49)
	Adjuvant	Parasite antigen[a] (1.5 mg protein/injection)	1×10^6 FVO infected rbc's
A328	None	+	+
*A329	B30-MDP[b]	+	+
A330	B30-MDP	+	+
A332	B30-MDP	+	+
A335	B30-MDP	+	+
A333	Freund's complete[c]	+	+

* A329 died of unrelated causes on day 33. [a]Parasite antigen was prepared from the FVO strain of *P. falciparum*. FVO has been maintained by coninuous *in vitro* cultivation since May 1977. [b]0.2 mg of B30-MDP was incorporated into liposomes. [c]0.5 ml of FCA (Difco) per dose.

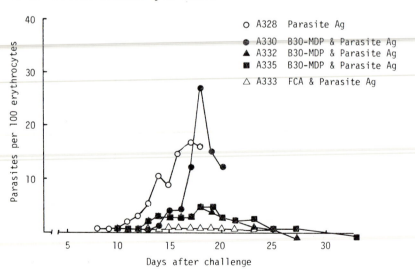

FIG. 8 *Course of infection of* Plasmodium falciparum *(FVO strain) in* Aotus *monkeys (Type VI): Vaccinated and controls.*

experiment. Figure 8 illustrates that a control monkey
(A328) and one of the monkeys vaccinated with B30-MDP as
adjuvant (A330) were killed by the challenge on days 18 and
20, respectively, while the other 3 vaccinated monkeys (A332
and A335 receiving B30-MDP as adjuvant, and A333 receiving
Freund's complete adjuvant (FCA)) survived the challenge and
no parasites could be found about 4 weeks after the challenge.
The experiment was a very preliminary one, but the result is
remarkable: (1) it is believed that there is some loss of
potency of protective antigens when parasites are kept in
continuous culture for a long period of time; (2) for the
further development of P. *falciparum* vaccine, *in vitro* cultured parasites have to be used for practical reasons; and
(3) in vaccination studies in which parasite antigens derived
from long-term *in vitro* cultures of P. *falciparum* are used,
the need for a specific adjuvant in achieving good protection
is quite evident as shown in Table 7, where it is shown that

TABLE 7

Role of adjuvants in vaccination studies with
Plasmodium falciparum *using parasite antigens*
derived from long-term (>1 y) culture parasites

Adjuvant	Survived/Vaccinated	References
FCA, FIA	0/3	Reese, R. *et al.* (1978)
MDP, FIA	1/3	Reese, R. *et al.* (1978)
FCA (2x)	3/3	Siddiqui, W. *et al.* (1978)
B30-MDP (2x)	2/3	Siddiqui, W. *et al.* (1978)

Quoted from W.A. Siddiqui (1982), and modified.

out of the 6 monkeys surviving vaccination, 3 received FCA
and one received MDP in water-in-mineral oil emulsion in conjunction with parasite antigen, and the other 2 received B30-
MDP in liposomes as adjuvant. We are hoping that the results
of our ongoing experiment will further prove that B30-MDP
can substitute FCA for effective immunization of *Aotus* monkeys against P. *falciparum*.

IV. ANTITUMOUR ACTIVITY OF BENZOQUINONYL DERIVATIVE OF N-ACETYLMURAMYL-L-VALYL-D-ISOGLUTAMINE METHYL ESTER (ABBREVIATED AS QUINONYL-MDP-66)

There are a great number of studies attempted to utilize immunostimulating activities of bacterial cell walls and related synthetic compounds in tumour immunotherapy (Azuma *et al.*, 1979a,b,c; Johnson *et al.*, 1978; Saiki *et al.*, 1981; Yamamura *et al.*, 1981). Among them, Azuma and his coworkers (1976a; 1978; 1979a,b,c) in the last few years have synthesized several hydrophobic derivatives of MDP in pursuing compounds having stronger antitumour activity than hitherto described MDP analogues, and have chosen quinonyl-MDP-66 as one of provisional leading compounds. This new compound was synthesized, in consideration of the fact that ubiquinone and its related compounds have manifold biological activities, such as activation of reticuloendothelial system, enhancement of host resistance to infections and tumours (Bliznakov, 1977; Imada *et al.*, 1972) and stimulation of humoral immune response (Imada *et al.*, 1972; Sugimura *et al.*, 1976).

Antitumour activity of quinonyl-MDP-66 is shown in Tables 8 to 10 using the combination of line-10 hepatoma and strain-2 guinea pigs. Groups of 6 animals were inoculated intradermally with 10^6 tumour cells, and received single intralesional injection on day 5 of 10% squalene- or squalane-treated quinonyl-MDP-66 (400 µg/animal) in test groups and 10% squalene alone in control groups. Quinonyl-MDP-66 caused tumour regression and inhibition of lymph node metastases in 3 of 6 guinea pigs in both test groups, while none of the animals in the 2 control groups escaped from extensive tumour growth and lymph node metastases (Table 8). Four injections of 100 µg or 400 µg of quinonyl-MDP-66 induced complete regression of tumour development in all 7 guinea pigs in test groups under the experimental conditions where practically all of the 7 to 10 animals in control groups suffered from extensive tumour growth. Two injections of 400 µg of quinonyl-MDP-66 was partially effective (Table 9).

Comparison of the antitumour activity of quinonyl-MDP-66 with those of 6-O-acyl-MDP analogues which were synthesized by introduction of natural nocardomycolic acid or mycolic acid-like synthetic fatty acids revealed that quinonyl-MDP-66 was significantly more effective than 6-O-"mycoloyl"-MDP derivatives in either regression of the original tumour or prevention of lymph node metastases (Table 10). In this connection, the suppressive activity of quinonyl-MDP-66 against Meth-A fibrosarcoma in syngeneic mice was reported previously (Azuma *et al.*, 1979c).

TABLE 8

Regression of line-10 hepatoma by single intralesional injection of quinonyl-MDP-66 in strain-2 guinea pigs

Dose (μg)	Vehicle	Injection (on day)	Regressed /treated	Metastases in regional lymph nodes
400	10% Squalene	5	3/6	3/6
400	10% Squalane	5	3/6	3/6
None	10% Squalene	5	0/6	6/6
None	10% Squalane	5	0/6	6/6

Groups of 6 strain-2 guinea pigs were inoculated intradermally with line-10 hepatoma cells (10^6), and injected intralesionally with squalene or squalane-treated quinonyl-MDP-66 (test groups) and squalene or squalane (10%) alone (control groups).

TABLE 9

Regression of line-10 hepatoma by repeated intralesional injections of quinonyl-MDP-66 in strain-2 guinea pigs.

Dose (μg)	Injection (on day)	Tumour-free/Treated (on day 60)
400 x 4	2, 5, 8, 15	7/7
100 x 4	2, 5, 8, 15	7/7
400 x 2	2, 5	4/7
None x 4	2, 5, 8, 15	1/7
None x 2	2, 5	0/10

Groups of 7 to 10 strain-2 guinea pigs were inoculated intradermally with line-10 hepatoma cells (10^6), and injected intralesionally with squalene (10%)-treated quinonyl-MDP-66 (test groups) or squalene (10%) alone (control groups).

TABLE 10

*Comparison of antitumour activity to regress line-10 hepatoma
in strain-2 guinea pigs of quinonyl-MDP-66 and MDP analogues
with muramic acid residue substituted with natural
or synthetic "mycolic acid" at C-6 position*

Test MDP analogue	Tumour-free /treated	Metastases in regional lymph nodes
6-O-QS-10-MurNAc-L-Val-D-Glu (OCH$_3$)-NH$_2$ (Quinonyl-MDP-66)	4/7	2/7
6-O-Nocardomycoloyl-MurNAc-L-Ser-D-isoGln	1/7	6/7
6-O-BH32-MurNAc-L-Ser-D-isoGln	2/7	5/7
6-O-BH48-MurNAc-L-Ser-D-isoGln	3/6	3/6
6-O-BH32-MurNAc-L-Val-D-isoGln	2/6	4/6
6-O-BH48-MurNAc-L-Val-D-isoGln	1/6	5/6
Control (10% Squalene)	0/6	6/6

A group of 6 to 7 strain-2 guinea pigs were inoculated intradermally with line-10 hepatoma cells (10^6). Squalene (10%)-
treated adjuvant (400 μg each) or squalene (10%) alone was
injected intralesionally, on days 2 and 5. BH32-: 3-hydroxy-
2-tetradecyloctadecanoyl- ; BH48-: 3-hydroxy-2-docosylhexa-
cosanoyl-.

V. SOME BIOLOGICAL ACTIVITIES OF B30-MDP AND MDP-LYS(L18) ON WHICH SHOULD BE CAUTIOUS IN CLINICAL APPLICATION

A. Pyrogenicity

Pyrogenicity of bacterial cell wall peptidoglycans was first
demonstrated with *Streptococcus pyogenes* (Roberson and Schwab,
1961; Rotta and Bednár, 1969). This was followed by the
demonstration that a water-soluble higher molecular weight
fraction obtained from an enzymatic digest of *S. pyogenes*
walls caused definite febrile response and leukopenia in
rabbits (Hamada *et al.*, 1971), and then by the discovery
that MDP was the minimum structure holding the pyrogenicity
of bacterial cell walls (Dinarello *et al.*, 1978; Kotani *et
al.*, 1976; Mašek *et al.*, 1978; Riveau *et al.*, 1980). Pyrogenicity of cell wall peptidoglycans and MDP may at least
partially explain fever response in bacterial infections,

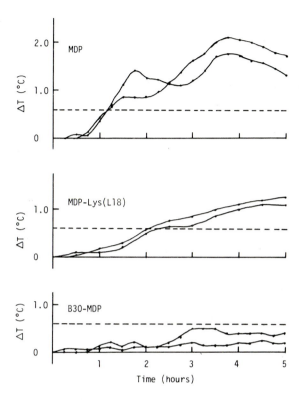

FIG. 9 *Changes in body temperature of rabbits receiving intravenous injection of MDP, MDP-Lys(L18) and B30-MDP in saline (0.1 mg/ml/kg of rabbit).*

particularly those by Gram-positive bacteria lacking endotoxin. But it may retard clinical use of peptidoglycan-
related synthetic compounds.

Figures 9 and 10 illustrate febrile response of rabbits
receiving i.v. injection of MDP, MDP-Lys(L18) and B30-MDP.
It is evident that pyrogenicity of MDP-Lys(L18) was far less
than that of MDP, and B30-MDP was practically nonpyrogenic.
Other MDP derivatives having no or less pyrogenicity have
been described (Audibert and Chedid, 1980; Azuma *et al.*,
1978; Parant *et al.*, 1980; Uemiya *et al.*, 1979). Quinonyl-
MDP-66, unlike 6-*O*-mycoloyl-MDP, was pyrogenic (Azuma *et al.*,
unpublished).

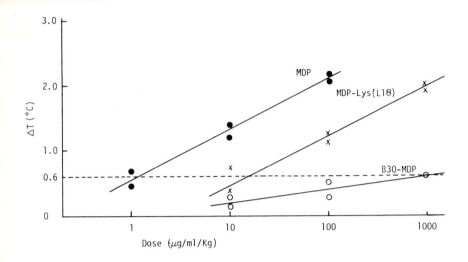

FIG. 10 *Dose-febrile response relationships in rabbits receiving intravenous injection of MDP, MDP-Lys(L18) and B30-MDP in saline.*

B. Extensive Haemorrhagic Necrosis Caused by Provocative Injection of MDP

This unexpected activity has been recently reported by Nagao *et al.* (1982) in the course of study on antigenicity of MDP. When guinea pigs receiving an intra-footpad injection of heat-killed tubercle bacilli (100 µg) suspended in water-in-mineral oil emulsion were injected intravenously or intracutaneously with MDP (100 µg) in PBS several weeks later, an extensive inflammation with a marked swelling, exudation, haemorrhage, necrosis and ulceration occurred within 24 h after the provocative injection of MDP at the site prepared with tubercle bacilli. Some animals were killed through a generalized shock. This phenomenon was observed clearly with guinea pigs of outbred Hartley strain, inbred strain JY-1, JY-2 and JY-7, and to a lesser extent with inbred strains 2 and 13. Neither rats (WKA) nor domestic albino rabbits were susceptible.

Several substances other than tubercle bacilli were tested for their ability to prepare animals for the necrotic

TABLE 11

Ability of heat-killed cells of various bacteria,
MDP, B30-MDP, and LPS to prepare for provocation of the
necrotic inflammation by MDP

Test specimen (100 µg)	Inflammatory swelling with necrosis
M. tuberculosis (H37Rv)	+
M. bovis (BCG)	+
N. corynebacteriodes (ATCC 14898)	(+)[a]
S. aureus (209P)	−
L. plantarum (ATCC 8014)	−
LPS (Difco, *E. coli*)	−
MDP	−
B30-MDP	−

Three outbred Hartley guinea pigs (male and female) per group
received a preparatory, intra-footpad injection of a test
specimen with water-in-mineral oil emulsion. Thirty days
later, each animal was examined for provocation by intracutaneous injection of MDP (400 µg) in PBS. [a]Weakly positive.

inflammation provoked by MDP. None of the whole cells of *S.
aureus* (209P) and *Lactobacillus plantarum* (ATCC 8014), LPS
(Difco, from *E. coli*), MDP and B30-MDP were active, while
whole cells of *Nocardia corynebacteriodes* (ATCC 14898) were
weakly active as a preparative agent (Table 11). Table 12
shows the result of an experiment to examine provocative
ability of various MDP analogues. While MDP, L18-MDP, MDP-
Lys, MDP-Lys(L18) and *N*-acetylglucosaminyl (GlcNAc)-*N*-acetyl-
muramyl (MurNAc)-dipeptide were active in provocation of the
necrotic inflammation, neither muramyl tetrapeptides nor
GlcNAc-MurNAc-tetrapeptide showed provocative activity.
MurNAc(-GlcNAc)-dipeptide and GlcNAc-MurNAc-tripeptide gave
marginally positive reactions. B30-MDP, on the other hand,
was inactive when it was injected as a suspension in PBS,
but was definitely provocative as a solution in Nikkol HCO-60,
a nonionic detergent for clinical use. The phenomenon described above seems to require a specific set of conditions
for its expression, but attempts to use some of the MDP analogues in humans should be done cautiously taking this phenomenon into consideration.

TABLE 12

Ability of muramyl peptides, their lipophilic derivatives and disaccharide peptides to provoke the necrotic inflammation at the site prepared with tubercle bacilli

Test compound (400 µg)	Route of administration	Inflammatory swelling with necrosis
MDP	i.c., i.v.	+ - 2+
MDP-L-Lys	i.c.	+
MDP-L-Lys-D-Ala[a]	i.c.	-
MDP-$meso$-A$_2$pm-D-Ala[a]	i.c.	-
GlcNAc-MurNAc-L-Ala-D-isoGln	i.v.	+
MurNAc(-GlcNAc)-L-Ala-D-isoGln	i.c.	±(1/3)[b]
GlcNAc-MurNAc-L-Ala-D-isoGln-$meso$-A$_2$pm[a]	i.v.	±(1/3)[b]
GlcNAc-MurNAc-L-Ala-D-isoGln-$meso$-A$_2$pm-D-Ala[a]	i.c., i.v.	-
L18-MDP	i.c.	+
MDP-L-Lys(L18)	i.c.	2+
B30-MDP	i.c.	-
B30-MDP (Nikkol HCO-60)[c]	i.v.	2+
LPS (Difco, $E. coli$)	i.v.	-

Three outbred Hartley guinea pigs (male and female) per group were prepared by intra-footpad injection of heat-killed human tubercle bacilli (H37Rv, 100 µg) with water-in-mineral oil emulsion. Thirty days later, each animal was submitted to provocation by intracutaneous or intravenous injection of a test compound in PBS. [a]Prepared from cell walls of *L. plantarum* (ATCC 8014) by use of peptidoglycan-hydrolysing enzyme and kindly supplied by Drs S. Kawata and K. Yokogawa (Research and Development Division, Dainippon Pharmaceutical Co., Osaka). [b]One out of 3 guinea pigs showed a positive reaction. [c]Dissolved in Nikkol HCO-60.

C. Polyarthritis-producing Activity

Various bacterial cell walls and their peptidoglycans were
proved to be capable of substituting for mycobacterial cells
in FCA to produce adjuvant arthritis (Koga *et al.*, 1973,
1976). Previous studies showed that neither a monomer of
peptidoglycan subunit nor synthetic MDP produced the arthritis, while a dimer of the subunits was highly arthritogenic
(Koga *et al.*, 1979; Kohashi *et al.*, 1976, 1977). Subsequent
studies, however, have revealed that MDP can produce moderate
to severe arthritis in WKA and Lewis rats with high incidence
when injected in a water-in-mineral oil emulsion made of some
FIA (Difco) (Kohashi *et al.*, 1980; Nagao and Tanaka, 1980).
     Keeping this somewhat complicated history of studies in
mind, B30-MDP and MDP-Lys(L18) have been examined for their
polyarthritis-inducing activity. It can be seen from Table
13 that intra-footpad or subcutaneous injection into SD/NCrj
rats of 1 mg/kg body weight of B30-MDP in Intralipid®, PBS or
Nikkol HCO-60 did not induce polyarthritis even after repeated
injections. However, 12 daily intravenous injections of
1 mg/kg of B30-MDP dissolved in Nikkol HCO-60, or daily subcutaneous administration of 5 mg/kg of B30-MDP in PBS over
2 to 3 weeks caused mild and moderate to severe polyarthritis,
respectively. Similarly, daily injections, subcutaneous or
intravenous, of 1 to 2 mg/kg of MDP-Lys(L18) in PBS for one
to several weeks resulted in a high incidence of induction of
moderate polyarthritis. Histological studies revealed that
arthritis induced by B30-MDP or MDP-Lys(L18) consisted of an
acute exudative synovitis, subsynovitis and tendinitis infiltrated with massive numbers of polymorphonuclear leucocytes
(Kohashi *et al.*, 1981). No information is available on
arthritogenicity of quinonyl-MDP-66 at this time.
     It is unknown whether humans are as susceptible to the
arthritogenic activity of B30-MDP or MDP-Lys(L18) as the rats
used in the present study. But the above finding should be
kept in mind in attempts to utilize the immunostimulating
activities of lipophilic derivatives of MDP in humans, particularly when repeated injections are needed to obtain satisfactorily beneficial effects.

VI. CONCLUSION

It is possible to conclude from the results given above that
some lipophilic derivatives of MDP could be utilized in
treatment and prevention of microbial infections and tumours
in humans. However, we must realize that these compounds,
which mainly exert their biological effects through modulation of various mechanisms of host defence, natural or

TABLE 13

Polyarthritis-inducing ability of lipophilic derivatives of MDP by single or daily injections in SD/NCrj rats

Test compound	Vehicle	Dose (mg/kg)	Route	Total injection (times, daily)	Sex[a]	Polyarthritis Incidence (attacked/total)	Polyarthritis Onset day	Polyarthritis Score (severity)
B30–MDP	PBS	1	s.c.	12	M	0/8		
	PBS	1	s.c.	12	F	0/9		
	PBS	5	s.c.	14	M	8/8	15–16	9.5 (moderate)
	PBS	5	s.c.	21	F	9/9	12–15	13.8 (severe)
	Nikkol HCO-60	1	i.v.	12	F	10/10	2– 9	5.3 (mild)
	Nikkol HCO-60	1	f.p.	1	F	0/7		
	Intralipid®	1	f.p.	1	F	0/7		
MDP-Lys(L18)	PBS	1	s.c.	46	M	7/7	6–14	7.6 (moderate)
	PBS	1	s.c.	46	F	9/9	7–15	7.8 (moderate)
	PBS	2	i.v.	7	M	6/6	2–10	8.8 (moderate)
	PBS	2	i.v.	7	F	5/5	3– 9	9.6 (moderate)
None	PBS	—	s.c.	12	F	0/5		
	PBS	—	i.v.	7	F	0/5		
	Nikkol HCO-60	—	i.v.	12	F	0/10		

[a] F: female; M: male.

acquired, are primarily a double-edged sword, because over-stimulation of immunological mechanisms sometimes brings about more deleterious than beneficial effects on the host. We should thus be cautious in clinical application of MDP-relating compounds of possible occurrence of adverse side-effects, even when using those compounds whose detrimental activity has been proved to be the minimum by basic experiments in laboratory animals.

VII. REFERENCES

Abdulla, E.M. and Schwab, J.H. (1965). Immunological properties of bacterial cell wall mucopeptides. *Proc. Soc. Exp. Biol. Med.* **118**, 359-362.

Adam, A., Souvannavong, V. and Lederer, E. (1978). Nonspecific MIF-like activity induced by the synthetic immunoadjuvant: *N*-acetyl muramyl-L-alanyl-D-isoglutamine (MDP). *Biochem. Biophys. Res. Commun.* **85**, 684-690.

Akagawa, K.S. and Tokunaga, T. (1980). Effect of synthetic muramyl dipeptide (MDP) on differentiation of myeloid leukaemic cells. *Microbiol. Immunol.* **24**, 1005-1011.

Audibert, F. and Chedid, L. (1980). Recent advances concerning the use of muramyl dipeptide derivatives as vaccine potentiators. *Prog. Clin. Biol. Res.* **47**, 325-338.

Audibert, F. and Jolivet, M. (1982). "Immunomodulation by Microbial Products and Related Synthetic Compounds" (Eds Y. Yamamura and S. Kotani), pp. 241-244. Excerpta Medica, Amsterdam.

Azuma, I., Sugimura, K., Taniyama, T., Yamawaki, M., Yamamura, Y., Kusumoto, S., Okada, S. and Shiba, T. (1976a). Adjuvant activity of mycobacterial fractions: adjuvant activity of synthetic *N*-acetylmuramyl-dipeptide and the related compounds. *Infect. Immun.* **14**, 18-27.

Azuma, I., Taniyama, T., Sugimura, K., Aladin, A.A. and Yamamura, Y. (1976b). Mitogenic activity of the cell walls of mycobacteria, nocardia, corynebacteria and anaerobic coryneforms. *Jpn. J. Microbiol.* **20**, 263-271.

Azuma, I., Sugimura, K., Yamawaki, M., Uemiya, M., Kusumoto, S., Okada, S., Shiba, T. and Yamamura, Y. (1978). Adjuvant activity of synthetic 6-*O*-"mycoloyl"-*N*-acetylmuramyl-L-alanyl-D-isoglutamine and related compounds. *Infect. Immun.* **20**, 600-607.

Azuma, I., Uemiya, M., Saiki, I., Yamawaki, M., Tanio, Y., Kusumoto, S., Shiba, T., Kusama, T., Tobe, K., Ogawa, H. and Yamamura, Y. (1979a). Synthetic immunoadjuvants — new immunotherapeutic agents. *Dev. Immunol.* **6**, 311-330.

Azuma, I. and Yamamura, Y. (1979b). Immunotherapy of cancer with BCG cell wall skeleton and related materials. *GANN Monogr. Cancer Res.* **24**, 121-141.

Azuma, I., Yamawaki, M., Uemiya, M., Saiki, I., Tanio, Y., Kobayashi, S., Fukuda, T., Imada, I. and Yamamura, Y. (1979c). Adjuvant and antitumour activities of quinonyl-*N*-acetylmuramyl-dipeptides. *GANN* **70**, 847-848.

Barker, L.A., Campbell, P.A. and Hollister, J.R. (1977). Chemotaxigenesis and complement fixation by *Listeria monocytogenes* cell wall fractions. *J. Immunol.* **119**, 1723-1726.

Barot-Ciorbaru, R., Wietzerbin, J., Petit, J.-F., Chedid, L., Falcoff, E. and Lederer, E. (1978). Induction of interferon synthesis in mice by fractions from *Nocardia*. *Infect. Immun.* **19**, 353-356.

Bliznakov, E.G. (1977). "Biomedical and Clinical Aspects of Coenzyme Q" (Eds K. Folkers and Y. Yamamura), pp. 73-83. Elsevier, Amsterdam.

Carelli, C., Audibert, F. and Chedid, L. (1981). Persistent enhancement of cell-mediated and antibody immune responses after administration of muramyl dipeptide derivatives with antigen in metabalizable oil. *Infect. Immun.* **33**, 312-314.

Chedid, L., Parant, M., Parant, F., Lefrancier, P., Choay, J. and Lederer, E. (1977). Enhancement of nonspecific immunity to *Klebsiella pneumoniae* infection by a synthetic immunoadjuvant (*N*-acetylmuramyl-L-alanyl-D-isoglutamine) and several analogues. *Proc. Natl. Acad. Sci. USA* **74**, 2089-2093.

Chedid, L., Audibert, F. and Johnson, A.G. (1978a). Biological activities of muramyl dipeptide, a synthetic glycopeptide analogues to bacterial immunoregulating agents. *Prog. Allergy* **25**, 63-105.

Chedid, L., Parant, M. and Parant, F. (1978b). "Immunology 1978" (Eds Gergely, Medgyesi and Hollán), pp. 449-462. Publishing House of the Hungarian Academy of Science, Budapest, Hungary.

Chedid, L., Parant, M., Parant, F., Audibert, F., Lefrancier, F., Choay, J. and Sela, M. (1979). Enhancement of certain biological activities of muramyl dipeptide derivatives after conjugation to a multi-poly(DL-alanine)--poly(L-lysine) carrier. *Proc. Natl. Acad. Sci. USA* **76**, 6557-6561.

Chedid, L., Audibert, F., Lefrancier, P., Choay, J. and Lederer, E. (1976). Modulation of the immune response by a synthetic adjuvant and analogues. *Proc. Natl. Acad. Sci. USA* **73**, 3472-2475.

Damais, C., Bona, C., Chedid, L., Fleck, J., Naucier, C. and
    Martin, J.P. (1975). Mitogenic effect of bacterial pep-
    tidoglycans possessing adjuvant activity. *J. Immunol.* **115**,
    268-271.
Damais, C., Parant, M. and Chedid, L. (1977). Nonspecific
    activation of murine spleen cells *in vitro* by a synthetic
    immunoadjuvant (*N*-acetylmuramyl-L-alanyl-D-isoglutamine).
    Cell. Immunol. **34**, 49-56.
Dewhirst, F.E. (1982). *N*-acetylmuramyl dipeptide stimulation
    of bone resorption in tissue culture. *Infect. Immun.* **35**,
    133-137.
Dinarello, C.A., Elin, R.J., Chedid, L. and Wolff, S.M.
    (1978). The pyrogenicity of the synthetic adjuvant muramyl
    dipeptide and two structural analogues. *J. Infect. Dis.*
    138, 760-767.
Dziarski, R. (1978). Immunosuppressive effect of *Staphylo-
    coccus aureus* peptidoglycan on antibody response in mice.
    Int. Arch. Allergy Appl. Immunol. **57**, 304-311.
Elin, R.J., Wolff, S.M. and Chedid, L. (1976). Nonspecific
    resistance to infection induced in mice by a water-soluble
    adjuvant derived from *Mycobacterium smegmatis*. *J. Infect.
    Dis.* **133**, 500-505.
Ellouz, F., Adam, A., Ciorbaru, R. and Lederer, E. (1974).
    Minimal structural requirements for adjuvant activity of
    bacterial peptidoglycan derivatives. *Biochem. Biophys.
    Res. Commun.* **59**, 1317-1325.
Emori, K. and Tanaka, A. (1978). Granuloma formation by syn-
    thetic bacterial cell wall fragment: muramyl dipeptide.
    Infect. Immun. **19**, 613-620.
Fidler, I.J., Sone, S., Fogler, W.E. and Barnes, Z.L. (1981).
    Eradication of spontaneous metastases and activation of
    alveolar macrophages by intravenous injection of liposomes
    containing muramyl dipeptide. *Proc. Natl. Acad. Sci. (USA)*
    78, 1680-1684.
Franser-Smith, E.B. and Mattews, T.R. (1981). Protective
    effect of muramyl dipeptide analogues against infections
    of *Pseudomonas aeruginosa* or *Candida albicans* in mice.
    Infect. Immun. **34**, 676-683.
Greenblatt, J., Boackle, R.J. and Schwab, J.H. (1978). Acti-
    vation of the alternate complement pathway by peptidogly-
    can from streptococcal cell wall. *Infect. Immun.* **19**, 296-
    303.
Hadden, J.W. (1978). Effects of isoprinosine, Levamisole,
    muramyl didpeptide, and SM1213 on lymphocyte and macrophage
    function *in vitro*. *Cancer Treat. Rep.* **62**, 1981-1985.
Hamada, S., Narita, T., Kotani, S. and Kato, K. (1971).
    Studies on cell walls of group A *Streptococcus pyogenes*,

type 12. II. Pyrogenic and related biological activities
of the higher molecular weight fraction of an enzymatic
digest of the cell walls. *Biken J.* **14**, 217-231.

Harada, K., Kotani, S., Takada, H., Tsujimoto, M., Hirashi, Y.
Kusumoto, S., Shiba, T., Kawata, S., Yokogawa, K., Nishi-
mura, H., Kitaura, T. and Nakajima, T. (1982). Liberation
of serotonin from rabbit blood platelets by bacterial cell
walls and some related compounds. *Infect. Immun.* (In
press).

Heymer, B., Finger, H. and Wirsing, C.H. (1978). Immunoadjuvant effects of the synthetic muramyl-dipeptide (MDP)
N-acetylmuramyl-L-alanyl-D-isoglutamine. *Z. Immunitaetsforsch. Immunobiol.* **155**, 87-92.

Igarashi, T., Okuda, M., Azuma, I. and Yamamura, Y. (1977).
Adjuvant activity of synthetic *N*-acetylmuramyl-L-alanyl-
D-isoglutamine and related compounds on cell-mediated
cytotoxicity in syngeneic mice. *Cell. Immunol.* **34**, 270-
278.

Imada, I., Azuma, I., Kishimoto, S., Yamamura, Y. and Morimoto, H. (1972). The effect of ubiquinone-7 and its metabolites on the immune response. *Int. Arch. Allergy Appl.
Immunol.* **43**, 898-907.

Imai, K., Tomioka, M., Nagao, S., Kushima, K. and Tanaka, A.
(1980). Biochemical evidence for activation of guinea pig
macrophages by muramyl dipeptide. *Biomed. Res.* **1**, 300-307.

Imai, K. and Tanaka, A. (1981). Effect of muramyldipeptide,
a synthetic bacterial adjuvant, on enzyme release from
cultured mouse macrophages.   *Microbiol. Immunol.* **25**,
51-62.

Iribe, H., Koga, T., Onoue, K., Kotani, S., Kusumoto, S. and
Shiba, T. (1981). Macrophage-stimulating effect of a synthetic muramyl dipeptide and its adjuvant-active and -in-
active analogues for the production of T-cell activating
monokines. *Cell. Immunol.* **64**, 73-83.

Ishihara, Y., Takada, H., Kotani, S., Kusumoto, S., Shiba, T.,
Kawata, S. and Yokogawa, K. (1982). "Immunomodulation by
microbial products and related synthetic compounds". (Eds
Y. Yamamura and S. Kotani), pp. 217-220. Excerpta Medica,
Amsterdam.

Johnson, A.G., Audibert, F. and Chedid, L. (1978). Synthetic
immunoregulating molecules: a potential bridge between
cytostatic chemotherapy and immunotherapy of cancer.
Cancer Immunol. Immunother. **3**, 219-227.

Jones, J.M. and Schwab, J.H. (1970). Effects of streptococcal
cell wall fragments on phagocytosis and tissue culture
cells. *Infect. Immun.* **1**, 232-242.

Juy, D. and Chedid, L. (1975). Comparison between macrophage activation and enhancement of nonspecific resistance to tumours by mycobacterial immunoadjuvants. *Proc. Natl. Acad. Sci. (USA)* **72**, 4105-4109.

Kato, K., Kotani, S., Kawano, K., Monodane, T., Kitamura, H., Kusumoto, S. and Shiba, T. (1982). "Immunomodulation by Microbial Products and Related Synthetic Compounds". (Eds Y. Yamamura and S. Kotani), pp. 181-184. Excerpta Medica, Amsterdam.

Kawasaki, A. (1982). Activation of human complement system by bacterial cell walls, their water-soluble enzymatic digests and related synthetic compounds. *J. Osaka Univ. Dent. Soc.* **27**, 46-61. (In Japanese with English summary).

Kierszenbaum, F. and Ferraresi, R.W. (1979). Enhancement of host resistance against *Trypanosoma cruzi* infection by the immunoregulatory agent muramyl dipeptide. *Infect. Immun.* **25**, 273-278.

Kimura, Y., Norose, Y., Kato, T., Furuya, M., Hida, M., Banno, Y., Kotani, S., Kusumoto, S. and Shiba, T. (1982). "Immuno-modulation by Microbial Products and Related Synthetic Compounds". (Eds Y. Yamamura and S. Kotani), pp. 225-228. Excerpta Medica, Amsterdam.

Kinoshita, F. (1978). Studies on the immunoadjuvant activities of bacterial cell wall components — with special reference to the effects of administration with various vehicles. *J. Osaka Univ. Dent. Soc.* **23**, 141-157. (In Japanese with English summary).

Kishimoto, T., Hirai, Y., Nakanishi, K., Azuma, I., Nagamatsu, A. and Yamamura, Y. (1979). Regulation of antibody response in different immunoglobulin classes. VI. Selective suppression of IgE response by administration of antigen conjugated muramylpeptides. *J. Immunol.* **123**, 2709-2715.

Koga, T., Pearson, C.M., Narita, T. and Kotani, S. (1973). Polyarthritis induced in the rat with cell walls from several bacteria and two streptomyces species. *Proc. Soc. Exp. Biol. Med.* **143**, 824-827.

Koga, T., Kotani, S., Narita, T. and Pearson, C.M. (1976). Induction of adjuvant arthritis in the rat by various bacterial cell walls and their water-soluble components. *Int. Arch. Allergy Appl. Immunol.* **51**, 206-213.

Koga, T., Maeda, K., Onoue, K., Kato, K. and Kotani, S. (1979). Chemical structure required for immunoadjuvant and arthritogenic activities of cell wall peptidoglycans. *Mol. Immunol.* **16**, 153-162.

Kohashi, O., Pearson, C.M., Watanabe, Y., Kotani, S. and Koga, T. (1976). Structural requirements for arthrito-genicity of peptidoglycans from *Staphylococcus aureus* and

Lactobacillus plantarum and analogous synthetic compounds.
J. Immunol. **116**, 1635-1639.

Kohashi, O., Pearson, C.M., Watanabe, Y. and Kotani, S.
(1977). Preparation of arthritogenic hydrosoluble peptidoglycans from both arthritogenic and nonarthritogenic bacterial cell walls. *Infect. Immun.* **16**, 861-866.

Kohashi, O., Tanaka, A., Kotani, S., Shiba, T., Kusumoto, S.,
Yokogawa, K., Kawata, S. and Ozawa, A. (1980). Arthritis-
inducing ability of a synthetic adjuvant, *N*-acetylmuramyl
peptides, and bacterial disaccharide peptides related to
different oil vehicles and their composition. *Infect.
Immun.* **29**, 70-75.

Kohashi, O., Kohashi, Y., Kotani, S. and Ozawa, A. (1981). A
new model of experimental arthritis induced by an aqueous
form of synthetic adjuvant in immunodeficient rats (SHR
and nude rats). *The Ryumachi* **21** (supp.), 149-156.

Kotani, S., Watanabe, Y., Kinoshita, F., Shimono, T., Morisaki, I., Shiba, T., Kusumoto, S., Tarumi, Y. and Ikenaka,
K. (1975). Immunoadjuvant activities of synthetic *N*-
acetylmuramyl-peptides or -amino acids. *Biken J.* **18**, 105-
111.

Kotani, S., Watanabe, Y., Shimono, T., Harada, K., Shiba, T.,
Kusumoto, S., Yokogawa, K. and Taniguchi, M. (1976).
Correlation between the immunoadjuvant activities and
pyrogenicities of synthetic *N*-acetylmuramyl-peptides or
-amino acids. *Biken J.* **19**, 9-13.

Kotani, S., Kinoshita, F., Morisaki, I., Shimono, T., Okunaga,
T., Takada, H., Tsujimoto, M., Watanabe, Y., Kato, K.,
Shiba, T., Kusumoto, S. and Okada, S. (1977a). Immunoadjuvant activities of synthetic 6-*O*-acyl-*N*-acetylmuramyl-
L-alanyl-D-isoglutamine with special reference to the
effect of its administration with liposomes. *Biken J.* **20**,
95-103.

Kotani, S., Watanabe, Y., Kinoshita, F., Kato, K., Harada, K.,
Shiba, T., Kusumoto, S., Tarumi, Y., Ikenaka, K., Okada,
S., Kawata, S. and Yokogawa, K. (1977b). Gelation of the
amoebocyte lysate of *Tachypleus tridentatus* by cell wall
digest of several Gram-positive bacteria and synthetic
peptidoglycan subunits of natural and unnatural configura-
tions. *Biken J.* **20**, 5-10.

Kotani, S., Tsujimoto, M., Morisaki, I., Okunaga, T., Kinoshita, F., Takada, H., Kato, K., Shiba, T., Kusumoto, S.,
Inage, M., Yamamura, Y. and Azuma, I. (1978). "Thirteenth
Joint Meeting Tuberculosis Panel" (Ed. Y. Yamamura), pp.
163-178. Osaka, Japan.

Kotani, S., Takada, H., Tsujimoto, M., Ogawa, T., Kato, K.,
Okunaga, T., Ishihara, Y., Kawasaki, A., Morisaki, I.,
Kono, N., Shimono, T., Shiba, T., Kusumoto, S., Inage, M.,

Harada, K., Kitaura, T., Kano, S., Inai, S., Nagai, K.,
Matsumoto, M., Kubo, T., Kato, M., Tada, Z., Yokogawa, K.,
Kawata, S. and Inoue, A. (1981). "Immunomodulation by
Bacteria and their Products". (Eds H. Friedman, T.W.
Klein and A. Szentivanyi), pp. 231-274. Plenum Press,
New York.
Krahenbuhl, J.L., Sharma, S.D., Ferraresi, R.W. and Remington,
J.S. (1981). Effects of muramyl dipeptide treatment on
resistance to infection with *Toxoplasma gondii* in mice.
Infect. Immun. **31**, 716-722.
Leclerc, C., Löwy, I. and Chedid, L. (1978). Influence of
MDP and of some analogous synthetic glycopeptides on the
in vitro mouse spleen cell viability and immune response
to sheep erythrocytes. *Cell. Immunol.* **38**, 286-293.
Leclerc, C., Bourgeois, E. and Chedid, L. (1979a). Enhancement by muramyl dipeptide of *in vitro* nude mice responses
to a T-dependent antigen. *Immunol. Commun.* **8**, 55-64.
Leclerc, C., Juy, D., Bourgeois, E. and Chedid, L. (1979b).
In vivo regulation of humoral and cellular immune responses
of mice by a synthetic adjuvant, *N*-acetylmuramyl-L-alanyl-
D-isoglutamine, muramyl dipeptide for MDP. *Cell. Immunol.*
45, 199-206.
Leclerc, C., Juy, D. and Chedid, L. (1979c). Inhibitory and
stimulatory effects of a synthetic glycopeptide (MDP) on
the *in vitro* PFC response: factors affecting the response.
Cell. Immunol. **42**, 336-343.
Lederer, E. (1980a). "Fourth International Congress of
Immunology. Immunology 80. Progress in Immunology IV".
(Eds M. Fougereau and J. Dausset), pp. 1194-1211. Aca-
demic Press, London.
Lederer, E. (1980b). Synthetic immunostimulants derived from
the bacterial cell wall. *J. Med. Chem.* **23**, 819-825.
Löwy, I., Bona, C. and Chedid, L. (1977). Target cells for
the activity of a synthetic adjuvant: muramyl dipeptide.
Cell. Immunol. **29**, 195-199.
Löwy, I., Leclerc, C., Bourgeois, E. and Chedid, L. (1980).
Inhibition of mitogen-induced polyclonal activation by a
synthetic adjuvant, muramyl dipeptide (MDP). *J. Immunol.*
124, 320-325.
Maeda, K., Koga, T., Onoue, K., Kotani, S. and Sumiyoshi, A.
(1980). Induction of delayed type hypersensitivity-like
skin reaction by peptidoglycans of bacterial cell walls.
Microbiol. Immunol. **24**, 335-348.
Mašek, K., Kadlecová, O. and Petrovický, P. (1978). "Toxins:
Animal, Plant and Microbial". (Ed. P. Rosenberg), pp. 991-
1003. Pergamon Press, Inc., Oxford and New York.
Matsumoto, K., Ogawa, H., Kusama, T., Nagase, O., Sawaki, N.,
Inage, M., Kusumoto, S., Shiba, T. and Azuma, I. (1981).
Stimulation of nonspecific resistance to infection induced

by 6-*O*-acyl muramyl dipeptide analogues in mice. *Infect. Immun.* **32**, 748-758.

Matsumoto, K., Otani, T., Une, T., Osada, Y., Ogawa, H. and Azuma, I. (1982). Stimulation of nonspecific resistance to infection induced by MDP analogues: structure-activity relationships of analogues substituted in the γ-carboxyl group, and bacteriological evaluation of MDP-Lys(L18). (Manuscript in preparation).

Misaki, A., Yukawa, S., Tsuchiya, K. and Yamasaki, T. (1966). Studies on cell walls of *Mycobacteria*. I. Chemical and biological properties of the cell walls and the mucopeptide of BCG. *J. Biochem.* **59**, 388-396.

Morisaki, I. (1980). Species and strain differences in responsiveness of laboratory animals to immunopotentiating activities of bacterial cell walls and their related adjuvants. *J. Osaka Univ. Dent. Soc.* **25**, 229-249. (In Japanese with English summary).

Nagai, Y., Akiyama, K., Kotani, S., Watanabe, Y., Shimono, T., Shiba, T. and Kusumoto, S. (1978a). Structural specificity of synthetic peptide adjuvant for induction of experimental allergic encephalomyelitis. *Cell. Immunol.* **35**, 168-172.

Nagai, Y., Akiyama, K., Suzuki, K., Kotani, S., Watanabe, Y. Shimono, T., Shiba, T., Kusumoto, S., Ikuta, F. and Takeda, S. (1978b). Minimum structural requirements for encephalitogen and for adjuvant in the induction of experimental allergic encephalomyelitis. *Cell. Immunol.* **35**, 158-167.

Nagao, S. and Tanaka, A. (1980). Muramyl dipeptide-induced adjuvant arthritis. *Infect. Immun.* **28**, 624-626.

Nagao, S., Iwata, Y. and Tanaka, A. (1982). "Immunomodulation by Microbial Products and Related Synthetic Compounds". (Eds Y. Yamamura and S. Kotani), pp. 189-192. Excerpta Medica, Amsterdam.

Namba, M., Ogura, T., Hirao, F., Azuma, I. and Yamamura, Y. (1978). Antitumour activity of peritoneal exudate cells inudced by cell-wall skeleton of *Mycobacterium bovis* BCG. *Gann* **69**, 831-834.

Nguyen-Huy, H., Nauciel, C. and Wermuth, C.-G. (1976). Immunochemical study of the peptidoglycan of gram-negative bacteria. *Eur. J. Biochem.* **66**, 79-84.

Nozawa, R.T., Sekiguchi, R. and Yokota, T. (1980). Stimulation by conditioned medium of L-929 fibroblasts, *E. coli* lipopolysaccharide, and muramyl dipeptide of candidacidal activity of mouse macrophages. *Cell. Immunol.* **53**, 116-124.

Ogawa, T., Kotani, S., Tsujimoto, M., Kusumoto, S., Shiba, T., Kawata, S. and Yokogawa, K. (1982a). Contractile effects of bacterial cell walls, their enzymatic digests, and muramyl dipeptides on ileal strips from guinea pigs. *Infect. Immun.* **35**, 612-619.

Ogawa, T., Takada, H., Kotani, S., Kusumoto, S., Shiba, T., Kawata, S., Yokogawa, K. and Inoue, A. (1982b). "Immuno-modulation by Microbial Products and Related Synthetic Compounds". (Eds Y. Yamamura and S. Kotani), pp. 221-224. Excerpta Medica, Amsterdam.

Ogawa, T., Kotani, S., Fukuda, K., Tsukamoto, Y., Mori, M., Kusumoto, S. and Shiba, T. (1982c). Stimulation of chemo-taxis of human monocytes by bacterial cell walls and mura-myl peptide. *Infect. Immun.* (Submitted).

Ohkuni, H. and Kimura, Y. (1976). Increased capillary perme-ability in guinea pigs and rats by peptidoglycan fraction extracted from group A streptococcal cell walls. *Exp. Cell Biol.* **44**, 83-94.

Ohkuni, H., Norose, Y., Hayama, M., Kimura, Y., Kotani, S., Shiba, T., Kusumoto, S., Yokogawa, K. and Kawata, S. (1977). Adjuvant activities in production of reaginic antibody in mice of bacterial cell wall peptidoglycans or peptidoglycan subunits and of synthetic *N*-acetylmuramyl dipeptides. *Biken J.* **20**, 131-136.

Ohkuni, H., Norose, Y., Ohta, M., Hayama, M., Kimura, Y., Tsujimoto, M., Kotani, S., Shiba, T., Kusumoto, S., Yoko-gawa, K. and Kawata, S. (1979). Adjuvant activities in production of reaginic antibody by bacterial cell wall peptidoglycan or synthetic *N*-acetylmuramyl dipeptides in mice. *Infect. Immun.* **24**, 313-318.

Oppenheim, J.J., Togawa, A., Chedid, L. and Mizel, S. (1980). Components of mycobacteria and muramyl dipeptide with adjuvant activity induce lymphocyte activating factor. *Cell. Immunol.* **50**, 71-81.

Osada, Y., Mitsuyama, M., Matsumoto, K., Une, T., Otani, T., Ogawa, H. and Nomoto, K. (1982a). Augmenting effect of a synthetic derivative of muramyl dipeptide, L18-MDP(Ala), on resistance against microbial infections of immuno-suppressed hosts. *Infect. Immun.* (submitted).

Osada, Y., Mitsuyama, M., Une, T., Matsumoto, K., Otani, T., Satoh, H., Ogawa, H. and Nomoto, K. (1982b). Effect of L18-MDP(Ala), a synthetic derivative of muramyl dipeptide, on nonspecific resistance against infections with various species of bacteria. *Infect. Immun.* (submitted).

Pabst, M.J. and Johnston, R.B. Jr. (1980). Increased produc-tion of superoxide anion by macrophages exposed *in vitro* to muramyl dipeptide or lipopolysaccharide. *J. Exp. Med.* **151**, 101-114.

Pabst, M.J., Cummings,N.P., Shiba, T., Kusumoto, S. and Kotani, S. (1980). Lipophilic derivative of muramyl di-peptide is more active than muramyl dipeptide in priming macrophages to release superoxide anion. *Infect. Immun.* **29**, 617-622.

Parant, M. (1979). Biologic properties of a new synthetic adjuvant, muramyl dipeptide (MDP). *Springer Semin. Immunopathol.* **2**, 101-118.

Parant, M., Damais, C., Audibert, F., Parant, F., Chedid, L., Sache, E., Lefrancier, P., Choay, J. and Lederer, E. (1978). *In vivo* and *in vitro* stimulation of nonspecific immunity by the β-D-p-aminophenyl glycoside of *N*-acetylmuramyl-L-alanyl-D-isoglutamine and an oligomer prepared by cross-linking with glutaraldehyde. *J. Infect. Dis.* **138**, 378-386.

Parant, M.A., Audibert, F.M., Chedid, L.A., Level, M.R., Lefrancier, P.L., Choay, J.P. and Lederer, E. (1980). Immunostimulant activities of a lipophilic muramyl dipeptide derivative and of desmuramyl peptidolipid analogues. *Infect. Immun.* **27**, 826-831.

Prunet, J., Birrien, J.L., Panijel, J. and Liacopoulos, P. (1978). On the mechanism of early recovery of specifically depleted lymphoid cell populations by nonspecific activation of T cells. *Cell. Immunol.* **37**, 151-161.

Reese, R.T., Trager, W., Jensen, J.B., Miller, D.A. and Tantravahi, R. (1978). Immunization against malaria with antigen from *Plasmodium flaciparum* cultivated *in vitro*. *Proc. Natl. Acad. Sci. USA* **75**, 5665-5668.

Reichert, C.M., Carelli, C., Jolivet, M., Audibert, F., Lefrancier, P. and Chedid, L. (1980). Synthesis of conjugates containing *N*-acetylmuramyl-L-alanyl-D-isoglutaminyl (MDP). Their use as hapten-carrier system. *Mol. Immunol.* **17**, 357-363.

Riveau, G., Mašek, K., Parant, M. and Chedid, L. (1980). Central pyrogenic activity of muramyl dipeptide. *J. Exp. Med.* **152**, 869-877.

Roberson, B.S. and Schwab, J.H. (1961). Endotoxic properties associated with cell walls of group A streptococci. *J. Infect. Dis.* **108**, 25-34.

Rook, G.A.W. and Stewart-Tull, D.E.S. (1976). The dissociation of adjuvant properties of mycobacterial components from mitogenicity, and from the ability to induce the release of mediators from macrophages. *Immunology* **31**, 389-396.

Rotta, J. and Bednář, B.B. (1969). Biological properties of cell wall mucopeptides of hemolytic streptococci. *J. Exp. Med.* **130**, 31-47.

Saiki, I., Tanio, Y., Yamawaki, M., Uemiya, M., Kobayashi, S., Fukuda, T., Yukimasa, H., Yamamura, Y. and Azuma, I. (1981). Adjuvant activities of quinonyl-*N*-acetylmuramyl dipeptides in mice and guinea pigs. *Infect. Immun.* **31**, 114-121.

Saito -Taki, T., Tanabe, M.J., Mochizuki, H., Matsumoto, T.,
    Nakano, M., Takada, H., Tsujimoto, M., Kotani, S., Kusu-
    moto, S., Shiba, T., Yokogawa, K. and Kawata, S. (1980).
    Polyclonal B cell activation by cell wall preparations of
    Gram-positive bacteria. *In vitro* responses of spleen
    cells obtained from Balb/c, nu/nu, nu/+, C3H/He, C3H/HeJ
    and hybrid (CBA/N x Balb/c)F1 mice. *Microbiol. Immunol.*
    24, 209-218.
Schwab, J.H. (1979). "Microbiology — 1979". (Ed. D. Schles-
    singer), pp. 209-214. American Society for Microbiology,
    Washington, D.C.
Sharma, B., Kohashi, O., Mickey, M.R. and Terasaki, P.I.
    (1975). Effect of water-soluble adjuvants on *in vitro*
    lymphocyte immunization. *Cancer Res.* **35**, 666-669.
Sharma, S.D., Tsai, V., Krahenbuhl, J.L. and Remington, J.S.
    (1981). Augmentation of mouse natural killer cell activity
    by muramyl dipeptide and its analogues. *Cell. Immunol.*
    62, 101-109.
Siddiqui, W.A. (1982). "Immunomodulation by Microbial Pro-
    ducts and Related Synthetic Compounds". (Eds Y. Yamamura
    and S. Kotani), pp. 245-248. Excerpta Medica, Amsterdam.
Siddiqui, W.A., Taylor, D.W., Kan, S.-C., Kramer, K., Rich-
    mond-Crum, S.M., Kotani, S., Shiba, T. and Kusumoto, S.
    (1978). Vaccination of experimental monkeys against *Plas-
    modium falciparum*: a possible safe adjuvant. *Science* **201**,
    1237-1239.
Smialowicz, R.J. and Schwab, J.H. (1978). Inhibition of
    macrophage phagocytic activity by group A streptococcal
    cell walls. *Infect. Immun.* **20**, 258-261.
Sone, S. and Fidler, I.J. (1980). Synergistic activation by
    lymphokines and muramyl dipeptide of tumoricidal proper-
    ties in rat alveolar macrophages. *J. Immunol.* **125**, 2454-
    2460.
Specter, S., Friedman, H. and Chedid, L. (1977). Dissocia-
    tion between the adjuvant vs mitogenic activity of a
    synthetic muramyl dipeptide for murine splenocytes. *Proc.
    Soc. Exp. Biol. Med.* **155**, 349-352.
Stewart-Tull, D.E.S., Shimono, T., Kotani, S., Kato, M.,
    Ogawa, Y., Yamamura, Y., Koga, T. and Pearson, C.M. (1975).
    The adjuvant activity of a nontoxic, water-soluble glyco-
    peptide present in large quantities in culture filtrate
    of *Mycobacterium tuberculosis* strain DT. *Immunology* **29**,
    1-15.
Stewart-Tull, D.E.S., Davies, M. and Jackson, D.M. (1978).
    The binding of adjuvant-active mycobacterial peptidogly-
    colipids and glycopeptides to mammalian membranes
    and their effect on artificial lipid bilayers. *Immunology*
    34, 57-66.

Sugimoto, M., Germain, R.N., Chedid, L. and Benacerraf, B.
    (1978). Enhancement of carrier-specific helper T cell
    function by the synthetic adjuvant, *N*-acetylmuramyl-L-
    alanyl-D-isoglutamine (MDP). *J. Immunol.* **120**, 980-982.
Sugimura, K., Azuma, I., Yamamura, Y., Imada, I. and Morimoto,
    H. (1976). The effect of ubiquinone-7 and its metabolites
    on the immune response. III. The effect on the immune
    response to sheep erythrocytes and DNP-Lys-Ficoll in mice.
    Int. J. Vitam. Nutr. Res. **46**, 464-471.
Sugimura, K., Uemiya, M., Saiki, I., Azuma, I. and Yamamura,
    Y. (1979). The adjuvant activity of synthetic *N*-acetyl-
    muramyl-dipeptide: evidence of initial target cells for
    the adjuvant activity. *Cell. Immunol.* **43**, 137-149.
Takada, H., Kotani, S., Kusumoto, S., Tarumi, Y., Ikenaka, K.
    and Shiba, T. (1977). Mitogenic activity of adjuvant-
    active *N*-acetylmuramyl-L-alanyl-D-isoglutamine and its
    analogues. *Biken J.* **20**, 81-85.
Takada, H., Tsujimoto, M., Kato, K., Kotani, S., Kusumoto, S.,
    Inage, M., Shiba, T., Yano, I., Kawata, S. and Yokogawa,
    K. (1979a). Macrophage activation by bacterial cell walls
    and related synthetic compounds. *Infect. Immun.* **25**, 48-53.
Takada, H., Tsujimoto, M., Kotani, S., Kusumoto, S., Inage,
    M., Shiba, T., Nagao, S., Yano, I., Kawata, S. and Yoko-
    gawa, K. (1979b). Mitogenic effects of bacterial cell
    walls, their fragments, and related synthetic compounds on
    thymocytes and splenocytes of guinea pigs. *Infect. Immun.*
    25, 645-652.
Tanaka, A. (1982). "Immunomodulation by Microbial Products
    and Related Synthetic Compounds". (Eds Y. Yamamura and
    S. Kotani), pp. 72-83. Excerpta Medica, Amsterdam.
Tanaka, A., Nagao, S., Nagao, R., Kotani, S., Shiba, T. and
    Kusumoto, S. (1979). Stimulation of the reticuloendothel-
    ial system of mice by muramyl dipeptide. *Infect. Immun.*
    24, 302-307.
Tanaka, A., Nagao, S., Imai, K. and Mori, R. (1980). Macro-
    phage activation by muramyl dipeptide as measured by
    macrophage spreading and attachment. *Microbiol. Immunol.*
    24, 547-557.
Tanaka, A., Nagao, S., Ikegami, S., Shiba, T. and Kotani, S.
    (1982). "Immunomodulation by Microbial Products and
    Related Synthetic Compounds". (Eds Y. Yamamura and S.
    Kotani), pp. 201-204. Excerpta Medica, Amsterdam.
Taniyama, T., Azuma, I., Aladin, A.A. and Yamamura, Y. (1975).
    Effect of cell-wall skeleton of *Mycobacterium bovis* on
    cell-mediated cytotoxicity in tumour-bearing mice. *Gann*
    66, 705-709.

Taniyama, T. and Holden, H.T. (1979). Direct augmentation of
   cytolytic activity of tumour-derived macrophages and macrophage cell lines by muramyl dipeptide. *Cell. Immunol.* **48**,
   369-374.
Tenu, J.-P., Lederer, E. and Petit, J.-F. (1980). Stimulation
   of thymocyte mitogenic protein secretion and of cytostatic
   activity of mouse peritoneal macrophages by trehalose dimycolate and muramyl dipeptide. *Eur. J. Immunol.* **10**, 647-
   653.
Tsujimoto, M. (1981). Fundamental studies to utilize the
   adjuvant activity of 6-*O*-acyl-muramyl dipeptides to potentiate cellular and humoral immune responses for practical
   purpose. *J. Osaka Univ. Dent. Soc.* **26**, 63-83. (In Japanese with English summary).
Uemiya, M., Sugimura, K., Kusama, T., Saiki, I., Yamawaki, M.,
   Azuma, I. and Yamamura, Y. (1979). Adjuvant activity of
   6-*O*-mycoloyl derivatives of *N*-acetylmuramyl-L-seryl-D-iso-
   glutamine and related compounds in mice and guinea pigs.
   Infect. Immun. **24**, 83-89.
Wahl, S.M., Wahl, L.M., MaCarthy, J.B., Chedid, L. and
   Mergenhagen, S.E. (1979). Macrophage activation by mycobacterial water soluble components and synthetic muramyl
   dipeptide. *J. Immunol.* **122**, 2226-2231.
Waters, R.V. and Ferraresi, R.W. (1980). Muramyl dipeptide
   stimulation of particle clearance in several animal
   species. *J. Reticuloendothel. Soc.* **28**, 457-471.
Watson, J. and Whitlock, C. (1978). Effect of a synthetic
   adjuvant on the induction of primary immune responses in
   T cell-depleted spleen cultures. *J. Immunol.* **121**, 383-389.
Yamamoto, Y., Nagao, S., Tanaka, A., Koga, T. and Onoue, K.
   (1978). Inhibition of macrophage migration by synthetic
   muramyl dipeptide. *Biochem. Biophys. Res. Commun.* **80**,
   923-928.
Yamamoto, K., Kakinuma, M., Kato, K., Okuyama, H. and Azuma,
   I. (1980). Relationship of anti-tuberculous protection to
   lung granuloma produced by intravenous injection of synthetic 6-*O*-mycoloyl-*N*-acetylmuramyl-L-alanyl-D-isogluta-
   mine with or without specific antigens. *Immunology* **40**,
   557-564.
Yamamura, Y., Yasamoto, K., Ogura, T. and Azuma, I. (1981).
   "Agents for Augmentation of Cancer Immunology". (Eds
   E.M. Hersh, M.A. Chirigos, and M.J. Mastrangelo), pp. 71-
   90. Raven Press, New York.

DISCUSSION I
CHAPTERS 1-4

Dr BALDWIN: Many of the trials giving Coley's toxins to patients with either operable or nonoperable tumours show variable response rates. For example, inoperable colorectal cancer patients show a 5-year survival of 46%. How does this response compare to other control trials? Our clinical trial on the use of intrapleural BCG post-operatively in treating lung cancer does not show any clinical benefit after 5 years (Lowe *et al.*, 1980).

This has been the finding more recently in other trials. Do you consider this to be due to the use of a less aggressive form of treatment?

Helen NAUTS: We have not compared results of Coley's toxins with other control trials. Answering your second question, I cannot say which form of immunotherapy is less aggressive: BCG or Coley's toxins.

Dr SEDLACEK: Are there any epidemiological studies showing an inverse correlation between people who respond to infection with a high fever rate and their tumour occurrence rate? Are there any experimental data showing that exogenous fever induction in infected animals might help to overcome the disease?

Helen NAUTS: No.

Dr SEDLACEK: So you have no basis for your assumption, that fever induction might help in overcoming tumour disease?

Dr CHEDID: Yes, with respect to parasitic or bacterial infection, this is the case.

Helen NAUTS: Fever is not the only effect the Coley's preparation induces.

Dr KONDO: It is interesting that fever may play a role in the antitumour effect of Coley's toxin. But, from a different point of view, is it possible that those patients with fever have an ability to respond to their tumour, since fever is a result of defence mechanisms against foreign substance?

Dr ROSZKOWSKI: Fever and hyperthermia must be distinguished, as fever involves much more complex mechanisms than hyperthermia. It was shown that cancer cells, particularly sarcoma cells, are highly thermosensitive. Elevation of the body temperature can facilitate killing of cancer cells without the participation of the immune system. On the other hand, prolonged elevation of body temperature may result in immunosuppression. I think that we have to consider this biphasic effect of elevated temperature when we are discussing the beneficial effects of fever.

Dr CHEDID: Fever has been shown to play an important role in the control of bacterial infections. Some recent observations have established very clearly that several bacterial agents produce hyperthermia through the release of a monokine called endogenous pyrogen (E.P.). E.P. has also been shown to activate lymphocytes and to lower iron levels via its effect on the hepatocytes. The combination of elevation of temperature with lowering of iron levels can inhibit *in vivo* and *in vitro* the proliferation of pathogens via blockade of the bacterial "iron-pump" enzymes.

Dr KIRN: I would like to stress the fact that the effect of fever may not be the same on both virus infections and cancer. Hyperthermia may act directly on virus multiplication. We were able to show 15 years ago that when you cut down the fever in rabbits infected with vaccinia virus (with amidopyrine), they die. In this case, fever directly inhibits the virus multiplication in the lung. In cancer cells, at least in cancers induced by viruses, the situation is completely different: at nonpermissive temperature (high temperatures) normal cells may be transformed by oncogenic viruses, at permissive temperatures (low temperatures) transformed cells may lose their transformed status.

Dr CHEDID: We are in complete agreement with Dr Kirn's comments. Nevertheless, it is most exciting to note that thermosensitivity is not the only factor, and that the host secretes a monokine which elevates temperature, which *also* stimulates lymphocytes (LAF) and via the liver cell lowers iron concentrations. These 3 factors build up the host's resistance to bacterial invasiveness considerably.

Dr SCHWAB: Dr Chedid, do you postulate that bacterial-derived adjuvants are a normal part of immune regulation? Normal digestion of bacterial cell walls by macrophages yield peptidoglycan derivatives which work synergistically with monokines. Therefore, this modulation of immune response by MDP may reflect a normal process of regulation by bacteria.

Dr CHEDID: Bacterial components could indeed play a part in normal immune regulation, as increasing evidence seems to show.

Dr ROSZKOWSKI: Dr Chedid, did you observe increased suppressive activity of monocytes after MDP treatment? Is MDP safe in the treatment of such diseases (like cancer or tuberculosis), where we expect that at least part of suppression comes from monocyte function?

Dr CHEDID: Let me start by answering your second question. Yes indeed, MDP like any other immunostimulating agent may suppress the immune response and this has been shown in experimental models. Studies of the mechanism of such effects have until now been correlated to T-suppressor cells and to my knowledge no evidence of an MDP-effect via a suppressor macrophage population has been shown.

Dr GOLDIN: Does MDP decrease the lethal toxicity of cyclophosphamide or does it permit usage of a higher dosage of cyclophosphamide? Could advantage be taken of this for therapy?

Dr CHEDID: Dr Masek from Prague has found evidence that MDP allows the use of smaller amounts of cyclophosphamide. He believes that this effect is mediated through a decreased catabolism in the liver. As with BCG, barbiturate sleep is prolonged by MDP-pretreatment. I think that advantage of this observation could be taken for therapy.

Dr SEDLACEK: MDP seems to be effective in prophylaxis of infections. But what about giving MDP to animals which are already infected? Does it aggravate the disease or increase the cure rate?

Dr KOTANI: Sorry, I cannot answer your question, because we have never done systematic studies on this subject.

Dr CHEDID: Most agents (including MDP) have been used preventively. However, a therapeutic effect has been shown by initial control injection (with hepatocarcinoma) or with mild *Klebsiella* challenge. However, the efficacy is smaller in the case of infections if MDP is given *after* the challenge. That is why we believe that administration of adjuvants *before* cytostatic therapy (such as cyclophosphamide) may be a fruitful strategy.

Dr MITCHELL: What is the role of T-cell stimulation in explaining the effectiveness of lipid-containing MDPs systemically? Our work several years ago suggested that lipid-free materials fail to stimulate T-cells (mitogenically) *in vitro*.

Dr CHEDID: Hydrophilic MDP is not a T-cell mitogen, although certain lipophilic MDP-derivatives have been shown to stimulate T-lymphocytes. However, there exists a lot of data on *in vitro* activity of MDP. Thus MDP was active on several systems measuring macrophage activation. Activity has been shown on NK-cells by Remington *et al*. and MDP was found also to be a weak B-mitogen. Adoptive transfer experiments have shown that *in vivo* there is a cascade of events leading probably through the production of ILy to a stimulation of a T-helper carrier specific cell which in turn accounts for the increased antibody synthesis by the B-cell.

Dr SINKOVICS: In regard to necrotic inflammation by MDP: how is this phenomenon related to Shwartzman reaction? Is it possible to counteract this reaction without loss of immunostimulating activity?

Dr KOTANI: We do not believe that this phenomenon is related to Shwartzman reaction because rabbits, unlike Hartley guinea pigs, are not susceptible to provocation of necrotic inflammation by MDP. We also know that even among guinea pigs, there are big strain differences in regard to susceptibility. Muramyltetrapeptides, either L-Lys or meso-Asp type, are inactive in provocation of necrotic inflammation, unlike MDP. So, it is possible.

Dr SCHWAB: What is the histology of lymph node swelling induced by MDP and derivatives?

Dr KOTANI: Dr Tanaka has shown that certain MDP derivatives could induce granulomas. Although MDP did not induce granulomas, it stimulated the draining lymph nodes especially when administered in an oily medium.

Dr KIRN: Is there an increase in the number of macrophages? Is it possible in MDP-treated mice to demonstrate that macrophages from a treated animal are more resistant *in vitro* to intracellular bacteria or viruses?

Dr CHEDID: As stated by Dr Kotani there does indeed appear a significant monocytosis which remains elevated for several hours. Such *in vivo* results have been studied recently *in vitro* by Dr Hadden, who showed that MDP was strongly synergistic with MMF (macrophage mitogenic factor), a lymphokine which stimulates macrophage proliferation (thymidine uptake). He also found a strong potentiation of another lymphokine (MAF), which enabled these cells to kill tumour targets, and also to kill engulfed *Listeria monocytogenes*.

Dr PRAGER: An inexpensive, readily available, water soluble lipoidal material - dimethyldioctadecylammonium bromide (DDA)

— is a potent enhancer of immune responses. By selection of conditions of immunization, it is possible to enhance either an antibody response or delayed hypersensitivity. DDA with tumour antigen has been effective in both immunoprophylaxis and chemoimmunotherapy models, and I commend it to your attention for use with tumour systems.

Dr SINKOVICS: Dr Hill, which are the clostridia that produce cocarcinogens from bile acids? Is *Clostridium difficile* one of them? Would ingestion of lactobacilli suppress these clostridia? Can you give specific references (publications) for increased incidence of gastric and colonic carcinoma after gastric resection or ureterosigmoid anastomosis after cystectomy? What is the exact chemical mechanism for this carcinogenesis: is it simply nitrosamine production? Let me have a comment on iron. Excess iron may block RES and thus permit infections which are usually controlled by the macrophage system. The example is *Listeria monocytogenes* infection (sepsis and meningitis) in patients with haemochromatosis.

Dr HILL: The clostridia do not belong to a particular species, but are lecithinase-negative organisms producing butyric acid as a major end-product of fermentation; the main species involved are *Clostridium butyricum*, *Cl. indolis*, *Cl. tartium* and *Cl. paraputrifficum* but do not include *Cl. difficile*. The work has been described by Dasa *et al.* (1975) in *Journal of Medical Microbiology* and has been reviewed recently in my chapter in "Gastrointestinal Cancer" (Eds DeCorse and Sherlock), Martinus Nijhoff, 1982, pp. 187-226. Reference to the work on partial gastrectomy and ureterosigmoidostomy has been made in a recent review paper that I wrote in *The Practitioner* (1981) or *British Medical Bulletin* (1980).

PRODUCTION AND STANDARDIZATION OF
C. PARVUM SUSPENSIONS:
A REVIEW OF 12 YEARS' EXPERIENCE

C. Adlam[*], P.A. Knight[+] and R.N. Lucken[+]

*Departments of Bacteriology R & D[*] and Quality Control[+],
The Wellcome Research Laboratories,
Langley Court, Beckenham, Kent BR3 3BS, UK*

I. INTRODUCTION

It is now 18 years since Halpern *et al.* (1964) first published
their paper on the lympho-reticular stimulatory properties of
an organism which they called *Corynebacterium parvum*. They
showed that mice injected intravenously or intraperitoneally
with killed suspensions of these bacteria were able to clear
intravenously administered carbon more rapidly from their
circulation than control animals. In addition, *C. parvum*-
treated mice went on to develop enlarged livers and spleens
which reached a peak size at about 14 days after injection
and thereafter gradually returned to normal (Figs. 1 and 2).
Two years later, in 1966, Halpern's group (Halpern *et al.*,
1966) and Woodruff and Boak (1966), independently showed that
mice treated with *C. parvum* were protected against tumour cell
challenge.

II. TAXONOMY OF *C. PARVUM*

Our involvement in *C. parvum* began towards the end of 1969
when research workers interested in the organism were experi-
encing problems in reproducing these early results. Cultures
lodged in various collections as *C. parvum* proved to be inactive and the original French strain 936B had been lost. By
testing a large number of anaerobic diphtheroids for their
spleen weight increasing ability, we managed to find several
active organisms which at that time carried a variety of
names (Adlam and Scott, 1973). The most active of these
organisms, at that time called *Corynebacterium anaerobium* was
numbered CN6134 and continues to be the strain from which we
manufacture suspensions for clinical and experimental use.

FIG. 1 *Spleen and liver enlargement 14 days after a single i.v. injection of 1.4 mg dry weight of killed* C. parvum *CN6134 organisms.*

Due to advances in our understanding of the taxonomy of these organisms, we are now able to appreciate that the reason for the failure of certain strains of *C. parvum* to stimulate the lymphoreticular system lies in the fact that they belong to a separate inactive species. Indeed, the organism name *C. parvum* is no longer recognized by taxonomists as a separate entity. Using the taxonomic scheme proposed by Johnson and Cummins (1972) it is clear that active strains belong to one of 3 species named *Propionibacterium acnes*, *P. avidum*, and *P. lymphophilum*. The inactive species is named *P. granulosum* (Table 1).

III. PRODUCTION OF *C. PARVUM* SUSPENSIONS

Having selected a strain, the next problem was to grow it in bulk for production. This turned out to be reasonably straightforward, but as with any bacterial vaccine, certain precautions need to be taken during growth. Since anaerobic diphtheroids are present on the skin they may also contaminate growth media and be difficult to distinguish from the desired production culture (Cummins and Johnson, 1974). For this reason, in process biochemical fermentation tests are

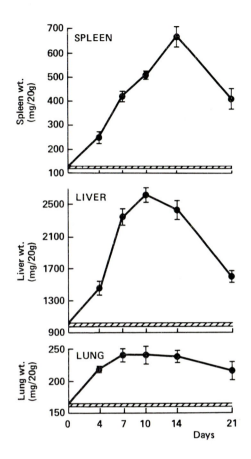

FIG. 2 *Kinetics of the increase in spleen, liver and lung weights observed following a single intravenous injection of 1.4 mg dry weight of killed* C. parvum *CN6134 organisms. Reproduced from Adlam and Scott (1973), by kind permission of the editors and publishers, Longmans.*

carried out and every batch is routinely tested for potency and toxicity.

Another problem which we have experienced relates to the tendency of "wet-filled" *C. parvum*, i.e. washed organisms resuspended in saline containing preservative, to aggregate when stored for long periods of time in small ampoules.

TABLE 1

Taxonomy of the anaerobic diphtheroids
(Johnson and Cummins, 1972) with reference to their
lymphoreticular stimulatory properties.
(This table has been abstracted from Adlam and Reid (1978)
by kind permission of the
editors and publishers, S. Karger AG, Basel.)

		Activity
P. *acnes* I	Wellcome "*C. parvum*"[1] CN6134	+++
	Pasteur Institute "*C. granulosum*"	+++
P. *acnes* II	NCTC[2] 10390 (ATCC[3] 12930) "*C. parvum*"	++[4]
P. *granulosum*	NCTC 10387 (ATCC 11829 "*C. parvum*"	-[5]
P. *avidum* I	Merieux *C. parvum* IM 1585	+++
P. *avidum* II		+++
P. *lymphophilum*[6]		+++

(1) Registered Trade Mark. "Coparvax".
(2) National Collection of Type Cultures.
(3) American Type Culture Collection.
(4) Other strains of this species (3/6 tested) were fully
    active.
(5) All 6 strains of this species were inactive.
(6) Of 4 strains tested 3 were fully active and the fourth
    strain inactive. These strains form a separable sero-
    logical group and have a different peptidoglycan amino
    acid make up from the other species.

For this reason we now produce clinical *C. parvum* as a freeze-
dried product. This material resuspends well and shows no
deterioration on storage even at elevated temperatures.

IV. QUALITY CONTROL

A. Potency

There are many assays which can be used for assessing the biological activity of batches of *C. parvum* suspension. Clearly one needs to select tests which reflect activity in the human clinical context. The mechanisms by which *C. parvum* exerts its antitumour effects are many and are beyond the scope of this paper. In brief, in laboratory animals, *C. parvum* may produce nonspecific antitumour effects, for example when injected systemically in advance of a tumour cell challenge. Alternatively, effects on immunologically specific antitumour responses may be demonstrated when *C. parvum* is used as an adjuvant and either administered mixed with irradiated tumour cells or injected directly into a localized tumour lesion. For these specific effects, resistance may only be observed against the tumour line against which the animal was immunized. The reader is referred to the review by Milas and Scott (1977) for further information concerning the mechanisms of these antitumour effects.

Among the many assays for potency of *C. parvum* which we have investigated are: (1) antitumour effects (using a variety of tumour cell lines and a variety of inbred strains of mice); (2) adjuvant activity (using a variety of antigens and injection regimens); (3) histamine sensitizing activity (certain strains of mice pretreated with *C. parvum* become sensitized to subsequent challenge with histamine (Adlam, 1973)); (4) spleen and liver weight increases (see above). We have found that all of these parameters appear to go hand in hand. Thus batches of *C. parvum* and related organisms which produce a large spleen weight increase will also possess good antitumour properties (Fig. 3), adjuvant effects and histamine sensitizing effects.

Of these many assays, we have for reasons of simplicity and reproducibility now selected only 2 for routine batch testing. These are the spleen weight increase assay and an antitumour assay which employs T3 fibrosarcoma cells in CBA mice. These 2 assays and typical results achieved using them will now be considered.

1) Spleen weight assay. Although a wide variety of immunostimulators produce some increase in spleen weight, the magnitude of the increase observed is peculiar to *C. parvum*.

For quality control purposes we have examined 2 aspects of the spleen weight response, (a) its absolute magnitude, and (b) the quantity of *C. parvum* required to equal the response evoked by a reference preparation.

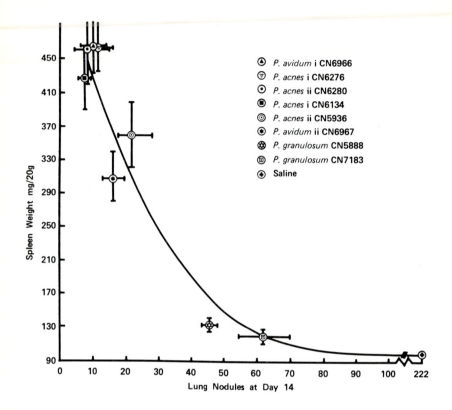

FIG. 3 *Spleen weight enhancement and antitumour properties of several different diphtheroid strains showing a correlation between ability to increase spleen weight and ability to depress tumour nodule formation. Reproduced from Adlam and Reid (1978). Dev. Biol.* Standard **38**, *115, by kind permission of the editors and publishers, S. Karger AG, Basel.*

In both instances the actual spleen weights are corrected for a standard mouse body weight of 20 g. The spleen weight 14 days after intravenous administration of *C. parvum* has been chosen because it was found that the discrimination between strongly and weakly active suspensions was greatest at that interval. Some typical early results are shown (Table 2).

It is evident that all 4 test batches and the reference batch on each occasion have evoked the very large increases in spleen weight that characterize active suspensions of

TABLE 2

Comparison of spleen weights resulting from injection of diluted test and early reference C. parvum batches using random bred MFI mice

| Batch | Mean corrected spleen weights | | | | | | Ratio: Neat *C.parvum* / saline | | Relative potency (95% fiducial limits) |
| | Dilution of *C. parvum* under test | | | Dilution of reference | | | | | |
	1/1	1/3	1/9	1/1	1/3	1/9	Saline	Test vaccine	Reference vaccine
A	601	480	406	660	419	374	73	8.2	9.0	96.5% (57 – 164)
B	488	180	207	489	401	245	79	6.2	6.2	58.5% (32 – 99)
C	510	397	255	513	495	295	98	5.2	5.2	71.7% (27 – 161)
D	557	407	231	571	475	246	92	6.1	6.2	82.1% (54 – 123)

* Groups of 10 mice received dilutions of *C. parvum* (0.2 ml) intravenously. Spleens were weighed 14 days later. Results are expressed as mg/20 g mouse.

C. parvum. It is also evident that the variation in ratio for the reference "between occasions" is at least as large as the variation between the 4 batches. For this reason we do not regard the absolute increase in spleen weight as an adequate quantitative index of activity for *C. parvum*, and more recently this view has been strengthened by the finding that the size of increase varies markedly between strains of mice.

The absolute magnitude in spleen weight however still serves as a useful check to eliminate from further processing any batches which are seriously defective, and we therefore ensure that all batches will evoke a spleen weight increase of at least 4-fold in Olac MF1 mice before we commit them for filling and freeze drying.

The relative potency assays quoted in Table 2 were performed using small numbers of outbred mice and yielded results with confidence limits extending over a range of 3-fold or more. In later assays, a new reference preparation has been introduced which is more homogenous with current batches of suspension and larger numbers of mice of the inbred NIH strain are used, reducing the range of confidence limits to 1.5-fold (Table 3). In this form the spleen weight assay is used as the principle test for immunostimulatory activity and only batches with a potency more than 50% of the new reference preparation are released for issue.

2) Antitumour assay. Intravenous inoculation of CBA mice with *C. parvum* 4 days before intravenous challenge with 10^5 autologous T$_3$ fibrosarcoma cells almost completely prevents the subsequent formation of tumour nodules in the lung (Fig. 4). This effect has been adopted as the second index of potency for *C. parvum* suspensions. Initially the percentage reduction of tumour numbers over control was used as an index of activity but it was found that the "take" of challenge suspension in unvaccinated controls was highly variable, and that this distorted the value found for percentage reduction in tumour count.

We are therefore obliged to use a parallel line assay similar to that used for the spleen weight test. However, the precision of tumour nodule assays is far lower than for spleen weight assays and many assays are statistically unsatisfactory (Table 4). The lung fibrosarcoma assay should therefore be regarded only as semiquantitative index of antitumour activity.

TABLE 3

Comparison of spleen weights resulting from injection of diluted test and new reference C. parvum batches using inbred NIH mice*.

Batch	Mean corrected spleen weights (mg)						Relative potency (95% fiducial limits)
	Dilution of *C. parvum* under test			Dilution of reference			
	1/1.5	1/3	1/6	1/1.5	1/3	1/6	
A	427	351	301	428	367	309	90.5% (78 – 104)
B	368	310	276	402	355	280	71.7% (53 – 93)
C	409	381	332	457	408	343	66.6% (48 – 88)
D	454	358	311	492	402	327	74.8% (60 – 91)

* Note as for Table 2.

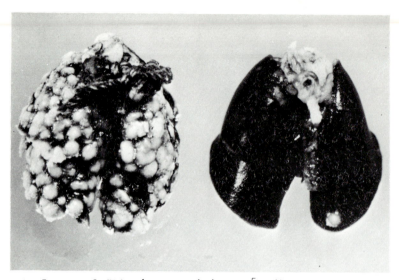

FIG. 4 *Lungs of CBA mice receiving 10⁵ fibrosarcoma cells.
The lungs on the right are taken from a mouse which received
1.4 mg of* C. parvum *intravenously 4 days prior to intravenous
challenge with tumour. The lungs on the left are taken from
a control mouse which received saline prior to tumour challenge. Lungs are inflated by injection with carbon to facilitate nodule counting.*

B. Toxicity

As with the assays for potency, those for toxicity are selected for their representation of effects which have been
observed in man following treatment with *C. parvum*. Since
pyrexia is a frequently reported side-effect, batches are
checked for this in rabbits.

An additional assay which has proved useful for the assessment of toxicity of other human vaccines (e.g. whooping cough
vaccine, Pittman and Cox, 1965) is the mouse weight gain
assay.

1) Rabbit pyrogenicity. Since the most frequently reported
side-effects following administration of *C. parvum* have been
pyrexial, each batch of suspension is tested for pyrogenicity
in rabbits. *C. parvum* suspension is pyrogenic in rabbits,
although it contains little or no endotoxin as measurable by
the Limulus Amoebocyte Lysate test and exhibits pyrexial
response kinetics quite different from those produced by
injection of Gram-negative organisms (Lucken *et al.*, 1977).
In order to obtain an intermediate response level where

TABLE 4

Comparison of T3 fibrosarcoma nodule numbers resulting in the lungs of CBA mice after the prior injection of test and reference C. parvum batches*

		Geometric mean tumour nodule counts								
Batch	Controls	Dilution of test C. parvum			% reduction	Dilution of reference			% reduction	Relative potency (95% fiducial limits)
		1/100	1/20	1/4		1/100	1/20	1/4		
A	183	76	30	18	90%	72	27	6.1	97%	50.2% (11 – 158)
B	446	430	54	31	93%	500	86	44.0	90%	93.5% (100 – 229)
C	183	95	17	15	92%	72	27	6.1	97%	67.0% (19 – 205)
D	161	58	44	9.9	94%	51	36	7.6	96%	119.0% (54 – 268)

* Groups of 10 mice received dilutions of *C. parvum* intravenously followed 4 days later by 10^5 tumour cells intravenously. Tumour nodules were counted 14 days later.

differences between vaccine batches are detectable, it is necessary to reduce the dose level to the equivalent of 0.025 ml of reconstituted suspension. The mean response to this dose level for batches of *C. parvum* used in trial was 0.8°C — more or less the middle of the dose response range — with one or 2 batches giving responses as high as 1.0°C. On the basis of this experience we have decided to release for issue only those batches for which a dose of 1.6 ml of a 1:64 dilution of suspension produces a mean peak rise not exceeding 1.0°C.

2) Mouse weight gain. In this test the weight gain of mice 3 days after an intraperitoneal inoculation with 0.2 ml of *C. parvum* is compared with that of mice receiving a dose of reference suspension with an effect on mouse weight gain equivalent to the mean of a group of batches used in clinical trials. Batches giving a significantly lower weight gain than the reference would be rejected.

Table 5 shows the mean weight gain of groups of mice 3 days after receiving a series of batches of *C. parvum* compared to a reference batch.

TABLE 5

Weight gain of mice 3 days after
intraperitoneal inoculation of C. parvum*

Batch	Dose (ml)	Mean weight gain (g)	Standard error
Reference	0.2	1.56	0.51
	0.1	1.97	0.37
	0.05	2.90	0.54
A	0.2	2.14	0.32
B	0.2	1.62	0.56
C	0.2	2.06	0.58
D	0.2	2.28	0.27
E	0.2	1.31	0.45
F	0.2	2.53	0.27
G	0.2	1.92	0.46
H	0.2	2.62	0.40
I	0.2	2.29	0.43
J	0.2	1.47	0.68
K	0.2	2.13	0.55

* Groups of 10 male mice of starting weight 14-16 g were used.

V. CLINICAL CONSIDERATIONS

Although a large number of clinical trials have been carried out using *C. parvum* (Wellcome Coparvax®) administered by several different routes against a wide range of different human malignancies, in general, it is true to say that the striking results obtained in animals have not, so far, been borne out in man. There may be several reasons for this, not least of which is the fact that experiments with inbred mice may not reflect the situation occurring in human beings.

In those conditions where malignant effusions and ascites occur, however, Coparvax® does provide clinical benefit. Clinical results from different groups of workers have consistently shown positive results (Webb *et al.*, 1978; Millar *et al.*, 1980; Mantovani *et al.*, 1981).

It is hoped that these promising results may reactivate some of the clinical interest in this novel immunostimulant and that new indications for its use may arise from controlled clinical studies.

VI. REFERENCES

Adlam, C. (1973). *J. Med. Microbiol.* **6**, 527-538.

Adlam, C. and Reid, D.E. (1977). *Develop. Biol. Standard* **38**, 115-120.

Adlam, C. and Scott, M.T. (1973). *J. Med. Microbiol.* **6**, 261-274.

Cummins, C.S. and Johnson, J.L. (1974). *J. Gen. Microbiol.* **80**, 433-442.

Halpern, B.N., Prevot, A.R., Biozzi, G., Stiffel, C., Mouton, D., Morard, J.C., Bouthillier, Y. and Decreusefond, C. (1964). *J. Reticuloendothelial Soc.* **1**, 77-96.

Halpern, B.N., Biozzi, G., Stiffel, C. and Mouton, D. (1966). *Nature, Lond.* **212**, 853-854.

Johnson, J.L. and Cummins, C.S. (1972). *J. Bact.* **109**, 1047-1066.

Lucken, R.N., Adlam, C. and Knight, P.A. (1977). *Develop. Biol. Standard.* **34**, 135-141.

Mantovani, A., Sessa, C. and Peri, G. (1981). *Int. J. Cancer* **27**, 437-446.

Milas, L. and Scott, M.T. (1977). *Adv. Cancer Res.* **26**, 257-306.

Millar, J.W., Hunter, A.M. and Horner, N.W. (1980). *Thorax* **35**, 856-858.

Pittman, M. and Cox, C.B. (1965). *Appl. Microbiol.* **13**, 447-456.

Webb, H.E., Oaten, S.W. and Pike, C.P. (1978). *Br. Med. J.* **1**, 338-340.

Woodruff, M.F.A. and Boak, J.L. (1966). *Br. J. Cancer* **20**, 345-355.

EXPERIMENTAL IMMUNOSTIMULATION BY PROPIONIBACTERIA

S. Szmigielski[1], W. Roszkowski[2], K. Roszkowski[3],
H.L. Ko[4], J. Jeljaszewicz[5], and G. Pulverer[4]

[1]*Centre for Radiobiology and Radioprotection, Warsaw, Poland;*

[2]*National Institute of Tuberculosis, Warsaw, Poland;*

[3]*Postgraduate Medical Centre, Warsaw, Poland;*

[4]*Institute of Hygiene, University of Cologne, Cologne, FRG;*

[5]*National Institute of Hygiene, Warsaw, Poland*

Halpern *et al.* (1964) and Woodruff and Boak (1966) showed that injection of killed *Corynebacterium parvum* into mice resulted in potent and quick stimulation of the reticulo-endothelial system and inhibition of transplantable tumour growth. Since that time, many experimental studies have been done to explain immunomodulatory, antineoplastic and anti-viral activities of these microorganisms which were classi-fied by Johnson and Cummins (1972) as propionibacteria (for review see Baldwin and Byers, 1979; Roszkowski *et al.*, 1982).

For the most part, the studies were performed on *C. parvum* strain CN 6134 (Adlam and Scott, 1973; Halpern *et al.*, 1964). In later studies other strains of propionibacteria were tested.

In 1973, Pulverer and Ko examined 72 strains of propioni-bacteria and chose 2 of those with potent immunomodulatory and antitumour properties for further experimental and clini-cal studies (Roszkowski *et al.*, 1980a-d; Ko *et al.*, 1981; Lipski *et al.*, 1981; Szmigielski *et al.*, 1980; Gil *et al.*, 1980; Roszkowski *et al.*, 1981). In the present paper we summarize our studies on immunomodulatory, antineoplastic and antiviral activity of *Propionibacterium granulosum* strain KP-45 (P. KP-45) which encouraged us to undertake clinical trials with these bacteria (Roszkowski *et al.*, this volume).

I. ACTIVATION OF MONOCYTE-MACROPHAGE SYSTEM

Single intravenous or intraperitoneal injections of killed propionibacteria in mice result in marked enlargement of spleen, liver (Fig. 1) and also lymph nodes and lungs (Adlam

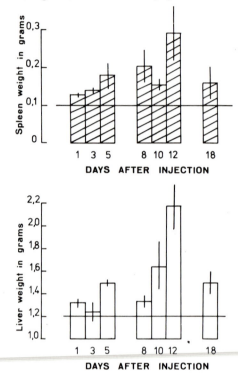

FIG. 1 *Spleen and liver enlargement after* Propionibacterium KP-45 *injection.*

and Scott, 1973; Milas *et al.*, 1975; Roszkowski *et al.*, 1980a) in the animals. Increase of these organ weights is significantly pronounced a few days after injection of the bacteria, reaching the highest values after 2 weeks. Histological studies showed that proliferation of macrophages, histiocytes and haemopoetic cells occurs in enlarged spleens (Milas *et al.*, 1974) whereas hepatomegaly seems to be related to infiltration of lymphocytes, monocytes and macrophages without change in number and location of Kupffer cells (Milas *et al.*, 1974; Gil *et al.*, 1982b). The degree of splenomegaly caused by different strains of propionibacteria correlates with their immunomodulatory and antitumour activity (Roszkowski *et al.*, 1980c). Thus spleen enlargement can be used as

a simple parameter of reticuloendothelial system stimulation.
P. KP-45 labelled with ^{51}Cr and injected intravenously is
quickly eliminated from circulating blood and is trapped in
the liver, spleen and lungs (Gil *et al.*, 1982b). In these
organs the bacteria are phagocytized by cells of the reticuloendothelial system. All strains of propionibacteria are
relatively susceptible to phagocytosis by macrophages. Only
strains which show potent immunomodulatory and antitumour
activity are resistant to intracellular killing whereas
others are quickly killed and degraded (Roszkowski *et al.*,
1980c). It is assumed that resistance of propionibacteria to
intracellular degradation and their immunomodulatory activity
are mainly related to cell wall composition, particularly to
the properties of peptidoglycans and carbohydrates (Adlam and
Reid, 1978).

Active strains of propionibacteria cause activation of
macrophages which is expressed, among other ways, by labiliza-
tion of lysosomal membranes and activation of lysosomal
enzymes. The activated cells show better and quicker adherence, increased vacuolization and changed ultrastructure of
their membranes (Puvion *et al.*, 1976). Such stimulation of
macrophages is combined with appearance of their cytostatic
and/or cytolytic activity directed against tumour or virus
infected cells (Basic *et al.*, 1974, 1975; Kirchner *et al.*,
1977) and with increased phagocytic abilities both *in vitro*
(Fig. 2) and *in vivo* (Fig. 3). Increased phagocytosis caused
by injection of different strains of propionibacteria was
measured by peripheral blood clearance of coloidal carbon
(Raynaud *et al.*, 1972; Warr and Sljivic, 1974), bovine albumin (McBride *et al.*, 1974) and different bacteria (Adlam *et*

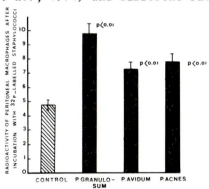

FIG. 2 *Phagocytosis of* ^{32}*P-labelled staphylococci by peri-*
toneal macrophages from rabbits treated with Propionibacterium *KP-45 (5 mg/Kg b.w.). Results are expressed in mean*
cpm × 10^3 ± *S.D.*

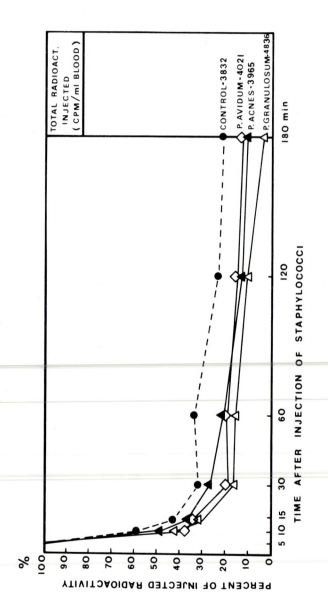

FIG. 3 *Clearance of* ^{32}P*-labelled staphylococci from peripheral blood in rabbits treated with propionibacteriae (5 mg/kg b.w.).*

al., 1972; Collins and Scott, 1974). In our studies this
effect was examined using ^{32}P labelled staphylococci. The
labelled bacteria were markedly more quickly eliminated from
circulating blood of rabbits which had previously received
different strains of propionibacteria than from blood of
control animals (Fig. 3). It seems that dynamics of blood
clearance after propionibacteria injection is directly related
to the degree of hepato- and splenomegaly caused by these
microorganisms and to their immunomodulatory activity.

It is commonly believed (Baldwin and Byers, 1979) that
activation of the monocyte—macrophage system is the most
important factor in the mechanisms of immunostimulatory, antibacterial, antiviral and antitumour activity of propionibacteria. A whole series of events, such as interferon induction, enhancement of specific immune response and activation
of natural cytolytic mechanisms (NK), is mainly the consequence of reticuloendothelial system stimulation (Baum and
Breese, 1976).

II. HUMORAL AND CELLULAR IMMUNITY

Propionibàcteria being immunogenic themselves also possess
ability to enhance humoral response directed against thymus
dependent antigens (Biozzi *et al.*, 1968; James *et al.*, 1974)
as well as thymus independent ones (Howard *et al.*, 1973),
thus having an adjuvant property. This property seems to be
unrelated to *prolonged* activation of the reticuloendothelial
system (as measured by spleen and liver enlargement), as the
enhancement of primary and secondary humoral response can
also be observed after application of "nonactive" strains
(Pincard *et al.*, 1967, 1968). Also results of our studies
(Szmigielski *et al.*, 1982) with different strains of propionibacteria (KP-40, KB, Pa 0162) confirmed lack of relationship
between enhancement of humoral response and enhancement of
macrophage phagocytosis *in vitro*, enlargement of spleen
weight and inhibition of transplantable lung tumour growth.
Although the process of macrophage activation is involved in
the enhancement of humoral immune response, at least in the
case of response to thymus dependent antigens (Wiener and
Bandieri, 1975), there are experimental data showing that
direct stimulation of B-cells by these bacteria may also
occur without involvement of the helper function of T-lymphocytes and macrophages (Halpern, 1975).

Simultaneously with enhancement of the humoral response,
suppression of the cellular immune response can be observed
after intravenous injection of propionibacteria. This was
shown using different models of specific cellular reactions
such as delayed skin hypersensitivity to oxazolone (Allwood

and Asherson, 1972), cell-mediated reaction to sheep red
blood cells (Scott, 1974) or skin graft rejection test in
mice (Castro, 1974). All these reactions were significantly
depressed in propionibacteria treated animals. This was
accompanied by inhibition of lymphocyte proliferation *in
vitro* induced by nonspecific mitogens (Scott, 1972; Howard *et
al*., 1973). The latter phenomenon could be reversed by removal of adherent cells which suggests the significance of
monocytes and macrophages in the mechanism of the above mentioned suppression. However, changed distribution of circulating lymphocytes has to be taken into account as a possible
reason for cellular immunity suppression (Roszkowski *et al*.,
1980b). The suppression may occur only when propionibacteria
are injected intravenously or intraperitoneally (systemic
treatment). On the other hand, subcutaneous application of
these bacteria together with antigen results in the opposite
effect, that is, a strong enhancement of cellular immune
response to the antigen (Table 1) (Bomford, 1975; Scott,
1975). In the case of usage of tumour cells as an antigen,
it leads to highly pronounced induction of cytotoxic effect
directed against these cells.

III. ANTITUMOUR ACTIVITY

Special interest has been shown in the antitumour activity of
propionibacteria, as well as the possibility of their application in cancer immunotherapy (for review see Roszkowski *et
al*., 1982). Antitumour properties of propionibacteria can be
reduced to 3 basic effects: immunoprophylaxis (prevention of
tumour induction or recurrence), inhibition of tumour growth
and total or partial tumour regression. The term "immunotherapy" can be referred to the 2 latter effects as well as
to prevention of recurrences after radical treatment (mainly
surgery). Immunoprophylactic effects in experimental animals
have been shown for various transplantable and spontaneously
grown tumours such as mastocytoma (Scott, 1974), mammary
adenocarcinoma (Woodruff and Inchley, 1971), osterosarcoma,
plasmocytoma (Smith and Scott, 1972) and lymphoma (Halpern,
1975). Although similar doses of various strains of propionibacteria were used (0.5 — 1.5 mg of lyophilized, killed bacteria per mouse) in the majority of the above quoted studies,
marked differences in intensity of the immunoprophylaxis
were observed. These differences may result not only from
different sensitivity of individual tumours to immunotherapy
or activity of applied vaccine but also from various routes
of propionibacteria administration. The immunoprophylactic
effect, in contrast to immunotherapy, is most pronounced after
systemic (intravenous or intraperitoneal) treatment of animals
(Table 1).

TABLE 1

*Effects of systemic versus local treatment
with bacterial immunomodulators*

	Systemic administration (intravenous or intraperitoneal)	Local administration (subcutaneous or intratumoural)
CELL-MEDIATED IMMUNITY (delayed hypersensitivity)	suppressed	enhanced
HUMORAL IMMUNITY (adjuvant reaction)	enhanced	enhanced
LIVER, SPLEEN, AND LUNG MASS	elevated	normal or slightly elevated
ACTIVATION OF THE RETICULO-ENDOTHELIAL (RES) SYSTEM	potent	slight
INDUCTION OF ALPHA-TYPE (VIRAL) INTERFERON	potent	no
NATURAL KILLER (NK) CELL ACTIVITY	enhanced or depressed	enhanced
STIMULATION OF HAEMOPOIESIS	potent	no
REACTION OF LYMPHOCYTES TO NONSPECIFIC MITOGENS *IN VITRO*	depressed	normal or slightly enhanced
ANTINEOPLASTIC ACTIVITY	slight or absent	potent
ANTIBACTERIAL ACTIVITY	potent	absent
ANTIVIRAL ACTIVITY	potent	absent

Woodruff and Boak (1966) were the first to demonstrate the immunotherapeutic effect of propionibacteria. These authors observed marked inhibition of syngeneic, transplantable mammary adenocarcinoma growth in mice. In further studies similar effects have been shown for other tumours in various experimental animals. Therapeutic results of propionibacteria treatment are dissimilar, the causes being different immuno-

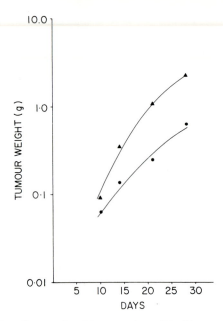

FIG. 4 *Effect of* Propionibacterium *KP-45 treatment on L-1 transplantable tumour growth.* (▲) *untreated control tumours;* (●) *tumour-bearing mice given i.p. injection of* Propionibacterium *KP-45.*

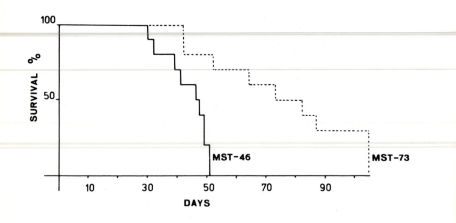

FIG. 5 *Effect of* Propionibacterium *KP-45 on death from 1 × 10⁵ L-1 tumour cells given s.c. (* —— *) untreated animals; (---)* Propionibacterium *KP-45 treated animals. MST — mean survival time.*

genicity of tumours and their susceptibility to this form of
therapy as well as the route of vaccine administration and
time of its application.

Intraperitoneal injection of *P. KP-45* in mice on day 7 or
14 after sarcoma L-1 implantation resulted only in inhibition
of tumour growth (Fig. 4) and prolongation of animal survival
time (Fig. 5) but did not lead to tumour regression (Janik
et al., 1980; Roszkowski *et al.*, 1981). No effects of immunotherapy were seen when therapy was started during advanced
growth of tumours but local therapy applied at the same time
was effective (Roszkowski *et al.*, 1981). Local immunotherapy
could even lead to total regression of tumours in some
animals (Milas *et al.*, 1975; Woodruff and Dunbar, 1975). All
these observations suggest different mechanisms of antitumour
action of propionibacteria given systemically and locally.
As yet these investigators (Baum and Breese, 1976; Baldwin
and Byers, 1979; Scott, 1975; Roszkowski *et al.*, 1982) consider nonspecifically activated macrophages to be the main
effector cells directed against tumours. The study performed
by Peters *et al.* (1977) provided convincing evidence supporting the above hypothesis. These authors observed significant
inhibition of murine fibrosarcoma growth when the tumour
cells were implanted subcutaneously or intraperitoneally
together with macrophages obtained from propionibacteria
treated animals. Moreover, it has been shown that propionibacteria activated macrophages possess cytostatic and cytolytic activity directed against tumour cells (Halpern, 1975).
These authors observed that the inhibitory effect of macrophages obtained from propionibacteria treated mice on [3]H-
thymidine incorporation into tumour cells is not abolished by
previous X-ray irradiation of animals. Cytotoxic effect of
such activated macrophages was visualized by morphological
techniques, showing the appearance of "holes" in the tumour
cell membranes after their direct contact with effectors.
Cytotoxic and/or cytostatic effect can also be demonstrated
in *in vivo* conditions, after administration of propionibacteria to tumour bearing animals (Ando *et al.*, 1978; Janik *et
al.*, 1980). An analysis of tumour cell cycle time performed
on the basis of a ^{3}H-thymidine labelled mitosis number showed
a significant decrease in the growth cell fraction during
systemic immunotherapy with *P. KP-45* (Janik *et al.*, 1980;
Figs. 6 and 7). Inhibition of tumour growth under the influence of therapy with propionibacteria may also be caused by
activation of specific antitumour immune mechanisms. Although
presence of a specific cellular immune function does not seem
to be a necessary condition for expression of the propionibacteria antitumour effect (Bomford and Olivotto, 1974),

138   SZMIGIELSKI *et al.*

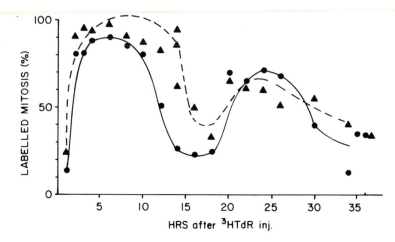

FIG. 6 *Labelled mitoses curves from (* ● *) untreated and*
(▲ *)* Propionibacterium KP-45 *treated L-1 sarcoma.*

	Normal [in hrs]	Propionibacterium treated [in hrs]
TD	132	156
TC	19	19
G_2	1,5	1
S	10,5	14
G_1	6	3
L.I.	36%	22%
G.F.	0,65	0,3
cell loss factor	0,85	0,72

FIG. 7 *Cell cycle tumour growth parameters in animals treated and not treated with* Propionibacterium KP-45. *TD = doubling time; TC = cell cycle time; L.I. = labelling index; G.F. = growth fraction.*

there are experimental data suggesting this relationship,
e.g. lymph node lymphocytes from propionibacteria treated
leukaemic mice show enhanced specific cytotoxic effect against
transformed cells.

The majority of investigators postulate that systemic
treatment with propionibacteria does not decisively influence
specific cellular immunity whereas in local therapy this

mechanism seems to play a significant role. At present the
significance of natural killer cells is emphasized as the
mechanism of antitumour immunity. Nonspecific cytotoxic
activity of these cells was observed against various types of
tumours both transplantable and spontaneous (Lotzova and
McCredie, 1978). Moreover, their activity correlates with
resistance of animals to implantation or induction of experimental tumours (Ojo, 1979; Ojo *et al.*, 1978). The influence
of propionibacteria on natural cell activity is dependent on
the route of the immunomodulator administration. Increased
activity was observed only after local or intraperitoneal
injection of these microorganisms, whereas intravenous administration could lead to significant diminishment of their
cytotoxic effect (Table 1).

IV. ANTIVIRAL ACTIVITY

Intravenous administration of propionibacteria induces a
protective effect against infection of mice with herpes virus
(Kirchner *et al.*, 1977). This phenomenon can also be evoked
in immunosuppressed animals (Kirchner *et al.*, 1978). Intraperitoneal injection of *P. KP-45* 3 days before HSV_1 virus
inoculation resulted in decreases in mortality by encepha-
litis, from 90% in controls to 30% in animals undergoing
immunoprophylaxis (Szmigielski *et al.*, 1980, Fig. 8). A protective effect was also observed against varicella and murine
hepatitis virus infection, in mice strains susceptible to
these viruses (Szmigielski *et al.*, 1980). C3H mice with
chronic viral hepatitis which were treated with *P. KP-45*
showed a marked decrease in virus titre in the liver on the
fourth month after infection. Recent studies (Kobus and
Szmigielski, in prep.) demonstrated that subcutaneous injection of *P. KP-45* together with live or attenuated viral
vaccines caused significant enhancement of immune response
which was expressed by significantly higher titre of serum
antiviral antibodies compared with animals immunized with
viral antigen alone. This points to the possibility of applying propionibacteria as a safe and effective adjuvant in
antiviral vaccination procedures. The above-mentioned antiviral activity of propionibacteria is related to activation
of the monocyte—macrophage system, interferon induction
(Hirt *et al.*, 1978) and/or activation of natural killer cells
(Baldwin and Byers, 1979). A high level of interferon has
been observed in serum of mice on days 5-12 after intravenous
or intraperitoneal administration of propionibacteria (Hirt
et al., 1978; Kirchner *et al.*, 1978). Thus it is assumed
that several days are needed to elicit the antiviral effect
of propionibacteria. In various acute experimental models

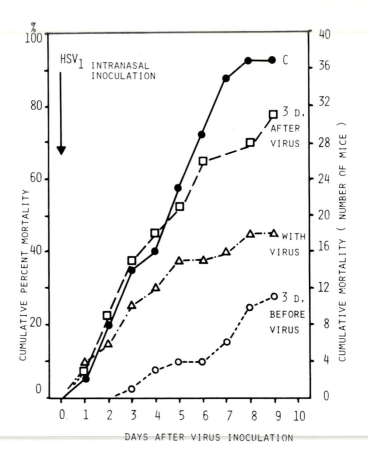

FIG. 8 *Effect of* Propionibacterium KP-45 *on death of mice from infection with herpes simplex virus.*

viral disease develops much more quickly, so this effect cannot be observed. The propionibacteria's ability to induce inferferon was also found in *in vitro* cultures of human and murine lymphocytes (Hirt *et al.*, 1978). The titer of inter-feron in these cultures was markedly higher than in cultures stimulated with nonspecific mitogens.

V. THE EFFECT OF PROPIONIBACTERIA ON THE HAEMOPOIETIC SYSTEM

An additive or even synergistic action of propionibacteria was observed when the bacteria were applied together with experimental tumour radiotherapy (Milas *et al.*, 1975; Suit *et al.*, 1976). But the most potent antitumour effect was noted

when propionibacteria were administered a few days before
tumour irradiation. It seems that this is due to the fact
that enhanced proliferation of reticuloendothelial system by
propionibacteria decreases the risk of local destruction of
this system during irradiation (Suit *et al.*, 1976). Intravenous or intraperitoneal administration of the bacteria
results in quick and potent stimulation of bone marrow (Baum
and Breese, 1976; Wrembel, 1982) with accompanying enhanced
proliferation of monocyte precursors and release of these
cells into the circulating blood (Chare and Baum, 1978; Wolmark and Fisher, 1974). Increased proliferation of granulo-
poietic cells is also observed as well as a higher percentage
of their nondifferentiated forms in bone marrow (Wrembel,
1982). The number of stem cells drops after its brief
increase in bone marrow. Simultaneously, these cells appear
in circulating blood and settle in the spleen (Roszkowski *et
al.*, 1980b). Stimulatory action of propionibacteria on
haemopoietic system causes its accelerated spontaneous
regeneration in mice irradiated with sublethal doses of X-
rays (Lipski *et al.*, 1981) or treated with cyclophosphamide
(Wrembel, 1982). Administration of propionibacteria to mice
4 hours before lethal irradiation with X-rays (850 R) resulted
in significant prolongation of animal survival time. The
strength of this effect was dependent on the bacterial strain
and correlated to its immunostimulatory and antitumour potency
(Roszkowski *et al.*, 1980d).

Stimulation of the haemopoietic system and its enhanced
postirradiation recovery points to advantages of combination
of immunotherapy with intensive radio- and/or chemotherapy.
In this case enhanced haemopoiesis can prevent the dangerous
risk of bacterial or viral infections as complications of the
therapy.

VI. SUMMARY

The immunoprophylactic and immunotherapeutic effect of propionibacteria is highly dependent on bacterial strain activity as well as on the route of immunomodulator administration and the time of commencement of therapy (Table 1). In
advanced experimental tumours, local (intralesional) application of propionibacteria is significantly more effective
than general treatment. However, the latter case is advis-
able to obtain immunoprophylactic effect or interferon induction. A key mechanism in the above phenomena is activation
of the monocyte-macrophage system, which triggers subsequent
events. Immunomodulatory and antineoplastic activities of
propionibacteria are determined by their cell wall composition particularly by peptidoglycan-teichoic acid complex

property. It seems that further progress in the application
of propionibacteria in cancer or viral disease immunotherapy
is dependent upon precise characterization of this active
complex. This would make possible the precise establishment
of the effective doses. So far the doses used in experimental
studies are noncomparable with those in clinical trials.
These differences could be the reason for discrepancies between the results of experimental and clinical studies.

VII. REFERENCES

Adlam, C. and Reid, D.E. (1978). Comparative studies on cell
wall composition of some anaerobic coryneforms of varying
lymphoreticular stimulatory activity. *Develop. Biol. Standard.* **38**, 115-120.

Adlam, C., Broughton, E.S. and Scott, M.T. (1972). Enhanced
resistance of mice to infection with bacteria following
pretreatment with *Corynebacterium parvum*. *Nature New
Biol.* **235**, 219-220.

Adlam, C. and Scott, M.T. (1973). Lymphoreticular stimulatory
properties of *Corynebacterium parvum* and related bacteria.
J. Med. Microbiol. **6**, 261-274.

Allwood, G.G. and Asherson, G.L. (1972). Depression of
delayed hypersensitivity by pretreatment with Freund-type
adjuvants. III. Depressed arrival of lymphoid cells at
recently immunized lymph nodes in mice pretreated with
adjuvants. *Clin. Exp. Immunol.* **11**, 579-584.

Ando, K., Urano, M. and Koike, S. (1978). Cytocidal and
cytostatic ability of *Corynebacterium liquefaciens* in mouse
squamous cell carcinoma *in vivo*. *Cancer Res.* **38**, 1769-
1773.

Baldwin, R.W. and Byers, V.S. (1979). Immunoregulation by
bacterial organisms and their role in the immunotherapy of
cancer. *Springers Sem. Immunopathol.* **2**, 79-100.

Basic, I., Milas, L., Grdina, D.J. and Withers, H.R. (1974).
Destruction of hamster ovarian cell cultures by peritoneal
macrophages from mice treated with *Corynebacterium granulosum*. *J. Nat. Cancer Inst.* **52**, 1839-1842.

Basic, I., Milas, L., Grdina, D.J. and Withers, H.R. (1975).
In vitro destruction of tumour cells by macrophages from
mice treated with *C. granulosum*. *J. Nat. Cancer Inst.* **55**,
589-596.

Baum, M. and Breese, M. (1976). Antitumour effect of *Corynebacterium parvum*. Possible mode of action. *Br. J. Cancer*
33, 468-473.

Biozzi, G., Stiffel, C., Mouton, D., Bouthillier, Y. and
Decreusefond, C. (1968). A kinetic study of antibody

producing cells in the spleen of mice immunized intravenously with sheep erythrocytes. *Immunology* **14**, 7-20.

Bomford, R. (1975). Active specific immunotherapy of mouse
methylcholanthrene induced tumours with *Corynebacterium
parvum* and irradiated tumour cells. *Br. J. Cancer* **32**,
551-557.

Bomford, R. and Olivotto, M. (1974). The mechanism of inhibition by *Corynebacterium parvum* of the growth of lung
nodules from intravenously injected tumorous cells. *Int.
J. Cancer* **14**, 226-235.

Castro, J.E. (1974). Antitumour effect of *Corynebacterium
parvum* in mice. *Eur. J. Cancer* **10**, 115-120.

Chare, M.J.B. and Baum, M. (1978). The effect of *Corynebacterium parvum* on the proliferation of monocyte precursors
in bone marrow of mice. *Develop. Biol. Standard.* **38**, 195-
200.

Collins, F.M. and Scott, M.T. (1974). Effect of *Corynebacterium parvum* treatment on the growth of Salmonella
enteritidis in mice. *Infect. Immun.* **9**, 863-869.

Gil, J., Orlowski, T., Nowakowski, W., Ko, H.L., Roszkowski,
K., Roszkowski, W., Szmigielski, S., Pulverer, G. and
Jejjaszewicz, J. (1980). Local immunotherapy of stomach
and intestinal carcinoma by *Propionibacterium granulosum*.
Dis. Rectum Colon **23**, 536-543.

Gil, J., Szmigielski, S., Jeljaszewicz, J. and Pulverer, G.
(1982a). 2.5-year follow-up of local immunotherapy of
advanced gastric and colorectal cancer. *J. Cancer Res.*
(in press).

Gil, J., Wegiel, J., Badowski, A., Szmigielski, S., Pulverer,
G. and Jeljaszewicz, J. (1982b). Morphology and function
of liver cells in animals treated with *Propionibacterium
granulosum*. *Exp. Path.* (in press).

Halpern, B. (1975). *Corynebacterium parvum*. *In* "Application
in Experimental and Clinical Oncology". (Ed. B. Halpern),
pp. 181-186. Plenum Press, New York.

Halpern, B., Prevot, A.R., Biozzi, G., Stiffel, C., Mouton,
D., Morard, J.C., Bouthillier, Y. and Decreusefond, C.
(1964). Stimulation de l'activité phagocytaire du système
reticuloendothelial provoquée par *Corynebacterium parvum*.
J. Reticuloendothel. Soc. **1**, 77-96.

Hirt, H.M., Becker, H. and Kirchner, H. (1978). Induction of
interferon production in mouse spleen cell cultures by
Corynebacterium parvum. *Cell. Immunol.* **38**, 168-175.

Howard, J.G., Christie, G.H. and Scott, M.T. (1973). Biological effects of *Corynebacterium parvum*. IV. Adjuvant
and inhibitory activities on B-lymphocytes. *Cell. Immunol.*
7, 290-301.

James, K., Ghaffar, A. and Milne, I. (1974). The effect of transplanted methylocholanthrene induced fibrosarcoma and *Corynebacterium parvum* on the immune response of CBA and A/HeJ mice to thymus dependent and independent antigens. *Br. J. Cancer* **29**, 11-20.

Janik, P., Roszkowski, W., Ko, H.L., Szmigielski, J., Pulverer, G. and Jeljaszewicz, J. (1980). The *in vivo* cytostatic effect induced by *Propionibacterium granulosum* on murine tumour cells. *J. Cancer Res. Clin. Oncol.* **98**, 51-58.

Johnson, J.M. and Cummins, C.S. (1972). Cell wall composition and deoxyribonucleic acid similarities among the anaerobic coryneforms, classical Propionibacteria and strains of *Arachnia propionica*. *J. Bacteriol.* **109**, 1047-1066.

Kirchner, H., Hirt, H.M., Becker, H. and Munk, K. (1977). Production of an antiviral factor by murine spleen cells after treatment with *Corynebacterium parvum*. *Cell. Immunol.* **31**, 172-176.

Kirchner, H., Scott, M.T. and Munk, K. (1978). Protection of mice against viral infections by *Corynebacterium parvum* and *Bordetella pertussis*. *J. Gen. Virol.* **41**, 97-104.

Ko, H.L., Roszkowski, W., Jeljaszewicz, J. and Pulverer, G. (1981). Comparative study on the immunostimulatory potency of different Propionibacterium strains. *Med. Microbiol. Immunol.* **170**, 1-9.

Lipski, S., Roszkowski, W., Ko, H.L., Szmigielski, S., Pulverer, G. and Jeljaszewicz, J. (1981). Postirradiation recovery of haemopoiesis in mice treated with *Propionibacterium granulosum*. *Exp. Path.* **19**, 179-185.

Lotzova, E. and McCredie, K.B. (1978). Natural killer cells in mice and man and their possible biological significance. *Cancer Immunol. Immunother.* **4**, 215-221.

McBride, W.H., Jones, J.T. and Weir, D.M. (1974). Increased phagocytic cell activity and anaemia in *Corynebacterium parvum* treated mice. Br. J. Exp. Pathol. **55**, 38-46.

Milas, L., Basic, I., Kogelnik, H.D. and Withers, H.R. (1975). Effects of *Corynebacterium granulosum* on weight and histology of lymphoid organs, response to mitogens, skin allografts and a syngeneic fibrosarcoma in mice. *Cancer Res.* **35**, 2365-2374.

Milas,, L., Hunter, N. and Withers, H.R. (1974). *Corynebacterium granulosum* induced protection against artificial pulmonary metastases of syngeneic fibrosarcoma in mice. *Cancer Res.* **34**, 613-620.

Milas, L., Withers, H.R. and Hunter, N. (1975). *C. granulosum* and *C. parvum* induced augmentation of the radiocurability of a murine fibrosarcoma (FSA). *Proc. Am. Ass. Cancer Res.* **16**, 154-160.

Ojo, M. (1979). Positive correlation between the levels of natural killer cells and the *in vivo* resistance to syngeneic tumour transplants as influenced by various routes of administration of *C. parvum* bacteria. *Cell. Immunol.* **45**, 182-187.

Ojo, M., Haller, O., Kimura, A. and Wigzel, H. (1978). An analysis of conditions allowing *Corynebacterium parvum* to cause either augmentation or inhibition of natural killer cell activity against tumour cells in mice. *Int. J. Cancer* **21**, 444-452.

Peters, L.J., McBride, W.H., Mason, K.A., Hunter, N., Basic, I. and Milas, L. (1977). *In vivo* transfer of antitumour activity by peritoneal exude cells from mice treated with *Corynebacterium parvum*: reduced effect in irradiated recipients. *J. Nat. Cancer Inst.* **59**, 881-887.

Pincard, R.N., Weir, D.M. and McBride, W.H. (1967). Factors influencing the immune response. II. Effect of the physical state of the antigen and of lymphoreticular cell proliferation on the response to intraperitoneal injection of bovine serum albumin. *Clin. Exp. Immunol.* **2**, 343-350.

Pincard, R.N., Weir, D.M. and McBride, W.H. (1968). Factors influencing the immune response. III. The blocking effect of *C. parvum* upon the induction of acquired immunological unresponsiveness to bovine serum albumin in the adult rabbit. *Clin. Exp. Immunol.* **3**, 413-421.

Pulverer, G. and Ko, H.L. (1973). Fermentative and serological studies on *Propionibacterium acnes*. *Appl. Microbiol.* **25**, 222-229.

Putten, L.M., van Kram, L.K.J., van Dierendonik, H.H.C., Smink, T. and Füzi, M. (1975). Enhancement by drugs of metastatic lung nodule formation after intravenous tumour cell injection. *Int. J. Cancer* **15**, 588-595.

Puvion, F., Fray, A. and Halpern, B. (1976). A cytochemical study of the *in vitro* interaction between normal and activated mouse peritoneal macrophages and tumour cells. *J. Ultrastruct. Res.* **54**, 95-108.

Raynaud, M., Kouznetzova, B., Bizzini, B. and Cherman, J.C. (1972). Etude de l'effect immunostimulant de diverses éspèces de corynebacterie anaerobie et de leurs fractions. *Ann. Inst. Pasteur* **122**, 685-695.

Roszkowski, W., Kobus, M., Luczak, M., Ko, H.L., Szmigielski, S., Laskowska, B., Pulverer, G. and Jeljaszewicz, J. (1980a). Action of three stains of Propionibacteria on experimental tumour system in mice. *Zbl. Bact. Hyg. I. Abt. Orig. A.* **246**, 405-414.

Roszkowski, W., Szmigielski, S., Ko, H.L., Janiak, M., Wrembel, J.K., Pulverer, G. and Jeljaszewicz, J. (1980b).

Effect of three strains of Propionibacteria and cell wall preparations on lymphocytes and macrophages. *Zbl. Bact. Hyg. I. Abt. Orig. A.* **246**, 393-404.

Roszkowski, W., Ko, H.L., Szmigielski, S., Jeljaszewicz, J. and Pulverer, G. (1980c). The correlation of susceptibility of different Propionibacterium strains to macrophage killing and antitumour activity. *Med. Microbiol. Immunol.* **169**, 1-8.

Roszkowski, W., Lipski, S., Ko., H.L., Szmigielski, S., Jeljaszewicz, J. and Pulverer, G. (1980d). Recovery of haemopoiesis in lethaly irradiated mice after treatment with Propionibacteria. *Strahlentherapie* **156**, 729-733.

Roszkowski, K., Roszkowski, W., Ko, H.L., Pulverer, G. and Jeljaszewicz, J. (1981). Macrophage and T-lymphocyte content of tumours in mice treated with *Propionibacterium granulosum*. *Oncology* **38**, 334-339.

Roszkowski, W., Roszkowski, K., Szmigielski, S., Pulverer, G. and Jeljaszewicz, J. (1982). Immunostimulatory and antineoplastic properties of Propionibacterium. *In* "Medical Microbiology". (Eds J. Jeljaszewicz and C. Easmon), vol. 1, pp. 433-452, Academic Press, New York, London.

Scott, M.T. (1972). Biological effects of the adjuvant *Corynebacterium parvum*. II. Evidence for macrophage T-cell interaction. *Cell. Immunol.* **5**, 469-473.

Scott, M.T. (1974). *Corynebacterium parvum* as a therapeutic antitumour agent in mice. II. Local injection of *C. parvum*. *Cell. Immunol.* **13**, 251-263.

Scott, M.T. (1975). Potentiation of the tumour-specific immune response by *Corynebacterium parvum*. *J. Nat. Cancer Inst.* **55**, 65-72.

Smith, S.E. and Scott, M.T. (1972). Biological effect of *Corynebacterium parvum*. III. Amplification of resistance and impairment of active immunity to murine tumours. *Br. J. Cancer* **26**, 361-367.

Suit, H.D., Sedlacek, R., Wagner, M., Orsi, L., Silobric, V. and Rothman, J. (1976). Effect of *Corynebacterium parvum* on the response to irradiation of C3H fibrosarcoma. *Cancer Res.* **35**, 1305-1314.

Szmigielski, S., Kobus, M., Gil, J., Jeljaszewicz, J. and Pulverer, G. (1980). Protection and therapy of mice with acute and chronic experimental virus infections with *Propionibacterium granulosum KP-45*. *Zbl. Bact. Hyg. I. Abt. Orig. A.* **248**, 286-295.

Szmigielski, S., Janiak, M., Wrembel, J.K., Lipski, S., Kobus, M., Pulverer, G. and Jeljaszewicz, J. (1982). Effect of 3 strains of Propionibacteria on enhancement of humoral immunity, activation of RES and inhibition of experimental tumour growth. (In prep.)

Warr, G.W. and Sljivic, V.S. (1974). Enhancement and depression of the antibody response in mice caused by *Corynebacterium parvum*. *Clin. Exp. Immunol.* **17**, 519-532.

Wiener, E. and Bandieri, A. (1975). Modification in the handling *in vitro* of ^{125}I-labelled keyhole limpet haemocyanin by peritoneal macrophages from mice pretreated with the adjuvant *Corynebacterium parvum*. *Immunology* **29**, 265-274.

Wolmark, N. and Fisher, B. (1974). The effect of single and repeated administration of *Corynebacterium parvum* on bone marrow macrophage colony production in syngeneic tumour-bearing mice. *Cancer Res.* **34**, 2869-2872.

Woodruff, M.F.A. and Boak, J.L. (1966). Inhibitory effect of *Corynebacterium parvum* on the growth of tumour transplants in isogeneic host. *Br. J. Cancer* **20**, 345-355.

Woodruff, M.F.A. and Dunbar, N. (1975). Effect of local injection of *Corynebacterium parvum* on the growth of a murine fibrosarcoma. *Br. J. Cancer* **32**, 32-41.

Woodruff, M.F.A. and Inchley, M.P. (1971). Synergistic inhibition of mammary carcinoma transplants in A-strain mice by antitumour globulin and *C. parvum*. *Br. J. Cancer* **25**, 584-593.

Wrembel, J.K. (1982). Modification of experimental post-irradiation or cyclophosphamide evoked injuries of the haemopoietic system by *Propionibacterium granulosum KP-45*. Ph.D. Thesis, pp. 1-104, WIHE, Warsaw.

A MODEL SYSTEM FOR ANTIMETASTATIC EFFICACY OF NONSPECIFIC IMMUNOMODULATING AGENTS

Ivan Bašić

Department of Animal Physiology,
Faculty of Natural Sciences and Mathematics,
University of Zagreb, Zagreb, Yugoslavia

I. INTRODUCTION AND RATIONALE

Development and assessment of methods for successful control of metastasizing tumours are very much dependent on the use of experimental tumour model systems. The similar pathogenesis of cancer in different mammalian species gives scope to the experimental cancer researcher to study the different mechanism(s) of metastasizing tumours and reactions of the disease to different approaches to cancer therapy. Furthermore, the majority of findings of such experimental work could easily and confidentially be extrapolated to clinical practice. Many animal tumour models used in cancer research do not fulfill necessary conditions to mimic clinical situations: a spontaneous origin, little or no capacity to effectively immunize animals syngeneic to the animal of origin and ability to spread from the tissue of origin to other sites in the host (Hewitt *et al.*, 1976). Experimental tumours possessing features or conditions which do not prevail in the clinical situations should be considered "artefact". Artefacts may be intrinsic to the model or can result from its technical handling. The commonest and most influential artefacts are avoidable by using experimental animal tumours of spontaneous origin in the inbred strain of animals providing transplant recipients, possessing low ability to immunize the host and having the ability to metastasize from primary to secondary sites. In this paper a suitable rat tumour model which accomplishes some of the requirements for mimicking clinical situations and its relevance to immunotherapy with nonspecific immunomodulating agents will be described.

II. IMMUNITY, IMMUNOMODULATION AND RESISTANCE TO TUMOUR

Ample data support the importance of immunological defence
mechanism in the aetiology and pathogenesis of malignant
tumours (Weiss, 1980). An increased incidence of malignant
tumours was found to be associated with immunological defici-
ency diseases (Waldmann *et al.*, 1972) and with the impairment
of the immune system caused by immunosuppressive treatments
(Starzl *et al.*, 1970). On the other hand, the infiltration
of tumours by lymphocytes is correlated with good prognoses
(Black *et al.*, 1954). It is now well established that the
growth of experimental animal tumours can sometimes be restricted by specific or nonspecific general host immunostimulation (Weiss, 1980). Many materials have been used to
stimulate the host immune responsiveness. Nonspecific immunomodifiers, particularly *Bacillus Calmette — Guérin* (Baldwin
and Pimm, 1980), *Corynebacterium parvum* (Milas and Scott,
1978), OK-432 (Aoki *et al.*, 1976) and Levamisole (LMS)
(Fisher *et al.*, 1977) have been shown to cause quite efficient
restriction of the growth and spread of experimental animal
malignant tumours. The extent to which malignant tumours
respond to the treatment with nonspecific immunomodifiers
depends on many factors comprising the stage, type and location of the tumour, the level of antigenicity of the target
cells, the status of the host immune system, and the route
of application of the agent. Using different mouse tumour
models we have shown that the antitumour activity of *C. parvum* (CP) mostly depended on immunogenicity of the tumour and
the time and route of immunomodifier administration (Bašić
et al., 1979; Bašić and Milas, 1980). Although the tumour
model systems in these studies have accomplished many purposes in studying *in vivo* antitumour activity of different
agents, they do not mimic clinical situations; in spite of
their spontaneous origin they are more or less immunogenic
and not metastasizing tumours.

III. A NEW METASTASIZING CARCINOMA MODEL IN THE RAT: MODEL CHARACTERISTICS AND APPLICATIONS

In the experimental study of the formation of tumour metastases and the influence of surgery, radio- and chemotherapy,
it would be desirable to use an animal model which resembles
the clinical tumour as closely as possible (Martin *et al.*,
1975). Many of the reported cancer models consist of induced
tumours which usually display marked antigenicity, in contrast
with tumours of spontaneous origin (Hewitt *et al.*, 1976).
Moreover, tumours which metastasize in a similar way as in
man are rarely used.

The main and primary route of spread of carcinomas in man is by way of the regional lymph nodes, and the main cause of death is from haematogenous metastases. In animals few tumours show both lymph node and blood borne metastases. Only a few model systems are known in mice which give lymphatic dissemination in a reproducible pattern, and these only rarely produce parenchimatous secondaris (van Putten *et al.*, 1975).

An animal tumour used in these studies is transplantable in a syngeneic host. The tumour metastasizes in the regional lymph node and the lungs, which serve as model organs. The staging with regard to tumour spread is easily followed. The tumour, although of low level of immunogenicity, responds to nonspecific immunotherapy. The tumour is an anaplastic carcinoma which arises spontaneously in the inguinal subcutaneous tissue in an old multiparous female of highly inbred strains of rats Y59 (Štark and Hauptfeld, 1969), produced in our own conventional breeding colony. The tumour has been maintained by serial passages in syngeneic recipients and it was in its 9th to 12th isotransplant generation when used for experiments.

A. Tumour Cell Suspension

Single cell suspension was prepared by digestion with trypsin of tumour tissues containing no visible necroses or haemorrhage (Milas *et al.*, 1974). Suspensions were passed through a stainless steel mesh (80 wires/cm) and washed out 3 times by centrifugation at 1200 r.p.m. for 5 min in Hanks solution supplemented with 5% normal syngeneic serum. Cells were counted with a haemacytometer. Viability, determined by trypan blue exclusion, was more than 90%.

B. Determination of Lung Metastases

The number of lung metastases was determined 22 days after tumour cell inoculation into footpad. Animals were killed, lungs removed, separated into lobes and fixed in Bouin's solution. Metastases appeared as round, white nodules which differed in size and were counted with the naked eye.

C. Immunization Procedure

Rats were given 3 intraperitoneal injections of irradiated (300 Gy) tumour cells (10^7 cells) at weekly intervals. Seven days after the last immunization they were challenged with viable tumour cells.

D. Nonspecific Esterase Activity

Tumour draining popliteal lymph node was removed, cellularity determined and nonspecific esterase activity examined by the method described in Bašić *et al.* (1980). Glass smears of the cells were fixed with phosphate-buffered 4% formaldehyde and incubated for 10 min with 1-naphthyl phosphate and Flast Blue B diazonium salt at pH 7.4.

E. Immunomodifiers

Four immunomodifiers were used: *Corynebacterium parvum* (CP) (Wellcome Research Laboratories, Beckenham, England), *Bacillus Calmette-Guérin* (BCG) (Torlak, Beograd, Yugoslavia), OK-432 (Picibanil, Chugai, Pharmaceutical Co. Ltd., Tokyo, Japan) and Levamisole (LMS) (Pliva, Zagreb, Yugoslavia). Agents were injected intravenously (i.v.), intrapleurally (i.p.) or into the tumour draining lymph node (i.t.d.n.). The doses were 0.35 mg for CP and BCG, and 7 mg for LMS.

F. Analysis of Results

The results were evaluated statistically by Student's *t*-test. Differences between groups were considered significant if the *p*-value of comparison was 0.05 or lower.

G. Model System

Figure 1 presents the model scheme used in these studies. Tumour in the footpad was generated by injecting 5×10^5 viable carcinoma cells resuspended in 0.05 ml of Hanks' solution.

Six days later the footpad with visible tumour was surgically removed. Surgery was performed under chloral hydrate general anaesthesia (300 mg/kg body weight). Subsequent to surgery animals received i.v. and i.t.d.n. treatment with immunomodifiers. In experiments with repeated i.t.d.n. injections modifiers were injected 7 and 14 days after first treatment. The antimetastatic activity of immunomodifiers was determined 22 days after tumour cell transplantation.

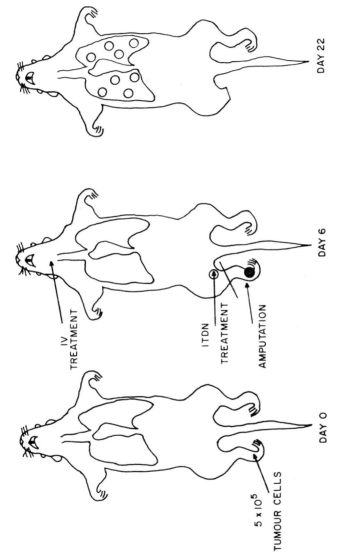

5 x 10⁵
TUMOUR CELLS

DAY O

IV
TREATMENT

ITDN
TREATMENT

AMPUTATION

DAY 6

DAY 22

FIG. 1 Schematic representation of the model.

IV. RESULTS

A. Effect of Specific Immunization on the Growth of Tumour Nodules in the Lung

Immunity to tumour was tested 7 days after the last immunization with irradiated tumour cells. Animals received i.v.
injection containing different numbers of tumour cells.
Table 1 shows that specific immunization slightly influenced
formation of tumour nodules in the lung.

TABLE 1

*Effect of specific immunization on the growth of
carcinoma nodules in the lung of Y59 rats*

Tumour dose	Number of lung nodules[a]		p (*t*-test)
	Nonimmunized	Immunized[b]	
10^3	2 (0 - 3)[c]	0	
5×10^3	19 (9 - 25)	12 (10 - 17)	NS[d]
10^4	77 (51 - 90)	62 (39 - 68)	<0.05
5×10^4	confluent	confluent	

[a] Nodules were counted 14 days after i.v. inoculation of carcinoma cells.

[b] Rats were immunized 3 times with 10^7 irradiated (30,000 R)
tumour cells given i.p. at weekly intervals.

[c] Mean (range). Groups contained 7–9 rats each.

[d] Not significant.

B. Therapy of Lung Metastases

Six days after tumour transplantation footpad with visible
tumour was amputated. Subsequent to surgery rats received
i.v. or i.p. injection of certain immunomodifiers. Tables 2
and 3 show the antimetastatic activity of the agents determined 22 days after tumour cell transplantation. Quite
efficient antimetastatic activity was achieved by CP and BCG
in both i.v. and i.p. treatments, while OK-432 and LMS seemed
to be ineffective. A number of immunomodifiers, when in
direct contact with tumour tissue, can excite a local host
response leading to suppression of tumour growth and spread.
To test this possibility immunomodifiers were given i.t.d.n.

TABLE 2

*Therapeutic effect of i.v. administered
immunomodifiers against pulmonary metastases
of anaplastic carcinoma of Y59 rats*

Treatment[a]	Lung metastases		P (t-test)
	Mean ± SE	Range	
None	26 ± 1.7	17 − 33	
CP	14 ± 0.8	9 − 17	<0.01
BCG	12 ± 1.8	3 − 18	<0.01
OK-432	31 ± 2.7	21 − 38	
LMS	27 ± 1.9	15 − 34	

[a] Amputation of the footpad with tumour was done 6 days after tumour cell transplantation. Treatments were subsequent to surgery; rats received 0.35 mg CP or BCG or OK-432 and 7.0 mg LMS, respectively. Groups contained 7 − 10 rats each.

TABLE 3

*Therapeutic effect of i.p. administered
immunomodifiers against pulmonary metastases
of anaplastic carcinoma of Y59 rats*

Treatment[a]	Lung metastases		P (t-test)
	Mean ± SE	Range	
None	18 ± 1.4	12 − 25	
CP	13 ± 0.4	9 − 14	<0.05
BCG	7 ± 0.7	4 − 11	<0.001
OK-432	−	−	−
LMS	17 ± 1.2	15 − 27	NS[b]

[a] Amputation of the footpad with tumour was done 6 days after tumour cell transplantation. Treatments were subsequent to surgery; rats received 0.35 mg CP or BCG and 7.0 mg LMS, respectively.
[b] Not significant. Groups contained 7 − 10 rats each.

TABLE 4

Therapeutic effect of i.t.d.n. administered
immunomodifiers against pulmonary metastases
of anaplastic carcinoma in Y59 rats

Treatment[a]	Lung metastases		P (t-test)
	Mean ± SE	Range	
None	22 ± 1.3	17 – 29	
CP	19 ± 0.9	16 – 22	NS[b]
BCG	18 ± 0.7	16 – 22	NS
OK-432	11 ± 1.1	7 – 15	<0.01
LMS	25 ± 2.7	14 – 33	NS

[a] Amputation of footpad with tumour was done 6 days after tumour cell transplantation. Treatments were subsequent to surgery; rats received 0.35 mg CP or BCG or OK-432, and 7.0 mg LMS, respectively.

[b] Not significant. Groups contained 7 – 10 rats each.

TABLE 5

Antimetastatic effect of various doses of i.v.
administered CP and BCG against lung metastases of
anaplastic carcinoma in Y59 rats

Treatment[a]	Dose (mg)	Lung metastases		P (t-test)
		Mean ± SE	Range	
None	–	24 ± 1.2	22 – 31	
CP	0.35	14 ± 0.9	9 – 17	<0.05
CP	100.00	16 ± 0.8	13 – 20	<0.05
BCG	0.35	11 ± 1.1	6 – 15	<0.01
BCG	100.00	10 ± 1.0	8 – 16	<0.001

[a] Amputation of the footpad with tumour was done 6 days after tumour cell transplantation. Treatments were subsequent to surgery. Groups contained 5 rats each.

subsequent to surgery. The number of lung metastases was
only reduced in animals receiving OK-432 (Table 4). The popliteal tumour growth, however, was equally controlled in
animals treated i.t.d.n. with either agent but LMS. Antimetastatic activity of CP or BCG was not improved with
increased doses of immunomodifiers in i.v. treatment (Table 5).

C. Antimetastic Activity of Combined Treatments

Single i.v. or i.p. treatments with CP or BCG efficiently
reduced the number of lung metastases (Tables 2 and 3). With
i.t.d.n. injection of these agents, however, local suppression of the tumour growth was achieved. It is likely that
this teratment also prevented further systemic distribution
of metastatic tumour cells. These findings and proposed possibilities suggested the combination of both systemic and
local treatment. Animals that underwent amputation of the
footpad with tumour received i.v. and/or i.p. (Tables 6 and 7).
Treatments with CP, BCG or OK-432 exerted strong antimetastatic effect while LMS was ineffective. Repeating i.t.d.n.
injection 3 times at weekly intervals the surviving time of
animals treated with BCG or CP was markedly prolonged; more

TABLE 6

*Combined i.v. and i.t.d.n. treatments and the growth
of carcinoma metastases in the lung of Y59 rats*

Treatment[a]	Lung metastases		P (t-test)
	Mean ± SE	Range	
None	19 ± 1.2	14 - 23	
CP	2 ± 0.3	0 - 4	<0.001
BCG	4 ± 0.4	1 - 7	<0.001
OK-432	7 ± 1.1	5 - 13	<0.05
LMS	17 ± 2.2	13 - 28	

[a] Amputation of the footpad with tumour was done 6 days after
tumour cell transplantation. Treatments were subsequent to
surgery; each injection contained 0.35 mg CP or BCG or
OK-432 and 7.0 mg LMS, respectively. Groups contained
5 — 7 rats each.

TABLE 7

Combination of i.p. and i.t.d.n. treatments and the number
of lung metastases of anaplastic carcinoma in Y59 rats

Treatment[a]	Lung metastases		P (t-test)
	Mean ± SE	Range	
None	19 ± 1.2		
CP	4 ± 0.3	2 - 6	<0.001
BCG	1 ± 0.2	0 - 2	<0.001

[a] Amputation of the footpad with tumour was done 6 days after
tumour cell transplantation. Treatments were subsequent to
surgery; each injection of immunomodifiers contained 0.35
mg. Groups contained 5 — 7 rats each.

TABLE 8

Effect of combined i.v. and i.t.d.n. treatments with BCG
against subcutaneous challenge with viable tumour cells

Treatment[a]	Tumour take		% of tumour rejected
	Initial	Final	
None	5/5	5/5	0
BCG	5/5	2/5	60

[a] Amputation of the footpad with tumour was done 6 days after
tumour cell transplantation; animals received BCG i.v. and
i.t.d.n. subsequent to surgery and i.t.d.n. therapy 7 and
14 days thereafter. Tumour-free animals were injected s.c.
with 10^4 viable tumour cells 90 days after first tumour
transplantation.

than 80% BCG and 40% CP treated animals survived 60 days
(Fig. 2). Mean survival time for control animals was 33 days.
Three of 5 rats, from the BCG treated group, that survived
90 days were specifically immune and rejected subcutaneous
challenge with 10^4 viable tumour cells (Table 8).

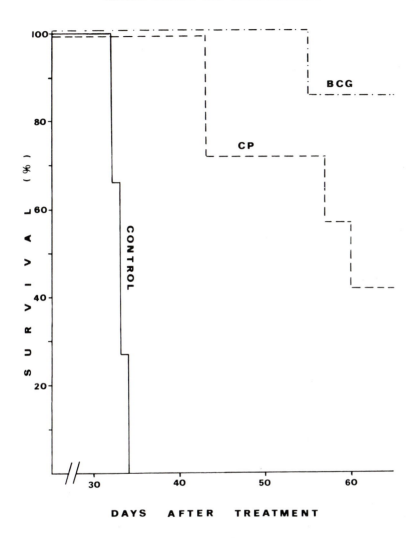

FIG. 2 *Effect on survival.   Viable tumour cells (5 × 10⁵) were injected into the footpad.   Six days after transplantation, the footpad with tumour was surgically removed, subsequent to amputation animals received i.v. and i.t.d.n. CP or BCG, treatments being repeated 7 and 14 days thereafter. Each injection contained 0.35 mg of immunomodifiers.   Groups contained 7 rats each.*

D. Nodal Tumour Growth Rate, Histology and Histo-
   chemistry of i.t.d.n. Treated Tumour Draining
   Lymph Node

Tumour draining popliteal lymph node was removed 3 days after
i.v. and i.t.d.n. injection of 0.35 mg CP or BCG, subsequent
to amputation of the footpad with the tumour. The nodes were
weighed and histopathological analyses were made. Popliteal
nodes from treated animals, in contrast to untreated ones,
showed that the infiltration of the node with tumour cells
occurred to a lesser degree in treated animals than in
untreated ones (Table 9); BCG produced a more profound effect.

TABLE 9

*Changes in tumour draining popliteal lymph node
after i.t.d.n. treatments with CP or BCG*

Treatment[a]	No. of rats	Mean lymph node weight (mg)	Histopathological grade[b]			
			−	+	++	+++
None	6	46 ± 1.9	0	0	0	6
CP	7	15 ± 1.7	0	2	4	1
BCG	7	13 ± 0.9	0	5	1	1

[a] Amputation of the footpad with tumour was done 6 days after
tumour cell transplantation; i.t.d.n. treatments with 0.35
mg CP or BCG were subsequent to surgery.

[b] Histopathological grading scale: − = no detectable tumour
cells; + = tumour cells confined to subcapsular marginal
sinus; ++ = 25% of node infiltrated with tumour cells;
+++ = more than 50% of node infiltrated with tumour cells.

Table 10 shows the effect of combined i.v. and i.t.d.n.
treatments on the number of nucleated cells in i.t.d.n.
treated lymph node and their histochemical characteristics
determined 3 days after treatment. Cellularity of treated
nodes was increased; plurality of cells in nodes from
untreated animals showed microscopical characteristics of
tumour cells. In contrast, the majority of cells from
treated animals were esterase-positive; we considered them
macrophages by comparing their nonspecific esterase activity
with that of peritoneal exudate cells induced in rats
treated intraperitoneally with BCG. The ability of draining
popliteal lymph node cells from animals subjected to combined

TABLE 10

*Cellularity of i.t.d.n. treated popliteal lymph node
and nonspecific esterase activity of respective cells*

Treatment[a]	Cellularity (x 10^6)	Esterase positive cells (%)
None	2.7 ± 0.4	27
CP	3.6 ± 0.5	67
BCG	3.9 ± 0.7	82

[a] Amputation of the footpad with tumour was done 6 days after
tumour cell transplantation; 0.35 mg CP or BCG was injected
i.v. and i.t.d.n. Parameters were determined 3 days after
treatment. Groups contained 7 rats each.

TABLE 11

*Induction of tumours in normal rats with tumour draining
"lymph node cells" from animals receiving combined
i.v. and i.t.d.n. therapy*

Donors[a]	Rats with tumour/rats injected with various numbers of node cells		
	10^4	10^5	5 x 10^5
Nontreated	7/7	5/5	5/5
CP-treated	0/7	3/7	5/7
BCG-treated	0/7	1/7	2/7

[a] Rats received 5 x 10^5 viable tumour cells on day 0. Six
days thereafter footpad with growing tumour was amputated
and animals were treated i.v. and i.t.d.n. with CP or BCG;
each injection contained 0.35 mg of immunomodifier. Tumour
draining popliteal lymph node was removed 3 days after
treatment, cell suspensions prepared and injected subcutaneously into normal recipients.

TABLE 12

Effect of repeated treatments with CP or BCG on tumour-bearing lymph nodes

Treatment[a]	No. of rats	Mean radii (mm) of nodes at days after surgery[b]				
		1	3	8	15	20
Control	6	5	7	10	18	27
CP	7	6	6	7	9	12
BCG	7	5	5	6	5	9

[a] Amputation of the footpad with tumour was done 6 days after tumour cell transplantation; i.t.d.n. treatments with 0.35 mg of CP or BCG were subsequent to surgery and 7 and 14 days thereafter.

[b] At intervals commencing one day after the first injection of immunomodifiers the surface area of tumour bearing lymph node (r^2) was determined by measurement through the skin.

i.v. and i.t.d.n. treatments to induce tumour in normal recipients was tested 3 days after treatment. Table 11 shows the results. For tumour induction many more "node cells" from treated donors were required than from untreated ones. Table 12 shows the effect of i.t.d.n. treatments given at weekly intervals on the size of the treated nodes. Treatments were administered on the day of surgery and 7 and 14 days thereafter. Mean radius (mm) of the treated node was measured on days 1, 3, 8, 15 and 20 after surgical removal of the footpad with tumour. Both CP and BCG treatments successfully controlled local tumour growth; BCG was more effective.

V. DISCUSSION AND CONCLUSION

Literature is abundant with the data describing successful treatments of malignant tumours with nonspecific immunomodulating agents (Milas and Scott, 1978; Baldwin and Pimm, 1980; Aoki *et al.*, 1976; Fisher *et al.*, 1977), mechanisms of their antitumour activity (Bašić *et al.*, 1975; Hibbs *et al.*, 1972) and their influence on haematopoietic activities (Bašić *et al.*, 1979). The main criticism to the results on antitumour activity obtained in different studies could be focussed just in one point: the animal tumour models used in most studies do not possess the characteristics of tumours in humans. This implies the obligation to the researchers to search for the animal tumour model system which will, as closely as possible, behave as malignant human tumours with special reference to metastases formation. The limitation of insight into the mechanism(s) of metastases formation and their sensitivity to therapy may be due, in part, to restrictions in experimental models used to approach this complex problem. With better models, a rational basis on which to found attempts at influencing metastases formation may be investigated. This relates especially to human tumours showing 2 modes of spread, via lymph node and subsequent haematogenous. Findings in these studies show that there is now a tumour model system which shows these 2 modes of spread in 100% of animals. The model could be used as a useful tool in testing the effectiveness of different types of therapy in various steps of metastases formation; these data showing the influence of different immunomodifiers on metastatic spread and growth of secondaries are a contribution to this proposal. It should be stressed, however, that no animal tumour can represent all the variations in growth potential and metastasizing characteristics of human tumours, but it is of significance in this system that the 2 modes of metastasizing are both 100% phenomena which may be of convenience for evaluation of therapy differences. In general, several conclusions could be drawn from the presented studies: (a) syngeneic rat anaplastic carcinoma system can be manipulated to more closely approximate to clinical experience; (b) the model offers the unique advantage of being a defined immunotherapy model which, besides surgery and immunotherapy, may also be treated by radiotherapy or chemotherapy; (c) to reach better goals in immunotherapy of tumour metastases the systemic application of an immunomodifier should be combined with local treatment associated with close contact of immunomodifier and tumour draining lymph node. Although no animal model can completely mimic malignant metastatic tumour in man, this model seems to be a very attractive experimental one.

VI. ACKNOWLEDGEMENTS

This investigation was supported by the Scientific Fund of S.R. of Croatia (SIZ IV).

The author is very grateful to Mrs Matilda Derikrava for technical assistance, Mrs Josipa Zake for the assistance with drawings, and Mrs Ivanka Rumora for secretarial work.

VII. REFERENCES

Aoki, T., Kveder, J.P., Hollis, V.M. Jr. and Bushar, G.S.
  (1976). *Streptococcus pyogenes* preparation OK-432.
  Immunoprophylactic and immunotherapeutic effects on the
  influence of spontaneous leukaemia. *J. Nat. Cancer Inst.*
  56, 687-690.

Bašić, I. and Milas, L. (1979). Effect of whole body irradiation in mice treated with *C. parvum*. *Eur. J. Cancer* **15**,
  901-908.

Bašić, I. and Milas, L. (1980). Role of macrophages in the
  antitumour activity of intrapleurally administered *Corynebacterium parvum*. *Period. Biol.* **82**, 119-125.

Bašić, I., Milas, L., Grdina, D.J. and Withers, R.H. (1975).
  In vitro destruction of tumour cells by macrophages from
  mice treated with *Corynebacterium granulosum*. *J. Nat.
  Cancer Inst.* **55**, 589-596.

Bašić, I., Malenica, B., Vujičić, N. and Milas, L. (1979).
  Antitumour activity of *Corynebacterium parvum* administered
  into pleural cavity of mice. *Cancer Immunol. Immunother.*
  7, 107-115.

Bašić, I., Rodé, B., Kaštelan, A. and Milas, L. (1980).
  Activation of pleural macrophages by intrapleural applica-
  tion of *Corynebacterium parvum*. *In* "Macrophages and
  Lymphocytes". (Eds M.R. Escobar and H. Friedman), Advances
  Exptl. Med. Biol., Vol. 121A, pp. 333-341. Plenum Press,
  New York and London.

Black, N.M., Opler, S.R. and Speer, F.D. (1954). Microscopic
  structure of gastric carcinomas and their regional lymph
  nodes in relation to survival. *Surg. Gynaecol. Obstet.*
  98, 725-734.

Baldwin, R.W. and Pimm, M.V. (1980). BCG in tumour immunotherapy. *Adv. Cancer Res.* **28**, 91-147.

Fisher, G.W., Crumrine, M.H., Balk, M.W., Chang, S.P. and
  Hokama, Y. (1977). Immunopotentiation in a herpes virus
  model: an overview. *In* "Control of Neoplasia by Modula-
  tion of the Immune System". (Ed. M.A. Chirigos), Vol. 2,
  Progress in Cancer Research and Therapy. pp. 107-114.
  Raven Press, New York.

Hewitt, H.B., Blake, A.S. and Walderer, S.A. (1976). A
  critique of the evidence for active host defence against
  cancer, based on personal studies of 27 murine tumours of
  spontaneous origin. *Br. J. Cancer* **33**, 241-249.

Hibbs, J.B., Jr., Lambert, L. Jr., and Remington, S.J. (1972).
  Possible role of macrophage mediated nonspecific cytotoxicity in tumour resistance. *Nature New Biol.* **235**, 48-
  50.

Martin, D.S., Fugman, R.A., Stolfi, R.L. and Hayworth, P.E. (1975). Solid tumour animal model therapeutically predictive for human breast cancer. *Cancer Chemother. Rep. Part 2* **5**, 89-94.

Milas, L., Hunter, N., Mason, K. and Withers, H.R. (1974). Immunological resistance to pulmonary metastases in C3Hf/Bu mice bearing syngeneic fibrosarcoma of different sizes. *Cancer Res.* **34**, 61-71.

Milas, L. and Scott, M.T. (1978). Antitumour activity of *Corynebacterium parvum*. *Adv. Cancer Res.* **26**, 257-306.

Putten, L.M., van Kram, L.K.J., van Dierendonck, H.H.C., Smink, T. and Fuzi, M. (1975). Enhancement by drugs of metastatic lung nodule formation after intravenous tumour cell injection. *Int. J. Cancer* **15**, 588-595.

Startzl, T.E., Porter, K.A., Andres, G., Halgrimson, C.G., Hurwitz, R., Giles, G., Terasaki, P.I., Penn, I., Schroter, G.T., Lilly, J., Starkie, S.J. and Putnam, C.W. (1970). Long term survival after renal transplantation in humans (with special reference to histocompatibility matching, thymectomy, homographt glomerulonephritis, heterologous ALG and recipient malignancy). *Ann. Surg.* **172**, 437-472.

Štark, O. and Hauptfeld, M. (1969). Serologically detected R+H-1 antigens of the A52, VM and Y59 rats. *Folia Biolog. (Praha)*, **15**, 35-40.

Waldmann, T.A., Strober, W. and Blaese, R.M. (1972). Immunodeficiency disease and malignancy: various immunological deficiencies of man and the role of immune processes in the control of malignant disease. *Ann. Intern. Med.* **77**, 605-628.

Weiss, D.W. (1980). Tumour antigenicity and approaches to tumour immunotherapy. *In* "Current Topics in Microbiology and Immunology" Vol. 89. Springer-Verlag, Berlin, Heidelberg, New York.

THE MECHANISM OF ANTITUMOUR ACTIVITY
OF THE BACTERIAL ENDOTOXINS

Masashi Kodama, Tomoyuki Mizukuro, Nozomi Yamaguchi,
Shunichi Yoshida and Yutaka Katayama

*Department of Surgery, Shiga University of Medical Science,
Ohtsu, Shiga;*

and

*Department of Public Health and Microbiology,
Kyoto Prefectural University of Medicine, Kyoto, Japan*

I. INTRODUCTION

The late Dr Coley used mixed bacterial vaccine consisting of
Streptococcus and *Serratia* for the therapy of human cancer
patients, and over 130 inoperable cancer patients were reported to have been cured completely (Nauts *et al.*, 1946).
Quite independently the phenomenon of haemorrhaging and necrosis of malignant tumour tissues by the i.p. injection of
bacterial endotoxin (LPS) was found by Gratia and Linz (1931).
The phenomenon was confirmed as the constant finding for
malignant tumours by many investigators, and is highly selective in that it affects only malignant tumour cells, suggesting the existence of a tumour-specific and common substance
in all malignant tumour cells. The study of this phenomenon
may help clarify the nature of the tumour cell and could lead
to the finding of an effective therapy for cancer patients.

II. TUMOUR NECROSIS BY THE LPS

A. Newly Established Tumour MC-B1 - BALB/c

By injecting BALB/c mice with methylcholanthrene, several
strains of fibrosarcoma were established in our laboratory
one of which was used in this experiment (MC-B1). The BALB/c
mice were inoculated with 5×10^5 MC-B1 cells intradermally
and 50 µg of LPS of *E. coli* 055 (Boivin) was injected intraperitoneally on days 3, 5, 7, 10, 12 or 22 respectively.

KODAMA *et al.*

TABLE 1

*Results of treatment of lipopolysaccharide
in MC-Bl fibrosarcoma-bearing mice*

The day of LPS after tumour inoculation	Tumour size in diameter (mm)	Haemorrhage (%)	Complete cure (%)
3	1.2 ± 0.1	0	0
5	2.1 ± 0.3	10	0
7	3.3 ± 0.3	80	20
10	6.3 ± 0.4	90	20
12	8.9 ± 0.4	100	30
22	14.2 ± 1.2	100	0

As shown in Table 1, the haemorrhage and necrosis, while not observed in 3 day old tumours, gradually increased as the tumour progressed and were observed in 100% of the mice on the 12th day. The cure rate of the mice with 12-day old tumours was 30%, and of those with the 22nd day old tumours 0%. All control mice died by the 65th day.

B. MH134 — C3H System

The C3H mice were inoculated with 10^6 ascitic tumour cells subcutaneously in the abdomen or hind region, and the tumour reached about 7 mm in diameter 7 days after inoculation. All the control mice died by the 40th day.

Fifty μg of *E. coli* 0127 LPS (Westphal) was injected intraperitoneally at day 7. Haemorrhage and necrosis were observed only at the tumour site without exception, and about 50% of the mice were cured completely.

The cured mice showed specific immunity to inoculation of MH134 tumour cells of 5×10^6 cells subcutaneously and this specific immunity was able to be adoptively transferred to the syngeneic mice by using spleen cell suspension.

When 200 μg of LPS was injected intratumourally one day before, on the same day, or one day after an intraperitoneal injection of LPS (50 μg), complete cure was obtained in about 80% of the mice. No statistically significant difference was observed between these groups.

OK-432, which is the penicillin-lysate of the haemolytic *streptococcus* and whose medical use is officially permitted

in Japan, did not affect the complete cure of MH134 tumours
7 mm in diameter by which 2 KE units were injected intraperitoneally, but did have a complete cure rate of about 20% when
the same dose was administered intratumourally. Intratumoural
inoculation of 2 KE units of OK-432 combined with intraperitoneal inoculation of 50 µg of *E. coli* LPS gave a complete
cure rate of about 80%.

III. THE MECHANISM : ENDOTOXIN SIDE

A. Nude Mouse System

Athymic nude mice BALB/c nu/nu were inoculated with 5×10^5
cells of MC-B1 and 10 days later, when the tumour had grown
to about 10 mm in diameter, were injected with 25 µg of LPS
intraperitoneally. Haemorrhaging of tumour tissues was
observed in all mice thus treated, and necrosis was evident
in the central portion of tumours, not completely at the
periphery. From here tumour growth resumed and all the mice
eventually died (Table 2).

TABLE 2

Haemorrhaging caused by lipopolysaccharide in
tumour-bearing nude mice

Tumour	Tumour size (mm)	Haemorrhage	Complete cure
MC-B1	10.4 ± 0.4	10/10 (100%)	0/10 (0%)
HeLa	10.3 ± 0.3	6/10 (60%)	0/10 (0%)

Other nude mice were implanted with 10^6 cells of a human
carcinoma, HeLa, subcutaneously, and were injected with 25 µg
of *E. coli* LPS after the tumours had grown to about 10 mm in
diameter in 3 weeks. Haemorrhaging was observed in the 6
mice out of the 10 thus treated which had larger sized tumours
(Table 2).
     These data suggest that the participation of T-lymphocytes
is not essential for the haemorrhaging and necrosis of tumour
tissues after LPS injection. This is supported by the fact
that the cheekpouch-implanted tumours of hamsters are brought
to haemorrhaging after LPS injection (Steinbring and Stevens,
1977).

This experiment also shows that TNF (tumour necrotizing factor) is not the major factor in this phenomenon, because Old had stated that TNF is not produced by athymic nude mice primed with *C. parvum* (1976). Since in our trial the mice were primed neither with BCG nor with *C. parvum*, or zymosan, it is therefore difficult to assume any participation at all of the TNF in the tumour haemorrhaging and necrosis in nude mice.

B. Histological Observations

We observed platelet thrombus and haemorrhage at the periphery of the tumour mass at 1 h, haemorrhage and makred infiltration of neutrophils in the tumour tissues at 4 h, degeneration and necrosis of the tumour cells at 12 h, and infiltration of mononuclear cells at 24 h. These changes were essentially similar with the necrotic vasculitis in the local Shwartzman phenomenon and in the Arthus reaction, suggesting that the haemorrhaging and necrosis phenomenon share some pathogenetic processes in common.

C. Many Agents That Can Cause Tumour Haemorrhage

The haemorrhage and necrosis of tumour tissues after i.v. or i.p. injection of LPS is very similar to the local Shwartzman phenomenon, since the histological manifestation of both phenomena are the same, and both are provoked by the inoculation of bacterial LPS.

The tumour haemorrhage can also be induced by the injection of antigen-antibody complex, or of agar (Apitz, 1933), or of glycogen (Stetson, 1951), all of which are known to have the capacity to provoke the local Shwartzman phenomenon.

D. Common Mechanism:Complement Activation?

These substances have the common capacity to activate complement. We confirmed that endotoxin is not accumulated in the tumour tissues after *in vivo* administration (Seligman *et al.*, 1948; Nowotny *et al.*, 1971), using MH134 tumour and C3H system with the peroxidase antibody method. These facts suggest the participation of the activated complement fragment(s) in tumour haemorrhaging and necrosis.

E. The Effect of Zymosan

Because of the above reasoning, we examined the effect of the zymosan on tumour tissues. BALB/c mice were injected with 100 mg of zymosan (Sigma Chem. Co., USA) intraperitoneally

12 days after MC-B1 tumour implantation. Tumour haemorrhage
was not observed. The administration of zymosan suppressed
tumour haemorrhage after LPS injection in 4 mice out of 10,
and the haemorrhaging in the remaining 6 mice was weaker than
that in the control tumour-bearing mice injected with LPS
without zymosan pretreatment.

F. The Effect of the Cobra Venom Factor (CVF)

The cobra venom of *Naja naja kaouthia* (Sigma Chem. Co. St.
Louis, USA) was purified using DEAE-cellulose and lyophilized.
A 0.4 mg dosage of this induced severe anaphylactoid reaction
in normal mice. BALB/c mice inoculated with MC-B1 tumour
cells 12 days previously were inoculated with 0.2 mg of CVF.
With this dosage haemorrhaging of the tumour tissue was not
observed in any of the mice, but CVF inoculation suppressed
the tumour haemorrhaging induced by LPS in 5 mice out of 7
(Table 3).

TABLE 3

*Results of treatment with Cobra Venom Factor
in MC-B1 fibrosarcoma-bearing mice*

Cobra V.F.	Tumour size	Haemorrhaging
0.2 mg	6.7 ± 0.3 mm	0/7 (0%)
0.6 mg	10.6 ± 0.4 mm	8/8 (100%)

When 0.6 mg of CVF was administered to 9 tumour-bearing
mice, one mouse died within 4 h, and the remaining 8 animals
showed strong haemorrhaging only at the tumour site, but 5 of
them died within 24 h (Table 3).
     The haemorrhaging of Sarcoma 37 in mice inoculated with
moccasin venom was observed by Shimkin and Zon (1942), but
the meaning has not been discussed since then.

G. The Effect of Yeast Activated Serum

The fresh serum of normal healthy humans and of normal mice,
incubated at $56^{\circ}C$ for 30 min, was treated with boiled (30 min)
baker's yeast at the concentration of 20 g/litre, with and
without the addition of epsilon-amino caproic acid (EACA).
     As a preliminary test, these serum specimens were inocu-
lated intraperitoneally into C3H mice bearing MH134 tumours,

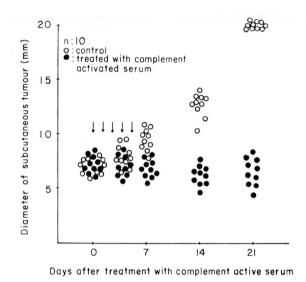

FIG. 1 *Results of treatment with the complement activated
serum in MH134 bearing mice. The complement activated serum
is administered from the 7th to the 11th day after tumour
inoculation.*

and slight hyperemea was observed. There was no marked difference in activity between mice serum and human serum, or
those sera activated in the presence or absence of EACA.
    The C3H mice carrying MH134 tumours about 7 mm in diameter
were inoculated with 3 ml of yeast-treated fresh human serum
for 5 consecutive days. As shown in Fig. 1, the growth of
the tumour was suppressed to the extent that the tumour dia-
meters were about 30% those in the control mice.

IV. THE MECHANISM: TUMOUR SIDE

A. A "Tumour-specific and Common Substance"

Though malignant tumour cell lines vary in susceptibility to
endotoxins (LPS) and though response to LPS is not observed
during the first few days after the inoculation of tumour
cells, the phenomenon of the haemorrhage and necrosis of
solid malignant tumour tissues following the i.v. or i.p.
injection of LPS is highly specific and common to all malignant tumour tissues. That injection of large amounts of the
cobra venom factor similarly causes tumour haemorrhaging and
necrosis, without there being any effect on normal tissues,
suggests the existence of a specific and common substance in
all malignant tumour cells.

B. The Shwartzman Active Substance in Human Cancer Tissues

The similarity of this phenomenon to the local Shwartzman reaction led Antopol to propose that Shwartzman active substance(s) may exist in the malignant tumour tissues of humans and mice (Antopol, 1937). Shear and his colleagues demonstrated that tryptic digestion of malignant tumour tissues resulted in a product having a strong Shwartzman activity in comparison to normal tissues in which only a weak reaction was demonstrated after tryptic digestion (Landy and Shear, 1957). Shear later reported that the latter was due to contamination by an endotoxin-containing bacteria (Merler *et al.*, 1960).

The authors have also demonstrated the existence of weak Shwartzman active substance(s) in malignant tumour cells of rats and of human cancer patients (Kodama *et al.*, 1973, 1977). Human gastric cancer tissues were extracted by saline, by 0.25 N trichloroacetic acid (TCA) or by 45% hot phenol, and these extracts were shown to have a very weak Shwartzman activity to prepare rabbits' skin for the local Shwartzman phenomenon without any exceptions in testing so far conducted. The strength of the activity was about one-thousandth that of the LPS of *E. coli* 055 prepared by the Westphal method by dry weight basis.

The human cancer extracts were also shown to have the ability to provoke the local Shwartzman phenomenon, to coagulate Limulus lysate, to have the pyrogenicity which was suppressed in endotoxin-tolerant rabbits, and to cause haemorrhage and necrosis of the solid sarcoma 180 tissues of dd strain mice.

The extracts of the healthier parts of the resected human tissues extracted in parallel with the malignant portions were shown to contain none of these activities.

C. The Shwartzman Active Substance in Yoshida Sarcoma Cells in Rats

The Yoshida sarcoma cells in saline, phenol and TCA extracts were shown to have the capability both to prepare and provocate the local Shwartzman phenomenon. A Shwartzman activity of the Yoshida sarcoma extract was about one-thousandth that of *E. coli* LPS by dry weight basis. A similar relationship was observed by Limulus lysate test.

These extracts were also shown to have the ability to cause pyrogenicity in rabbits, and to decrease neutrophils and platelets after i.v. injection, also at about one-thousandth the activity of that of *E. coli* LPS. They also caused

the haemorrhaging and necrosis of the sarcoma 180 tumour
tissues in dd strain mice after i.v. inoculation. Extracts
from rat normal tissues showed none of these activities.

D. The Shwartzman Active Substance(s) as Cell Surface Material

The antibody against TCA extract of the Yoshida sarcoma cell
was prepared from rabbits, and used to stain the Yoshida sarcoma cells followed by Fab fragment of sheep antirabbit IgG
conjugated with fluorescein isothiocyanate. Fluorescent
staining was observed only on the cellular surface of the
Yoshida sarcoma cells and not on the normal rat cells (Kodama
et al., 1973).

E. The Sensitization of Hosts by the Shwartzman Active Substance(s)

If our supposition about the Shwartzman active substance(s)
being contained in every cancer cell is correct, it follows
that cancer-bearing animals are sensitized by the Shwartzman
active substance liberated from dead tumour cells and may be
prepared in the Shwartzman sense. This altered behaviour of
the animals sensitized with the endotoxin-like substance may
be hypothesized from the observations of animals administered
with the endotoxin or endotoxin-containing living or dead
microorganisms. We named this the "Shwartzman model" (Katayama and Kodama, 1980).
     Our model of the cancer cells can be summarized as follows:

a) The cell membrane of the cancer cells is physically and
   chemically different from that of the normal cell, especially in that

b) the cell membrane of cancer cells contains a Shwartzman
   active substance which shows both preparatory and provocative activities;

c) the Shwartzman substance, liberated by occasional cell
   death, sensitizes not only the tumour site, but also the
   entire organisms and makes the host endotoxin-sensitive
   and prepared in the Shwartzman sense;

d) the endotoxin-sensitive state is expressed (i) either by a
   prepared state of the tumour site for a local Shwartzman
   phenomenon, in which endotoxin injection can cause haemorrhage and necrosis at the tumour site; (ii) by sensitization of the host for a generalized Shwartzman reaction,
   with occasional occurrence of an excessive increase of
   blood coagulation; (iii) by increased susceptibility of

the host to the LPS; (iv) or by increased sensitivities of
the hosts' skin to the intracutaneous injection of bacterial endotoxins.

A number of workers have reported the phenomenon of tumour
haemorrhaging and necrosis following the injection of LPS as
already described, and a number of clinical workers have
reported increased coagulation in cancer patients (Hjort and
Rapaport, 1965). An increase in susceptibility to the lethal
action of LPS has been repeatedly observed in tumour-bearing
animals (Shear and Perrault, 1944; Nowotny *et al.*, 1971). We
found this phenomon in solid Yoshida sarcoma-bearing rats, in
which all rats died following inoculation of 40 µg of *E. coli*
LPS, but no normal rats died after inoculation with 800 µg
(submitted for publication). We found also that the skin of
many cancer patients (48 in 125) shows an erythematous reaction of over 21 mm in diameter 4 h after the i.c. injection
of 40 ng of *E. coli* endotoxin to the centre line of the
flexor surface of the upper arm. The same dosage caused an
erythema of only 10—20 mm in diameter in 51 normal healthy
volunteers among 52 (manuscript in preparation).

F. "Shwartzman Gene" and "Cancer Gene": The Nature
   of the Malignant Tumour Cells

The phenomenon of the haemorrhaging and necrosis following
the LPS injection being highly selective and observed only in
tumour tissues suggests the existence of a tumour-specific
and common substance(s) in tumour cells. As one of the candidates for these substances, we assumed the existence of the
Shwartzman active substance in all malignant tumour cells.
To explain this existence, it has to be hypothesized that all
malignant tumour cells contain the Shwartzman gene which produces the Shwartzman active substance, that this gene is
closely linked to the malignancy-determining gene, and that
both of these genes must be strongly corepressed in normal
cells.

V. DISCUSSION

We have used 3 systems, chemically induced MH134 mouse hepatic
tumour cells, chemically induced MC-B1 tumour cells established in our laboratory, and human malignant tumour cells,
HeLa. In mice carrying each cell line, the inoculation of
20 to 50 µg of *E. coli* LPS intraperitoneally caused haemorr-
having and necrosis of tumour tissues, but this phenomenon
was not observed if the LPS was administered during the first
few days after tumour inoculation. Tumour haemorrhaging and

necrosis was also observed in tumour-inoculated nude mice
treated with LPS, suggesting that the T-lymphocyte is not an
essential component of this phenomenon. Although complete
regression was not obtained in nude mice, the necrosis of
major parts of the tumour tissue was evident.

There are several interesting features to the phenomenon
of tumour haemorrhaging and necrosis after LPS injection,
that is, it is observed only in solid tumour tissues, not in
free ascitic tumours; tumour tissues are not sensitive to LPS
injection during the first few days after tumour implantation;
the histological expression of tumour haemorrhaging and necrosis is very similar to that of the local Shwartzman phenomenon
and the Arthus reaction, and is essentially necrotic vasculitis, the main feature of which is the massive infiltration
of polymorphonuclear leukocytes (Apitz, 1933; Stetson, 1951);
weaker but similar phenomena with the same histological mani-
festations are induced in tumour-bearing animals by agar, by
antigen—antibody complexes (Apitz, 1933) or by glycogen
(Stetson, 1951) which are known to have the ability to provoke
the local Shwartzman phenomenon. Based on the similarity
between the tumour haemorrhaging and necrosis caused by LPS
and that of the local Shwartzman reaction, Antopol (1937),
Shear's group (Landy and Shear, 1957), and our own (Kodama
et al., 1973, 1977) have all tried to demonstrate Shwartzman
active substance(s) in malignant tumour tissues with varying
degrees of success.

Attempts have also been made to explain this LPS-instigated
phenomenon either as a direct effect of endotoxins on the
vascular system of the tumour, causing vacular collapse,
tumour anoxia, and subsequent tumour death, the angiogenesis
of which was augmented by substances released from tumour
cells such as "tumour angiogenesis factor" (TAF) (Algire *et
al*., 1952) or by the existence of tumour-specific cytotoxic
mediators, such as the "tumour necrotizing factor" (TNF)
(Carswell *et al*., 1975).

The theories based on the vulnerability of the newly formed
blood vessels in tumour tissues to direct endotoxin action
(Algire *et al*., 1952) or on the release of TNF (Carswell *et
al*., 1975) might be able to explain some, if not all, of the
features of the tumour haemorrhaging and necrosis after the
LPS injection, but interpretation of this phenomenon as a
kind of local Shwartzman phenomenon, assuming that the tumour
site was sensitized by the released Shwartzman active substance of cancer cells by occasional cell death, could
explain all the features easily.

It has been suggested recently that the *in vitro* tumour
cytotoxic activities (TNF) are multiple and are dissociable
from the *in vivo* tumour necrosis activity, and that tumour

tissue necrosis was observed without detectable TNF activity
in the mice sera after LPS injection (Kull and Cuatrecasas,
1981). We confirmed this latter finding using the BALB/c—
MC-B1 system. We also showed that the athymic nude mice
inoculated with the syngeneic MC-B1 tumour or with human carcinoma cell, HeLa, exhibited tumour haemorrhaging following
LPS injection. The nude mouse primed with *C. parvum* is
claimed to lack the ability to produce TNF (Old, 1976), making
it apparent that TNF is not the sole mechanism of the haemorrhaging and necrosis of malignant tumour tissues after LPS
injection. Of importance, however, is the fact that these
mechanisms are not mutually exclusive but could operate simultaneously.

Furthermore, a large amount of CVF was able to induce the
tumour haemorrhage in mice, and small amounts of CVF or zymosan suppressed the tumour haemorrhage by the LPS. These
phenomena may lead us to believe that activated complement
fragment(s) make some contribution to the antitumour activity.
Such an activity was reported for C3a by Ferluga *et al.*,
(1978). The supposed mechanisms of the LPS are illustrated
in Fig. 2.

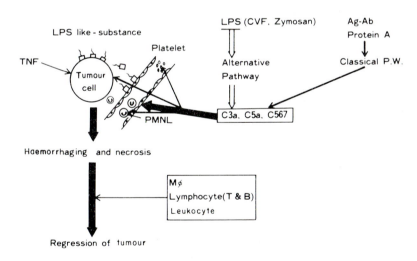

FIG. 2 *Mechanism of tumour haemorrhaging with the lipopolysaccharide.*

VI. SUMMARY

The bacterial endotoxin can cause haemorrhaging and necrosis in malignant tumour tissues only when the tumour size is greater than 3 mm in diameter.

The same phenomenon was also observed with athymic nude mice, indicating that the T-lymphocytes and the TNF are not essential components of this phenomenon. That cobra venom factor can cause a similar phenomenon, suggests the participation of activated complement component(s).

Since this phenomenon is highly selective for the malignant tumour tissues, there could be a tumour-specific and common substance. As one of the candidates, the existence of a Shwartzman active substance(s) in malignant tumour cells and a tumour model based on this supposition was proposed.

VII. REFERENCES

Algire, G.H., Legallais, F.Y. and Anderson, B.F. (1952). *J. Nat. Cancer Inst.* **12**, 1279-1295.

Antopol, W. (1937). *J. Infect. Dis.* **61**, 334-337.

Apitz, K. (1933). *Z. Krebsforsch.* **40**, 50-70.

Carswell, E.A., Old, L.J., Cassel, R.L., Green, S., Fiore, N. and Williamson, B. (1975). *Proc. Natl. Acad. Sci. USA* **72**, 3666-3670.

Ferluga, J., Schorlemmer, H.U., Baptista, L.C. and Assison, A.C. (1978). *Clin. Exp. Immunol.* **31**, 512-517.

Gratia, A. and Linz, R. (1931). *C.R. Soc. Biol.* **108**, 427-428.

Hjort, P.F. and Rappaport, S.I. (1965). *Ann. Rev. Med.* **16**, 135-168.

Katayama, Y. and Kodama, M. (1980). *Japan. J. Med. Sci. Biol.* **33**, 35-36.

Kodama, M., Yajima, H., Kubo, Y., Hashimoto, I. and Katayama, Y. (1973). *Japan. J. Med. Sci. Biol.* **26**, 3-6.

Kodama, M., Yajima, H., Hashimoto, I. and Katayama, Y. (1977). *Japan. J. Med. Sci. Biol.* **30**, 77-81.

Kull, F.C. and Cuatrecasas, P. (1981). *J. Immunol.* **126**, 1279-1283.

Landy, M. and Shear, M.J. (1957). *J. Exp. Med.* **106**, 77-97.

Merler, E., Perrault, A., Trapani, R.-J., Landy, M. and Shear, M.J. (1960). *Proc. Soc. Exp. Biol. Med.* **105**, 443-445.

Nauts, H.C., Swift, W.E. and Coley, B.L. (1946). *Cancer Res.* **6**, 205-216.

Nowotny, A., Golub, S. and Key, B. (1971). *Proc. Soc. Exp. Biol. Med.* **136**, 66-69.

Old, L.J. (1976). *Clin. Bull.* **6**, 118-120.

Seligman, A.M., Shear, M.J., Leiter, J. and Sweet, B. (1948). *J. Nat. Cancer Inst.* **9**, 13-18.

Shear, M.J. and Perrault, A. (1944). *J. Nat. Cancer Inst.*
 4, 461-476.
Shimkin, B. and Zon, L. (1942). *J. Nat. Cancer Inst.* **3**, 379-
 382.
Stetson, C.A. Jr. (1951). *J. Exp. Med.* **93**, 489-504.
Steinbring, W.R. and Stevens, D. (1977). *Proc. Soc. Exp.
Biol. Med.* **156**, 229-235.

DISCUSSION II
CHAPTERS 5-8

Dr BALDWIN: Wellcome's *C. parvum* is being used for control of malignant effusions — pleural and peritoneal. What are the clinical findings and is there any evidence as to the mechanism of action?

Dr ADLAM: The clinical findings show a beneficial effect but the mechanism of action is unknown at present. (References: Webb *et al.* (1978) *Brit. Med. J.* **I**, 338; Millar *et al.* (1980) *Thorax* **35**, 856; and Mantovani *et al.* (1981) *Int. J. Cancer* **27**, 437.)

Dr BALDWIN:The role of host cells stimulated by *C. parvum* in tumour rejection is complicated and will depend upon a number of factors, including the type (immunogeneity) of the tumour, its site and mode of treatment. For example, intralesional injection of *C. parvum* into tumours may function through stimulation of macrophages and so responses can be induced in immunodeprived hosts. Other modes of treatment may involve a series of pathways involving T-cells and NK cells. In this case immunosuppression will abrogate the therapeutic response.

Helen NAUTS: Dr Adlam, what site did you use in treating the ascites and pleural effusion cases with *C. parvum*? Coley's toxin was given intradermally and caused complete regression — no further ascites or effusion traced for several years.

Dr ADLAM: *C. parvum* was administered into the cavity involved, not intradermally.

Dr GOLDIN: Dr Adlam, what is the relation of lung nodule to survival time in your experimental studies?

Dr ADLAM: We dissect our mice 14 days after challenge with tumour and do not use "time to death" as a measure of activity. Presumably, there would be a relationship if "time to death" were used, but it is more quantitative to actually count the lung nodules.

Dr KIRN: How can you explain that nonactive strains of *C. parvum* are still quite active on tumour formation and not at all on spleen enlargement?

Dr ADLAM: *P. granulosum* strains do show some antitumour effect in the T3 fibrosarcoma lung model when compared to saline controls. But this is of an order of magnitude far less than the potency seen with active strains of *C. parvum*. This residual activity may just be due to the effects of injecting any large amount of particulate material which might cause slight macrophage activation and hence slight antitumour effects but this is just a speculation.

Dr EGGERS: Is there a relationship between active strains of *C. parvum* and their chemical structure (peptidoglycan)?

Dr ADLAM: This has been studied extensively. Active strains may have different peptidoglycan structures (Adlam and Reid (1978). *Develop. Biol. Standard.* **38**, 115). The important component for activity of these organisms is their carbohydrate but having said that, the purified carbohydrate does not increase spleen weight and broken organisms consisting of carbohydrate and peptidoglycan have greatly reduced or negative activity. It must be concluded therefore, that for full activity one needs the carbohydrate attached to a peptidoglycan backbone and it is the presentation of this carbohydrate in a particular configuration which is important.

Dr SCHWAB: Polysaccharide bound covalently to peptidoglycan is responsible for the persistence of peptidoglycan in macrophages. The capacity of peptidoglycan to activate macrophages to cytotoxicity is dependent upon this persistence (Smiałowicz, R. and Schwab, J.H. (1977). *Infect. Immun.*

Dr CHEDID: Is the histamine-sensitizing factor identified in *C. parvum*? Is it the same as in *Bordetella pertussis*?

Dr ADLAM: This histamine-sensitizing factor is similar to that seen with *B. pertussis* in its biological effects, but differs chemically. The component responsible for histamine sensitization is probably the same as that responsible for splenomegaly (Adlam (1973). *J. Med. Microbiol.* **6**, 527).

Dr KIRN: Dr Roszkowski, I would like to know if in the *in vitro* phagocytosis test you reported there is an increase in the adhesion and/or in the ingestion of the cells?

Dr ROSZKOWSKI: Both properties of macrophages are enhanced.

Dr KIRN: So one must admit that in *C. parvum*-treated animals there is a modification of the cytoskeleton of the macrophages?

Dr ROSZKOWSKI: The process of activation is complex, so we may say that the cytoskeleton of macrophages is involved in this phenomenon.

Dr PRAGER: Dr Roszkowski, I believe your data showed that after intratumoural injection of the bacteria there was an increased content of macrophages in the tumour. How does this compare to the natural history of the tumour?

Dr ROSZKOWSKI: The number of macrophages is increased as compared to nontreated tumours.

Dr BAŠIĆ: Dr Roszkowski, you have shown that *C. parvum* stimulates haematopoietic activity in mice submitted to whole-body irradiation. We have published work (Bašić and Milas (1979). *Eur. J. Cancer* **15**, 901-908) indicating that mice treated with *C. parvum* although having higher CFU activity in their bone marrow, blood and spleen, are much more sensitive to whole-body irradiation. Have you any comment on these findings?

Dr ROSZKOWSKI: The results depend on experimental design. In our experiments we applied propionibacteria after irradiation. It seems logical that after pretreatment of animals with propionibacteria, one can induce stimulation of bone marrow and irradiation applied during proliferative response of bone marrow may lead even to a greater damage of this system. So I think that our data are consistent.

Dr GOLDIN: Dr Bašić, have you made comparisons with the Lewis lung system and your system?

Dr BAŠIĆ: The growing characteristics of this tumour have not yet been compared with those of the Lewis lung tumour. We hope to be able to do these comparative studies pretty soon in collaboration with Professor Karrer from Vienna.

Dr KONDO: Dr Bašić, could you explain why OK-432 prevented the lung metastasis only when it was given intrapleurally? Was there any specific histological change?

Dr BAŠIĆ: OK-432 was effective against lung metastasis when given intrapleurally or in combined i.v. and intrapleural injections. Because of the lack of OK-432 in our laboratory, other studies have not been done, including histological studies of antimetastatic effects of OK-432.

Dr SINKOVIC: Dr Bašić, what human tumour does your system of spontaneous rat anaplastic carcinoma resemble most? Could it have arisen in an ectopic mammary gland of that old multiparous rat?

Dr BAŠIĆ: The main and primary route of mammary carcinoma in man is by way of the regional lymph nodes and the main cause

of death is from haematogenous metastases. In animals few
tumours show both lymph node and blood-borne metastases.
While blood-borne metastases are frequently observed, lymph
node metastases, if they occur, are seen only in a very low
percentage. This tumour, although having histological characteristics of an anaplastic carcinoma, arose spontaneously in
the area of a mammary complex in an old multiparous rat. The
possibility that it could have arisen in an ectopic mammary
gland is not excluded. Being of low antigenicity, spontaneous
tumour and spreading in lymph nodes and subsequently blood,
this tumour by this behaviour could mimic a human mammary
tumour.

IMMUNOSUPPRESSION BY BACTERIA

John H. Schwab and Dena L. Toffaletti

*Department of Bacteriology and Immunology,
University of North Carolina Medical School,
Chapel Hill, N.C. 27514, USA*

Numerous studies provide evidence that bacterial products can influence regulation of the immune system (Schwab, 1975, 1977). Furthermore, these products can be derived from bacterial species which are part of the normal microflora of animals or are in frequent association with the host. We postulate that because animals and their microflora have evolved together in intimate association, the normal functioning of the immune response may, in fact, depend upon such exogenous immunoregulators.

Why has there been reluctance to accept this important biological concept which has been proposed and reproposed since the days of Pasteur? This can be largely ascribed to the confusion produced by studies on germ-free animals which concluded that the immune response could be stimulated, albeit somewhat slower, in the absence of a microbial flora. However, in almost all studies germ-free animals have been maintained on autoclaved ordinary diet which contains large amounts of bacterial products and dead microorganisms. Thus, Wostman *et al.* (1970) have shown that by the criteria of lymphoid tissue and haematopoietic maturation the development of germ-free mice fed an ordinary autoclaved diet was slower than conventional animals; but germ-free mice fed a water-soluble, filtered diet developed even less lymphoid and haematopoietic function. It is also pertinent that absorption across the intestine is greater in germ-free animals, which would allow a greater effectiveness of bacterial products in the diet (Berg and Garlington, 1980). In addition, the antigen used to assess immune response has usually been incorporated with a bacterial adjuvant such as complete Freund's adjuvant; or the animals were immunized with antigens such as bovine serum albumin which are frequently contaminated with endotoxin or other bacterial products.

A recent report by Mattingly *et al.* (1979) provides some evidence to support the concept that the microflora is required for the effective regulation of the immune system. They have shown that the normal microbial flora of the rat is needed for the expression of suppressor macrophages in the spleen. They demonstrated that germ-free rats do not have active macrophage suppressor cells, but precursor cells can be activated by providing normal microflora. They propose that suppression is mediated by prostaglandin secretion and that microbial products provide the signal to release the prostaglandin. Pabst *et al.* (1982) have also concluded that human monocytes require stimulation from the microbial flora in order to maintain a primed state for generation of oxygen metabolites.

From this perspective the phenomenon of bacterial-induced immunosuppression reflects an exaggeration of the physiological role of bacterial products in normal regulatory mechanisms. Sudden exposure to a large dose of a bacterial product effecting immune regulation (Hunter *et al.*, 1981) or the accumulation of such products in the host during chronic infection (Bullock *et al.*, 1978; Katz *et al.*, 1979) appears to confound the regulatory system and leads to a variety of dysfunctions. This is particularly apparent with poorly biodegradable bacterial products such as the peptidoglycan-polysaccharide complexes of the cell wall (Schwab, 1982).

Most substances which can enhance the immune response can also be shown to suppress the immune response with appropriate manipulation of dose and timing of injections. Thus, all noncytotoxic potentiating or suppressive agents are modulators of normal immune regulation (Schwab, 1975). Reports showing that immunomodulation by microorganisms is subject to genetic control by the host add an exciting new parameter to these studies (Staruch and Wood, 1982).

This paper will be limited to an examination of immunosuppression by bacteria. Rather than cataloguing all of the bacterial products which have been described, the discussion is organized around some mechanisms of normal immunoregulation which bacteria may affect and thereby cause depression of immune functions. This is not intended to be an exhaustive review of the literature on this topic. Rather, we will use examples from other laboratories and some of our own studies to bring to your attention some speculations on possible methods by which bacteria suppress immunity.

I. SUPPRESSION BY STIMULATION OF MACROPHAGE AND/OR
T-SUPPRESSOR CELLS

Figure 1 is a model of macrophage—lymphocyte interactions which indicates how a variety of bacteria or their products may influence the regulation of this system. Macrophages can be stimulated directly or through activation of the alternate complement pathway by products such as lipopolysachharide (LPS), peptidoglycan, or synthetic muramyl dipeptide (MDP). Activated macrophages can secrete interleukin-1 (IL-1) which activates T-lymphocytes (Mizel *et al.*, 1978). Alternatively, macrophages can be activated to secrete prostaglandins or oxygen metabolites which suppress T-lymphocyte activity (Ellner and Spagnuolo, 1979; Mattingly *et al.*, 1979; Orme and Shand, 1981; Pabst and Johnston, 1980). The first obvious question is what determines which manifestation of macrophage activation will predominate? The second question is whether this is accomplished by effecting the balance of regulatory T-cells, that is, a selective activation of Ts or TH? Dose, route of injection, and physical state of the bacterial agent (particulate or soluble), as well as host genetic background are all important variables (Neta and Salvin, 1979; Turcotte, 1981; Dziarski, 1979). Interesting observations by Ishizaka and coworkers indicate that the nature of the bacterial activating agent can have a deciding role (Hirashima *et al.*, 1981a and b). These reports show that *Bordetella pertussis* vaccine or complete Freund's adjuvant (CFA) activate rat macrophages to release an IgE inducer factor which stimulates T-cells to release an IgE binding factor. However, a sub-population of T-cells must also be stimulated by *B. pertussis*

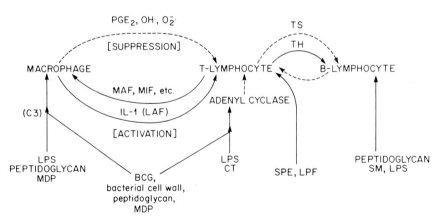

FIG. 1 *Macrophage—lymphocyte interactions which may be modulated by bacterial products. See text for explanation.*

to produce a soluble factor which facilitates the glycosylation of IgE binding factor, which in turn, results in the
formation of IgE potentiating factor. If the T-cell subset
is stimulated by CFA, however, soluble factors are produced
which prevent glycosylation of the IgE binding factor and
result in formation of an IgE suppressive factor. The suppressive effects of CFA on IgE response may be ascribed to
the mycoloyl-muramyl dipeptide according to the reports of
Kishimoto *et al.*, 1976, 1979). In any event, it is clear
that different bacterial products have unique effects upon
the lymphoid cells with which they interact.

Also illustrated in Fig. 1, bacterial products can effect
regulation through mechanisms not involving the macrophage.
Lipopolysaccharide and cholera toxin (CT) can activate adenyl
cyclase in T-cells and the resultant elevation of cyclic
nucleotides suppresses secretion and effector function of
cytotoxic T-cells (Henney *et al.*, 1973). Bacille Calmette
Guerin (BCG) can enhance all of these responses to LPS
(Rosenstreich and Vogel, 1981). Lymphocytosis promoting factor (LPF) from *B. pertussis* (Ho *et al.*, 1980) and streptococcal pyrogenic exotoxin (SPE; Schlievert, 1980) can act
directly on T-lymphocytes to stimulate T-cell suppressor (Ts)
activity. LPS (Uchiyama and Jacobs, 1978; Persson, 1977),
and a product associated with the cytoplasmic membrane of
group A streptococci (SM; Schwab *et al.*, 1981) can act
directly on B-cells to suppress immune function.

Of particular interest is the prolonged immunosuppression
associated with chronic infections, of mice or humans, in
which large amounts of bacterial debris accumulate. This
frequently results in specific anergy as well as nonspecific
suppression of T-cell response to other antigens and mitogens.
Suppression induced by mycobacterial infections has been
ascribed to Ts cells (Katz *et al.*, 1979; Bullock *et al.*, 1978;
Watson and Collins, 1981) or macrophage suppressor cells
(Klimpel and Henney, 1978) or both (Turcotte, 1981).

Collins and Watson (1979) showed that mice infected by
intravenous injection of a large number of *Mycobacterium bovis*
(BCG) organisms developed specific anergy to PPD both by *in
vitro* (^{3}H-thymidine incorporation) and *in vivo* (footpad
swelling) assays. Nonspecific response to PHA was also suppressed. Suppression was due to a Ts cell, first demonstrable
at 14 days. By 4 months suppressor activity had disappeared.
These authors also reported that *Mycobacterium avium* infection of mice induced Ts cells (Watson and Collins, 1981).
Turcotte (1981) has confirmed the reports by Neta and Salvin
(1979) that both macrophage suppressor cells and T-suppressor
cells coexist in the spleens of mice. Both develop early in
infection and persist for at least 40 days. Jyonouchi *et al.*

(1981) have shown that in mice injected with *C. parvum* (0.1 mg i.p.) B-cell development in the bone marrow was inhibited by an adherent suppressor cell, apparently mediated by production of prostaglandins. Complete Freund's adjuvant can also induce a suppression of cell mediated immunity if it is injected before antigen (Asherson, 1977).

It seems probable that most bacterial agents evoke an increase in nonspecific Ts through an influence upon macrophage function, rather than directly effecting the balance of T-cell subpopulations. It is pertinent to note here the studies of Benacerraf and Germain (1979) defining the IR gene control of the response of responder and nonresponder mice to a synthetic peptide. The major difference between these strains is the balance of Ts and T_H cells, which is determined by the inability of macrophages from nonresponder mice to present antigen in a way which can effectively stimulate T_H cells. The result is an increase in the balance of Ts cells. From this we can speculate that the evokation of Ts by bacteria could be due to interference with essential antigen presentation by macrophages, as well as to secretion of mediators such as prostaglandin E_2 and oxygen metabolites.

Current studies in our laboratory suggest that in the rat the extended suppression of cell mediated immunity, induced with either chronic infection or CFA, might be ascribed to the persistent peptidoglycan-polysaccharide complexes derived from the bacterial cell wall (Hunter *et al.*, 1980). Suppression induced with purified cell wall has features similar to that induced with infection or CFA, but has the advantage of utilizing a well defined, purified component of the bacterium and thus can be more precisely studied. In our model, anergy to peptidoglycan is induced by injection of cell wall isolated from group A streptococci. Anergy for peptidoglycan is dose related, and becomes apparent at 6 days with high doses and by 2 to 3 weeks with lower doses. This specific depression of T-cell function persists for at least 90 days. A nonspecific suppression of T-cell function, measured by *in vitro* stimulation of lymph node cells with PHA, is also observed but this response returns to normal by 30 days after injection of cell wall. Suppression of this model requires adherent cells (Regan and Schwab, unpublished).

II. SUPPRESSION BY INTERACTION WITH PRODUCTS OF THE MAJOR HISTOCOMPATIBILITY COMPLEX (MHC) ON CELL MEMBRANES

Ellner and Spagnuolo (1979) indicated that treatment of human monocytes with LPS induced suppression of T-cell proliferation and this was ascribed to secretion of prostaglandin E_2. Yem and Parmely (1981) also observed suppression of T-cell response to antigen stimulation associated with LPS activation

of macrophages. However, they could not demonstrate a suppressive factor or suppressive cells. They proposed that this
was because they used fewer macrophages in their cultures.
Of greater interest, Yem and Parmely went on to demonstrate
that LPS or zymosan A treatment of human monocytes resulted
in a significant reduction in the proportion of Ia$^+$ cells.
Since it is hypothesized (Benacerraf, 1981) that the presentation of antigen to T-cells requires the interaction of antigen with Ia glycoprotein on the surface of macrophages, Yem
and Parmely (1981) proposed the interesting idea that microbial products in the host microenvironment, such as LPS and
zymosan, can suppress specific immune responses by inter-
fering in this way with Ia expression.

Other studies (Geczy et al., 1980; Edmonds et al., 1981)
have also presented evidence that a bacterial factor can
modify an MHC product. In this system a component in culture
filtrates of Klebsiella pneumoniae, K43 will bind to human
peripheral blood lymphocytes from healthy individuals (without ankylosing spondylitis) that are HLA-B27 positive. Association with the B-27 gene product is detected by lysis of
lymphocytes by anti-K43 serum after incubation of cells with
Klebsiella K43 culture filtrate. This appears to be a
specific property of the K43 strain. Unfortunately, others
have failed to confirm this observation (Archer, 1981).

Another effect on Ia molecules has been described by
Parish and colleagues (1979). They reported that the level
of soluble Ia in serum can be modified by microbial infections. Thus, secretion of Ia by activated T-cells can be
first stimulated and then suppressed by LCM virus. Listeria
monocytogenes infection suppressed Ia levels, whereas Brucella abortus enhanced soluble Ia in serum.

Kronvall et al. (1978), Bjorck et al. (1981) and Schonbeck
et al. (1981) have described binding of surface protein structures on group A, B, C, and G streptococci to another molecule associated with human MHC products. They reported
receptor structures on these bacteria for β-2 microglobulin,
a light chain peptide of class I histocompatibility antigens.

Taken together, these reports support the concept that
some bacterial products may suppress a specific immune response at the level of IR gene control by nonspecifically
binding to an MHC product thereby blocking the appropriate
interaction with antigen.

III. SUPPRESSION BY ACTIVATION OF COMPLEMENT

According to Pepys (1976) C3 is necessary and sufficient for
localization of antigen in follicles. After treatment of
mice with cobra venom factor (CVF) Pepys showed a suppression

of T-dependent antibody response, particularly IgG. Klaus and Humphrey (1977) demonstrated that the predominant effect of COF treatment on the immune response of mice to DNP-KLH was to abrogate development of B-cell memory. C3 is involved in the binding of antigen on dendritic cells and is important for development of B memory cells. Romball *et al.* (1980) also showed that C3 is essential for antigen localization in follicles. They further showed that the primary T-dependent response of rabbits to aggregated human gamma globulin was not affected by COF treatment but the secondary IgG response was suppressed.

Evidence for the regulatory role of C3 is important in this discussion because numerous bacterial products can activate the alternate complement pathway and thus could modulate immune responses by this mechanism. Our studies have shown that peptidoglycan isolated from group A streptococcal cell walls is an extremely potent activator of the alternate complement pathway (Greenblatt *et al.*, 1978). Injection of the streptococcal cell wall fragments into rats induces a profound decrease in blood complement activity over the first 24 h, comparable to that obtained with COF treatment (Lambris *et al.*, 1982; Schwab *et al.*, 1982). Thus, suppression of some T-dependent antibody responses by prior injection of bacterial vaccines or products such as peptidoglycan, LPS, or cell wall preparations may involve depression of C3 levels. An equally important effect could be the depression of follicular localization and suppression of secondary responses. In our laboratory we have observed a selective reduction of IgG antibody response to cell wall polysaccharide and peptidoglycan haptens upon secondary injection of cell wall fragments (Esser and Schwab, in preparation).

It has been proposed that C3 can provide the second signal for antigen stimulated B-cells (Dukor *et al.*, 1974). It is also conceivable that activation of C3, by injection of activators of the alternate complement pathway, may initiate B-cell stimulation by reaction with C3 receptor. In the absence of specific antigen the clone would fail to differentiate to antibody forming cells. This could generate tolerant B-lymphocyte clones.

IV. SUPPRESSION THROUGH INTERACTION WITH B-LYMPHOCYTES

Previous reports have described immunosuppressive activity induced with extracts of group A streptococci. Injection of mice 24 h before sheep red blood cell antigen suppresses primary and secondary IgG and IgM plaque forming cells (Malakian and Schwab, 1968; 1971), while the delayed hypersensitivity response is enhanced (Toffaletti and Schwab,

1979). This activity is associated with cytoplasmic membranes isolated from osmotically disrupted protoplasts (Schwab *et al.* 1981).

Recent studies have shown that injection of mice with preparations of group A streptococcal membranes preferentially suppresses mitogenic responses of B-cells (Toffaletti and Schwab, 1979). Only the bone marrow cells stimulated by dextran sulfate are suppressed, suggesting that membranes preferentially affect immature B-cells. In contrast, T-cell mitogenic responses are unaltered or sometimes enhanced. The addition of membranes *in vitro* to cell cultures suppresses the proliferative response of bone marrow and spleen cells to dextran sulfate (DS) or lipopolysaccharide (LPS), but the DS response of bone marrow cultures are the most susceptible. This suppression is not mediated by adherent cells or Thy-1 bearing cells (Toffaletti and Schwab, 1979).

The studies we present here show that dialysed and diluted supernatants from bone marrow cell cultures, exposed to membrane preparations *in vitro*, selectively suppress the response of bone marrow cells to DS.

One million bone marrow cells from male A/J mice 7 to 10 weeks old were cultured in flat-bottomed microtitre plates with a suppressive dose of streptococcal membranes (25 µg) in the presence or absence of 20 µg DS. The total volume of the culture was 200 µl. Following a 40 h incubation period in a humidified 5% CO_2 in 95% air atmosphere at 37°C, the cultures were centrifuged for 20 min at 1000 x g and their supernatants collected and pooled into 2 groups, (+) DS and (-) DS. After centrifugation at 37,000 x g for 30 min to remove streptococcal membranes, the supernatants were filtered through a 0.22 µm Amicon filter. One half of the (+) DS and (-) DS supernatants were dialysed against RPMI-1640 medium (supplemented with, per ml: 5% heat-inactivated foetal calf serum, 2 mmoles l-glutamine, and 100 µg gentamucin sulfate reagent) for 24 h with 4 changes of medium. Either 10 µl (5%) or 100 µl (50%) of the nondialysed and dialysed supernatants were tested for their effect upon *de novo* DS stimulated bone marrow cells and for their effect upon LPS, PHA, and Con A stimulated spleen cells. The total volume of the test cultures was 200 µl, thus 10 µl or 100 µl of supernatants comprised 5% or 50% of the *de novo* cultures. Supernatants derived from bone marrow cell cultures not exposed to streptococcal membranes served as controls.

Supernatants derived from bone marrow cells cultured with streptococcal membranes (M) suppressed the net response of bone marrow cells to DS by 93%, at a level of significance of $P<0.001$, when compared to control (C) supernatants (Fig. 2a). This suppression was greater with the higher concentration of

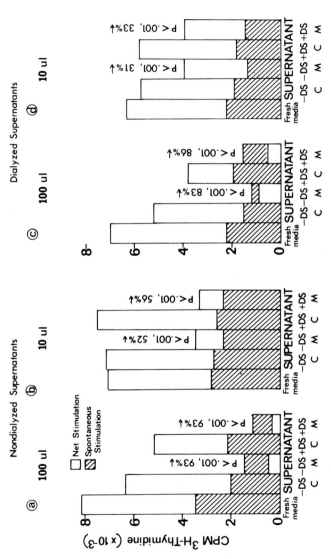

FIG. 2, a–d Effect of supernatants from bone marrow cells to dextran sulfate. Supernatants (dialysed or nondialysed) were derived from bone marrow cell cultures stimulated with: membranes only (−DS,M); membranes plus dextran sulfate (+DS,M); dextran sulfate only (+DS,C); neither dextran sulfate or membranes (−DS,C). Either 100 μl or 10 μl of the dialysed or nondialysed supernatant was added to bone marrow cells (1 × 10⁶ cells/200 μl culture) stimulated with DS (20 μg). The cultures were incubated for 2 days and pulsed with 0.5 μC ³H-thymidine during the last 8 h. Spontaneous stimulation = ³H-thymidine incorporated into cultures with no DS added. Net stimulation = ³H-thymidine incorporated into cultures stimulated with DS minus cultures without DS added.

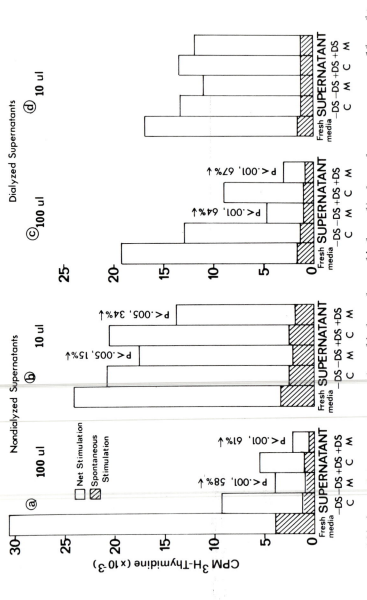

FIG. 3, a–d *Addition of supernatants (nondialysed or dialysed) from bone marrow cell cultures to LPS stimulated spleen cells. Same as Fig. 1, except bone marrow cell culture supernatants were added to LPS (3 µg) stimulated spleen cells (1 × 10⁶ cells/culture).*

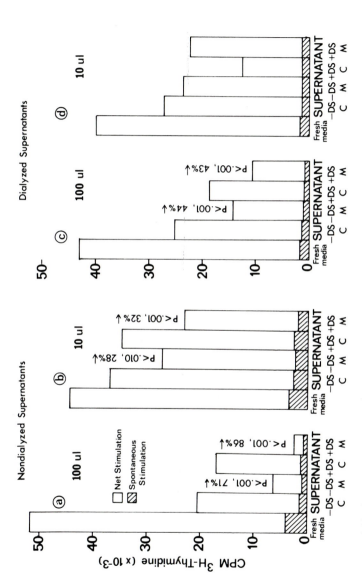

FIG. 4, a-d *Addition of supernatants (nondialysed or dialysed) from bone marrow cell cultures to PHA stimulated spleen cells. Same as Fig. 1, except bone marrow cell culture supernatants were added to PHA (3 µg) stimulated spleen cells (1 × 10⁶ cells/culture) and cultures were terminated on day 3.*

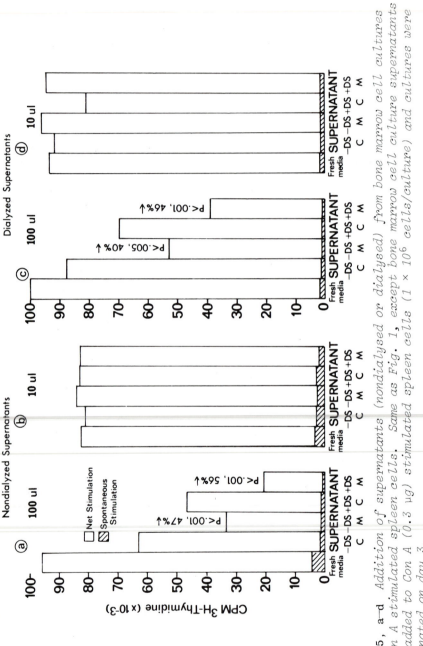

FIG. 5, a–d Addition of supernatants (nondialysed or dialysed) from bone marrow cell cultures to Con A stimulated spleen cells. Same as Fig. 1, except bone marrow cell culture supernatants were added to Con A (0.3 µg) stimulated spleen cells (1 × 10⁶ cells/culture) and cultures were terminated on day 3.

of supernatants (100 μl, 50%). In addition, neither the presence nor absence of DS in cultures which produced the supernatants had any significant effect upon the capacity to suppress the *de novo* DS response (Fig. 2a,b). Activity was still
present after dialysis and dilution of the supernatants (Fig.
2c,d).

As seen in Fig. 3—5 (a,b) stimulation of spleen cells by
LPS, PHA, or Con A was suppressed by nondialysed supernatants
derived from bone marrow cells cultured with streptococcal
membranes. Suppression of the LPS and PHA responses was
greater at the higher concentration of supernatants and suppression of the Con A response occurred only with the higher
concentration of supernatants. Again, the presence or
absence of DS in cultures which produced the supernatants had
no significant effect upon ability to suppress mitogenic responses. After dialysis of these supernatants, suppression of
LPS, PHA, and Con A responsive cells was still detected with
the higher concentration of supernatant but not with diluted
supernatants (Figs. 3—5, c,d).

These findings indicate that suppression of the response
to DS may be due to the generation of suppressor mediators
induced by streptococcal membranes. Exposure of bone marrow
cells to membranes in culture results in the generation of a
nondialysable suppressor molecule that selectively depresses
the response of bone marrow cells to DS. The specificity for
the DS stimulated cells is apparent only with dilution of the
dialysed supernatants. Both B-cell (DS and LPS) and T-cell
(PHA and Con A) mitogenic responses can be reduced by either
dialysed or nondialysed supernatants used at the higher concentration. This suggests that, in addition to the selective
suppression of DS-responsive bone marrow cells, there are
factors in the supernatants that depress other mitogen-respon-
sive cell populations.

The nonsusceptibility of T-cell mitogenic responses to the
suppressive activity of the diluted and dialysed supernatants
is consistent with the earlier finding that streptococcal membranes have no effect *in vitro* upon the response of lymphocytes stimulated with optimal doses of PHA or Con A (Toffaletti and Schwab, 1979).

Production of the postulated suppressor mediator is not a
general response of stimulated cells, since supernatants from
DS stimulated control cultures only marginally reduce the
responses to mitogens. The possibility of a digested product
of the streptococcal membrane, not removed by centrifugation
and filtration, being the active component in the supernatants cannot be excluded. Isolation of the nondialysable
suppressor molecule should resolve whether it is a component
of streptococcal membranes or bone marrow cells.

The cell that produces the suppressive factor is probably
not an adherent cell or a T-cell. These conclusions stem
from a previous report demonstrating that suppression of B-
cell mitogenic responses by membranes *in vitro* is not mediated
by adherent cells or Thy-1 bearing cells (Toffaletti and
Schwab, 1979). We speculate that the cells producing the
suppressor mediator(s) are a subpopulation of immature B-
cells. Persson (1977) has reported that LPS activates suppressor B-cells in the spleen which inhibit antibody formation
against a T-dependent antigen. Other studies have implicated
the existence of suppressor B-cells (Asherson, 1977; Neta and
Salvin, 1979).

V. CONCLUSION

Numerous bacterial products interact with regulatory cells or
soluble components of the immune system and modify the immune
response. Such exogenous modulators function through a
variety of mechanisms. Thus, the microbial environment of
the body is an important determinant in the relationship of
the host and developing tumour. This microbial influence can
undergo transient and subtle changes which could shift the
relationship if it were coincident with an oncogenic event.
Therefore, immune dysfunction induced by bacterial products
or persistent bacterial debris is a variable which must be
included in the analysis of carcinogenesis.

V. SUMMARY

1. Persistent, poorly biodegradable peptidoglycan-polysaccharide polymers derived from bacterial cell walls can induce
   prolonged immune dysfunction. This may be the basis of
   anergy observed in a variety of chronic infections or with
   agents such as *C. parvum* and BCG vaccines.

2. The dysfunction can be manifested as specific anergy of
   cell mediated immunity, and as a more transient nonspecific depression of T-cell response to mitogens. The dysfunction requires adherent cells.

3. The alternate complement pathway can be activated by
   several bacterial structures; among the most potent of
   these is peptidoglycan. Since complement components par-
   ticipate in many immunoregulatory steps, this could be an
   important mechanism by which bacteria modulate the immune
   response.

4. Another immunomodulating factor, associated with the cytoplasmic membrane of group A streptococci, can selectively

suppress B-cell responses, while enhancing certain T-cell functions. This can be measured as suppression of anti-body formation and as suppression of response of immature B-cells to the mitogen, dextran sulfate. The suppression of B-cells may be mediated by a soluble factor found in the culture medium of bone marrow cells exposed *in vitro* to the streptococcal membrane material.

5. Some bacterial factors may suppress immune responses by interaction with MHC gene products and thereby interfere with appropriate presentation of antigen to antigen-recognizing lymphocytes.

VI. ACKNOWLEDGEMENTS

*Supported by Public Health Service grant AM-25733 from the National Institute of Arthritis, Diabetes, and Digestive and Kidney Diseases.

VII. REFERENCES

Archer, J.R. (1981). Search for cross-reactivity between HLA B27 and *Klebsiella pneumoniae*. *Annal. Rheum. Dis.* **40**, 400-403.

Asherson, G.L. (1977). *In* "Microbiology — 1977" (Ed. D. Schlessigner), pp. 382-387. *Am. Soc. Microbiol.*, Wash. D.C.

Benacerraf, B. (1981). Role of MHC gene products in immune regulation. *Science* **212**, 1229-1238.

Benacerraf, B. and Germain, R.N. (1979). Specific suppressor responses to antigen under I region control. *Fed. Proc.* **38**, 2053-2057.

Berg, R.D. and Garlington, A.W. (1980). Translocation of *E. coli* from the gastrointestinal tract to the mesenteric lymph nodes in gnotobiotic mice receiving *E. coli* vaccines before colonization. *Infect. Immun.* **30**, 894-898.

Bjorck, L., Tylewska, S.K., Wadstrom, T. and Kronvall, C. (1981). Beta-α-microglobulin is bound to streptococcal M protein. *Scand. J. Immunol.* **13**, 391-394.

Bullock, W.E., Carlson, E. and Gershon, R.K. (1978). The evolution of immunosuppressive cell populations in experimental mycobacterial infection. *J. Immunol.* **120**, 1709-1716.

Collins, F. and Watson, S.R. (1979). Suppressor T-cells in BCG-infected mice. *Infect. Immun.* **25**, 491-496.

Dukor, P., Schumann, G., Gisler, R., Dierich, M., Konig, W., Hadding, U. and Bitter-Suermann, D. (1974). Complement-dependent B-cell activation by cobra venom factor and other mitogens. *J. Exp. Med.* **139**, 337-354.

Dziarski, R. (1979). Relationships between adjuvant, immuno-suppressive and mitogen activities of staphylococcal peptidoglycan. *Infect. Immun.* **26**, 508-514.

Edmonds, J., Macauley, D., Tyndall, A., Lieu, M., Alexander, K., Geczy, A. and Bashir, H. (1981). Lymphocytotoxicity of anti-Klebsiella antisera in ankylosing spondylitis and related arthropathies: patient and family studies. *Arthr. Rheum.* **24**, 1-7.

Ellner, J.J. and Spagnuolo, P.J. (1979). Suppression of antigen and mitogen induced human T-lymphocyte DNA synthesis by bacterial lipopolysaccharide: mediation by monocyte activation and production of prostaglandins. *J. Immunol.* **123**, 2689-2695.

Geczy, A.F., Alexander, K., Bashir, H.V. and Edmonds, J. (1980). A factor(s) in Klebsiella culture filtrates specifically modifies an HLA-B27-associated cell surface component. *Nature* **283**, 782-784.

Greenblatt, J., Boackle, R.J. and Schwab, J.H. (1978). Activation of the alternate complement pathway by peptidoglycan from streptococcal cell wall. *Infect. Immun.* **19**, 296-303.

Henney, C.S., Litchenstein, L.M., Gillespie, E. and Rolley, R.T. (1973). *In vivo* suppression of the immune response to alloantigen by cholera entertoxin. *J. Clin. Invest.* **52**, 2853-2857.

Hirashima, M., Yodoi, J. and Ishizaka, K. (1981a). Formation of IgE-binding factors by rat T-lymphocytes. II. Mechanisms of selective formation of IgE-potentiating factors by treatment with *Bordetella pertussis* vaccine. *J. Immunol.* **127**, 1804-1810.

Hirashima, M., Yodoi, J., Huff, T.F. and Ishizaka, K. (1981b). Formation of IgE-binding factors by rat T-lymphocytes. III. Mechanisms of selective formation of IgE-suppressive factors by treatment with complete Freund's adjuvant. *J. Immunol.* **127**, 1810-1816.

Ho, M.-K., Kong, A.S. and Morse, S.I. (1980). The *in vitro* effects of *Bordetella pertusis* lymphocytosis promoting factor on murine lymphocytes. V. Modulation of T-cell proliferation by helper and suppressor lymphocytes. *J. Immunol.* **124**, 362-369.

Hunter, N., Anderle, S.K., Brown, R.R., Dalldorf, F.G., Clark, R.L., Cromartie, W.J. and Schwab, J.H. (1980). Cell mediated immune response during experimental arthritis induced in rats with streptococcal cell walls. *Clin. Exp. Immunol.* **42**, 441-449.

Jyonouchi, H., Kincade, P.W. and Good, R.A. (1981). Immuno-suppression of marrow B-lymphocytes by administration of *Corynebacterium parvum* in mice. *J. Immunol.* **127**, 2502-2507.

Katz, D.H., Bargatze, R.F., Bogovitz, C.A. and Katz, L.R. (1979). Regulation of IgE antibody production by serum molecules. IV. Complete Freund's adjuvant induces both enhancing and suppressive activities detectable in the serum of low and high responder mice. *J. Immunol.* **122**, 2184-2190.

Katz, P., Goldstein, R.A. and Fauci, A.S. (1979). Immuno-regulation in infection caused by Mycobacterium tuberculosis: the presence of suppressor monocytes and the alteration of subpopulations of T-lymphocytes. *J. Infect. Dis.* **140**, 12-21.

Kishimoto, T., Hirai, Y., Suemura, M. and Yamamura, Y. (1976). Regulation of antibody response in different immunoglobulin classes. I. Selective suppression of anti-DNP IgE antibody response by preadministration of DNP-coupled mycobacterium. *J. Immunol.* **117**, 296-404.

Kishimoto, T., Hirai, Y., Nakanishi, K., Azuma, I., Nagamatsu, A. and Yamamura, Y. (1979). Regulation of antibody response in different Ig classes. VI. Selective suppression of IgE response by administration of antigen-conjugated muramylpeptides. *J. Immunol.* **123**, 2709-2715.

Klaus, G.G.B. and Humphrey, J.H. (1977). The generation of memory cells. I. The role of C3 in the generation of B memory cells. *J. Immunol.* **33**, 31-40.

Klimpel, G.R. and Henney, C.S. (1978). BCG-induced suppressor cells. I. Demonstration of a macrophage-like suppressor cell that inhibits cytotoxic T-cell generation *in vitro*. *J. Immunol.* **120**, 563-569.

Kronvall, G., Myhre, E.B., Bjorck, L. and Berggard, I. (1978). Binding of aggregated human B_2-microglobulin to surface protein structure in group A, C, and G streptococci. *Infect. Immun.* **22**, 136-142.

Lambris, J.D., Allen, J.B. and Schwab, J.H. (1982). *In vivo* changes in complement induced with peptidoglycan-polysaccharide polymers from streptococcal cell walls. *Infect. Immun.* **35**, 377-380.

Malakian, A. and Schwab, J.H. (1968). Immunosuppressant from group A streptococci. *Science* **159**, 880-881.

Malakian, A. and Schwab, J.H. (1971). Biological characterization of an immunosuppressant from group A streptococci. *J. Exp. Med.* **134**, 1253-1265.

Mattingly, J.A., Eardley, D.D., Kemp, J.D. and Gershon, R.K. (1979). Induction of suppressor cells in rat spleen: influence of microbial stimulation. *J. Immunol.* **122**, 787-790.

Mizel, S.B., Oppenheim, J.J. and Rosenstreich, D.L. (1978). Characterization of lymphocyte activating factor (LAF) produced by a macrophage cell line P388D. II. Biochemical characterization of LAD induced by activated T-cells and LPS. *J. Immunol.* **120**, 1504-1508.

Neta, R. and Salvin, S.B. (1979). Adjuvants in the induction of suppressor cells. *Infect. Immun.* **23**, 360-365.

Orme, I.M. and Shand, F.L. (1981). Inhibition of prosta-glandin synthetase blocks the generation of suppressor T-cells induced by Concanavalin A. *Int. J. Immunopharm.* **3**, 15-19.

Pabst, M.J. and Johnston, R.B. (1980). Increased production of superoxide anion by macrophage exposed *in vitro* to muramyl dipeptide or lipopolysaccharide. *J. Exp. Med.* **151**, 101-114.

Pabst, M.J., Hedegaard, H.B. and Johnston, R.B. Jr. (1982). Cultured human monocytes require exposure to bacterial products to maintain an optimal oxygen radical response. *J. Immunol.* **128**, 123-128.

Parish, C.R., Freeman, R., McKenzie, I., Cheers, C. and Cole, G.R. (1979). Ia antigens in serum during different murine infections. *Infect. Immun.* **26**, 422-426.

Pepys, M.B. (1976). Role of complement in induction of immunological responses. *Transplant. Rev.* **32**, 93-110.

Persson, U. (1977). Lipopolysaccharide-induced suppression of the primary immune response to a thymus-dependent anti-gen. *J. Immunol.* **118**, 789-796.

Romball, C.G., Ulevitch, R.J. and Weigle, W.O. (1980). Role of C3 in the regulation of a splenic PFC response in rab-bits. *J. Immunol.* **124**, 151-155.

Rosenstreich, D.L. and Vogel, S.N. (1981). Central role of macrophages in the host response to endotoxin. *In* "Immuno-modulation by Bacteria and Their Products". (Eds H. Friedman, T.W. Kline and A. Sventivanyi), Plenum Pub. Corp., New York.

Schonbeck, C., Bjorck, L. and Kronvall, G. (1981). Receptors for fibrinogen and aggregated B$_2$-microglobulin detected in strains of group B streptococci. *Infect. Immun.* **31**, 856-861.

Schwab, J.H. (1975). Suppression of the immune response by microorganisms. *Bact. Rev.* **39**, 121-143.

Schwab, J.H. (1977). *In* "Microbiology — 1977". (Ed. D. Schlessinger), pp. 366-373. *Am. Soc. Microbiol.* Wash. D.C.

Schwab, J.H. (1982). *In* "Immunomodulation by Microbial Pro-ducts and Related Synthetic Compounds". (Ed. S. Kotani), Excerpta Medica, Amsterdam. In press.

Schwab, J.H., Toffaletti, D.L. and Brown, R.R. (1981). *In* "Immunomodulation by Bacteria and their Products". (Eds H. Friedman, T.W. Kline and A. Sventivanyia), pp. 49-57. Plenum Publ. Corp. New York.

Schwab, J.H., Allen, J.B., Anderle, S.K., Dalldorf, F.G., Eisenberg, R.A. and Cromartie, W.J. (1982). Relationship of complement to experimental arthritis induced in rats with streptococcal cell walls. *Immunology* In press.

Schlievert, P.M. (1980). Activation of murine T-suppressor lymphocytes by group A streptococcal and staphylococcal pyrogenic exotoxins. *Infect. Immun.* **28**, 876-880.

Staruch, M.J. and Wood, D.W. (1982). Genetic influences on the adjuvanticity of muramyl dipeptide *in vivo*. *J. Immunol.* **128**(1), 155-160.

Toffaletti, D. and Schwab, J.H. (1979). Modulation of mouse lymphocyte functions by membranes from group A strepto-cocci. *Cell. Immunol.* **42**, 3-17.

Turcotte, R. (1981). Evidence for 2 distinct populations of suppressor cells in the spleen of *Mycobacterium bovis* BCG-sensitized mice. *Infect. Immun.* **34**, 315-322.

Uchiyama, T. and Jacobs, D.M. (1978). Modulation of immune response by bacteria lipopolysaccharides (LPS): multifocal effects of LPS-induced suppression of the primary antibody response to a T-dependent antigen. *J. Immunol.* **121**, 2340-2346.

Watson, S.R. and Collins, F.M. (1981). The specificity of suppressor T-cells induced by chronic *Mycobacterium avium* infection in mice. *Clin. Exp. Immunol.* **43**, 10-19.

Wostman, B.S., Pleasants, J.R., Bealmear, P. and Kincade, P.W. (1970). Serum proteins and lymphoid tissues in germ-free mice fed a chemically defined, water soluble, low molecular weight diet. *Immunology* **19**, 443-448.

Yem, A.W. and Parmely, M.J. (1981). Modulation of Ia-like antigen expression and antigen-presenting activity of human monocytes by endotoxin or zymosan A. *J. Immunol.* **127**, 2245-2251.

BCG-MEDIATED HOST RESPONSES IN TUMOUR IMMUNOTHERAPY

R.W. Baldwin and M.V. Pimm

Cancer Research Campaign Laboratories,
University of Nottingham,
University Park, Nottingham, UK

I. INTRODUCTION

Considerable progress has been made in various approaches to tumour immunotherapy (Baldwin, 1980) and in considering the agents being developed for this purpose it is evident that bacterial preparations, particularly BCG and *C. parvum* have proven to be effective in many animal tumour systems (Baldwin and Pimm, 1978; Milas and Scott, 1977). But these vaccines have not been very effective, if at all, when used clinically against a range of tumours (Goodnight and Morton, 1978; Terry and Windhorst, 1978). This has led to the development of synthetic immunomodulating agents such as muramyl dipeptide (Parant, 1980) and more recently to the use of factors such as interferon (Billiau, 1981) for immunotherapy.

It is argued that bacterial preparations as well as the more recently developed synthetic agents function through the stimulation of activated macrophages and/or natural killer (NK) cells (Baldwin and Byers, 1980). There is as yet, however, little understanding as to how these effector cells function in mediating tumour rejection as distinct from their *in vitro* effects upon cultured tumour cells. Related to this the role of sensitized T-lymphocytes, if any, in macrophage and/or NK cell mediated responses has not been resolved. These interactions have been investigated in the present studies which evaluate the therapeutic effectiveness of BCG and *C. parvum* on the growth of transplanted rat tumours derived initially from naturally arising mammary adenocarcinomas. These studies are concerned principally with the capacity of the bacterial preparations to produce tumour rejection when injected together with the tumour cells and how this regional immunotherapy can be enhanced by the

simultaneous generation of a local delayed type hypersensitivity response to nontumour antigens.

II. RESULTS

The 3 mammary carcinomas selected for these studies arose
naturally in the WAB/Not inbred rat strain. One of these
tumours Sp4 is immunogenic so that resistance to tumour challenge can be induced in rats preimmunized with tumour cells
prevented from progressive growth in various ways. This
includes total resection of a developing tumour graft or
immunization with γ-radiation attenuated cells, especially
when administered in conjunction with bacterial immunostimulants such as BCG or *C. parvum* (Baldwin and Embleton, 1969;
Pimm *et al.*, 1978a; Willmott *et al.*, 1979). However, the
most effective immunizing protocol has proved to be the
inoculation of viable mammary carcinoma Sp4 cells in admixture
with BCG or *C. parvum* under conditions where growth of the
immunizing inoculum is prevented. The immune response elicited by this means is sufficient to produce rejection of a
challenge with (2 x 10^5) Sp4 cells in preimmunized rats.
Moreover, as shown in Table 1, treatment with Sp4 tumour cells
admixed with BCG or *C. parvum* elicits a systemic tumour
immunity so that a contralateral challenge with Sp4 cells is
rejected.
    In contradistinction to the findings with tumour Sp4, the
2 other mammary carcinomas Sp15 and Sp22 have no, or only
marginal, immunogenicity. For example, rats treated with γ-
irradiated tumour Sp15 cells admixed with either BCG or *C.
parvum* did not reject a subsequent challenge with a dose of
viable Sp15 cells (10^3) which was just sufficient to produce
progressive growth in untreated controls. Immunity could be
induced to mammary carcinoma Sp22 following immunization with
vaccine containing irradiated tumour cells and either BCG or
C. parvum, but this was not very substantial and was inadequate in producing consistent systemic immunity (Table 1).

A. Regional Immunotherapy

Previous studies with carcinogen-induced rat tumours including 3-methylcholanthrene-induced sarcomas and aminoazo dye-
induced hepatomas (Baldwin and Pimm, 1973; Pimm and Baldwin,
1977) have demonstrated that bacterial vaccines administered
together with tumour cells suppress tumour growth. This
adjuvant contact suppression or "regional immunotherapy" has
been ascribed to nonspecific host responses elicited by the
bacterial preparations, but the role of effector cells including macrophages and natural killer (NK) cells has not been

TABLE 1

Active specific immunotherapy of subcutaneous growths of rat mammary carcinomas

Mammary carcinoma	Treatment Inoculum		Challenge Inoculum	Tumour takes in:	
	No. cells	Adjuvant[1]	No. cells	Treated rats	Controls
Sp4	5×10^6-IRR[2]	BCG	1×10^4	2/22	22/22
	5×10^6-IRR	BCG	1×10^4	9/16	17/17
	2×10^6-IRR	C. parvum	1×10^4	0/13	11/12
Sp15	5×10^6-IRR	BCG	1×10^3	3/6	3/6
	5×10^6-IRR	BCG	1×10^3	4/6	2/6
	2×10^6-IRR	C. parvum	1×10^3	7/7	7/6
Sp22	5×10^6-IRR	BCG	1×10^3	1/6	5/7
	5×10^6-IRR	BCG	1×10^3	6/6	4/7
	2×10^6-IRR	C. parvum	1×10^3	4/13	8/11

[1] BCG Glaxo Percutaneous 100–500 µg moist weight. C. parvum Wellcome CN6134, 200 µg dry weight.
[2] IRR-cells irradiated with 15,000 R ^{60}Co γ-irradiation.

Tumour cell suspensions were prepared by trypsin digestion of solid tumour tissue from s.c. grafts, washed and resuspended in Hank's solution. Tumour cells were mixed with bacterial suspensions immediately before s.c. injection into syngeneic female rats. Simultaneously animals received a challenge of tumour cells alone s.c. in the contralateral flank, control rats received only the challenge inocula. Tumour takes from these challenge inocula were scored after 30 to 40 days. No tumours developed following inoculation of tumour cells mixed with BCG or C. parvum.

TABLE 2

Inhibition of growth of rat mammary carcinoma cells injected subcutaneously in admixture with BCG *or* C. parvum

Mammary carcinoma	No. tumour cells	Tumour incidence from cells injected with		
		C. parvum[1]	BCG[2]	Control
Sp4	5×10^4	–	0/5	4/5
	1×10^5	0/5	3/5	4/4
	5×10^5	2/11	3/5	11/11
Sp15	1×10^3	–	2/14	8/12
	2×10^4	2/6	6/6	7/7
	1×10^5	11/12	12/12	13/13
Sp22	1×10^3	–	10/10	10/10
	2×10^4	0/5	6/6	5/5
	1×10^5	10/12	12/12	11/11

[1]*C. parvum* Wellcome CN6134 (100 µg dry weight);
[2]BCG Glaxo Percutaneous (100 µg dry weight).

Tumour cells prepared as described in Table 1 were mixed with bacterial suspensions immediately before injection and tumour takes scored after 30 to 40 days. Control rats were injected with tumour cells alone.

elucidated. Moreover, the role of T-cells is not clear, although previous studies showing that BCG or *C. parvum* will suppress tumour growth in athymic mice indicate that a T-cell response is not obligatory (Pimm and Baldwin, 1975,1976). These factors have been further analysed in the present studies by comparing the capacity of BCG and *C. parvum* to suppress growth of mammary carcinomas Sp4, Sp15 and Sp22 in normal immunocompetent rats (Table 2). These studies show that *C. parvum* can suppress growth of all 3 tumours, although it is clearly most effective against tumour Sp4 where challenges with up to 5×10^5 tumour cells are rejected. In comparison the maximum challenge doses of tumours Sp15 and Sp22 suppressed by *C. parvum* was 2×10^4 cells. Injection of BCG together with tumour Sp4 cells also controlled growth of this tumour but challenges with 1 or 5×10^5 Sp4 cells were only partially controlled. But in comparison, BCG was almost completely ineffective in suppressing growth of mammary carcinomas Sp15 and Sp22.

In comparing the data derived from studies upon the effect of BCG or *C. parvum* on growth of mammary tumours (Table 2) it is evident that the most pronounced response was achieved with tumour Sp4 which, as shown by the data in Table 1, is immunogenic. This suggests that the effects elicited in the local environment of tumours by BCG or *C. parvum* may be acting synergistically with host responses to the tumour. Since it has been established that sensitized lymphocytes, but not serum antibody, can transfer immunity to tumour Sp4 (Baldwin and Embleton, 1969), one possibility is that sensitized T-cells are involved in this synergistic response. But the function of these sensitized cells is not clear since the role of cytotoxic T-cells in tumour rejection has been brought into question (Baldwin, 1982). This is illustrated by the analysis of effector cells involved in the rejection of Moloney sarcoma virus-induced rat tumours (Fernandez-Cruz *et al.*, 1980). Expansion of spleen cells from rats in which the tumour had regressed by culturing them together with Mitomycin C-inactivated tumour cells generated a population which when injected intravenously produced tumour rejection. Fractionation of these stimulated spleen cells in a fluorescence activated cell sorter following reaction with anti-rat T-cell monoclonal antibody W3/25 yielded a subset (W3/25 positive) which was poorly cytotoxic *in vitro*, but nevertheless induced tumour rejection when transferred to tumour-bearing rats. Conversely, the W3/25 negative subset of splenic lymphocytes was cytotoxic *in vitro* but ineffective for transferring tumour immunity. In relation to these studies, it has been shown that lymphocytes infiltrating a transplanted rat sarcoma Mc7(TIL) are highly effective in suppressing tumour growth (Robins *et al.*, 1979). For example, injection of sarcoma Mc7 cell: TIL/mixtures in ratios as low as 1:6 completely suppressed tumour growth. Effector cells derived from lymph nodes of tumour-immune rats normally have to be injected in much higher ratios (1:200) to achieve a comparable tumour rejection response. When, however, Mc7-TIL were analysed for cytotoxicity for sarcoma Mc7 cells in culture, they did not display specificity and the response was not markedly better than that of normal splenic lymphocytes which do not suppress the *in vivo* growth of tumours (Flannery *et al.*, 1981).

These studies suggest that one population of T-cells arising following sensitization to immunogenic tumours such as mammary carcinoma Sp4 may function as "helper-like" cells promoting nonspecific effector cells such as NK cells and macrophages. This postulate has considerable implications since it suggests that tumour rejection may be enhanced through the promotion of local T-cell responses to irrelevant

TABLE 3

*Inhibition of mammary carcinomas by BCG
in BCG-sensitized rats*

Mammary carcinoma	BCG sensitization[1]	Tumour cell challenge[2]	Tumour incidence in rats receiving:	
			Tumour cells	Tumour cells + BCG[3]
Sp15	+	1×10^3	14/14	0/14
Sp15	−	1×10^3	9/14	8/14
Sp15	+	2×10^3	12/12	1/14
Sp15	−	2×10^3	14/14	12/14
Sp15	+	5×10^3	12/12	1/14
Sp15	−	5×10^3	14/14	12/14
Sp22	+	2×10^3	11/13	0/14
Sp22	−	2×10^3	12/13	10/13

[1]BCG Glaxo (1.0 mg moist weight), i.p., day − 14.
[2]Tumour challenge s.c., day 0.
[3]0.5 mg moist weight.

Tumour cells were injected either alone or in admixture with
BCG into untreated rats or rats sensitized to BCG.

antigens. This hypothesis has been explored with the spontaneous rat mammary carcinomas to evaluate whether the local
effects of BCG injected in admixture with tumour cells are
enhanced in rats presensitized to BCG. In these experiments
rats were sensitized to BCG by a single intraperitoneal
injection of vaccine (Glaxo percutaneous, 1 mg moist weight
organisms) 14 days before tumour challenge. Rats were then
inoculated subcutaneously with tumour cells with or without
BCG. The results summarized in Table 3 show that with mammary carcinoma Sp15, tumours grew progressively in normal
recipients whether or not the tumour cells were injected
together with BCG. The response was quite different, however, in recipients presensitized to BCG. Tumours grew
progressively in controls whether or not they were BCG sensitized whereas in the 3 experiments only 2/42 tumours developed when tumour cells were injected together with BCG.
Similarly with mammary carcinoma Sp22, tumours developed in
rats challenged with tumour cells alone and BCG sensitization had no effect. In comparison, BCG completely suppressed tumours in BCG sensitized but not in nonsensitized rats.

These experiments suggest that the delayed type hypersensitivity (DTH) response elicited by BCG enhances tumour rejection. Since BCG elicits a number of responses in addition to
the induction of DTH in sensitized recipients, a further
series of experiments was carried out to determine whether
the generation of a DTH response to a purified protein derivative of tuberculin (PPD) in BCG-sensitized rats would also
elicit tumour rejection. The PPD (PPD-CT68) isolated by tri-
chloracetic acid precipitation from the culture filtrate of
Mycobacterium tuberculosis var hominis Johnston Strain (Landi
and Held, 1980) was provided by Professor Landi (Tuberculosis
Division, Connaught Medical Research Laboratories, Canada).
This standardized preparation is highly effective for eliciting delayed type responses as measured by ear swelling tests
in BCG-sensitized rats (Fig. 1).

The experiments summarized in Table 4 show that tumour Sp4
grew progressively following inoculation of 2×10^4 cells
into either normal or BCG-sensitized rats. Also inoculation
of Sp4 tumour cells in medium containing PPD-CT68 (5 or 20 µg
$/2 \times 10^4$ cells) into normal rats did not materially inhibit
tumour growth. These findings differ from the effects observed when Sp4 tumour cells are injected in admixture with BCG
where complete suppression of tumour takes is obtained (Table
2).

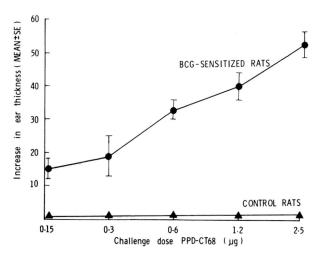

FIG. 1 *PPD-CT68 in 0.05 ml saline was injected intradermally
into the ears of normal rats and rats presensitized to BCG by
i.p. injection of 1 mg moist weight of vaccine 14 days before.
Ear thicknesses were measured 24 h after PPD injection and
compared with contralateral ears injected with saline alone
(6 to 12 rats per group).*

TABLE 4

Inhibition of mammary carcinomas by tuberculin purified
derivative (PPD) in BCG-sensitized rats

Mammary carcinoma	BCG sensitization[1]	Tumour cell challenge	PPD[2] dose (μg)	Tumour incidence in rats receiving Tumour	Tumour + PPD
SP4	+	2×10^4	5	7/7	0/7
	−	2×10^4	5	7/7	7/7
SP4	+	2×10^4	20	7/7	0/7
	−	2×10^4	20	7/7	7/7
Sp15	+	5×10^3	10	7/13	0/13
	−	5×10^3	10	10/14	9/13

[1]BCG Glaxo (1.0 mg moist weight) i.p. day − 14.
[2]Tuberculin purified protein derivation PPD-CT68.

Tumour cells were injected either alone or in admixture with
PPD into untreated rats or rats sensitized to BCG.

When however, Sp4 tumour cells are injected in medium containing PPD-CT68 (5 or 20 μg/2×10^4 cells) into rats presensitized to BCG there was a pronounced therapeutic response
and none of the rats developed tumours (Table 4). A similar
type of response was also obtained in comparable tests using
the nonimmunogenic mammary carcinoma Sp15. Again, PPD-CT68
did not modify growth of tumour Sp15 (5×10^3 cells) in normal rats, but in BCG-sensitized rats tumours were completely
rejected.

III. DISCUSSION

There is little doubt that bacterial vaccines particularly
BCG and *C. parvum* are highly effective for stimulating cell
populations involved in both specific and nonspecific antitumour responses (Baldwin, 1981; Baldwin and Byers, 1980).
The therapeutic effect is generally quite low, however, when
these vaccines are administered systemically and repeated
treatment frequently leads to adverse responses. In contrast,
promotion of local responses by regional administration of
bacterial vaccines frequently produces a strong antitumour
response. It may be argued that the local administration of

bacterial vaccines initially functions through activation of
host cells within tumour deposits leading to the generation
of activated macrophages and this may also result in the augmentation of NK cells, one pathway involving interferon production (Baldwin, 1981).

There is conclusive evidence that many experimental animal
tumours, particularly those induced with chemical carcinogens
or oncogenic viruses, contain infiltrating lymphocytes and
macrophages and these tumours respond to regional immunotherapy with bacterial preparations (Baldwin and Byers, 1980) as
well as subcellular products and synthetic analogues (Baldwin,
1981; Ogura *et al.*, 1979; Sone and Fidler, 1981). But some
animal tumours, including many naturally-arising (spontaneous)
tumours, have only low levels of infiltrating host cells.
For example, subcutaneous growths of the mammary carcinoma
SP15 used in the present studies contained only low levels of
infiltrating macrophages and lymphocytes (Pimm *et al.*, 1978b).
This has also been the experience in a number of recent investigations of tumour-infiltrating host cells in human tumours
(Moore and Vose, 1981; Nitsuma *et al.*, 1981; Eremin *et al.*,
1981). In these circumstances, regional immunotherapy would
not be expected to be very effective unless an additional
response is induced to increase extravasation of host cells
into tumours.

This is the basis of the present series of experiments
whose objective was to show that tumours which could not be
adequately controlled by direct contact with bacterial vaccines became sensitive to this form of immunotherapy following the induction of a local delayed type hypersensitivity
response. This is clearly demonstrated by the experiments
showing that BCG does not suppress the growth of 2 essentially
nonimmunogenic mammary carcinomas (Sp15 and Sp22) when inoculated in admixture into normal rats, although complete suppression of tumours was achieved in BCG-presensitized hosts.
Since BCG may elicit a number of other effects, it was also
important to show that this response could be obtained with
the mycobacterial component responsible for eliciting the
tuberculin response. This was conclusively established in
the series of experiments where suppression of growth of mammary carcinomas Sp15 and Sp4 was obtained when tumour cells
were inoculated together with PPD into BCG-sensitized rats.

The present view is that the host responses leading to the
rejection of rat mammary carcinomas is mediated by NK cells
and/or macrophages with the sensitized T-cells functioning as
accessory cells. In this respect it has been established
that tumour cells freshly derived from subcutaneous growth of
mammary carcinoma Sp4 and Sp22 are susceptible to NK-cell-
mediated lysis (Brooks *et al.*, 1981). It has also been

established that the susceptibility of rat tumour cells to NK
cells correlated with responses to BCG-activated peritoneal
macrophages (Brooks *et al.*, 1982). But the function of sensitized T-cells in tumour deposits remains to be elucidated.
They may function to recruit infiltration of host cells, perhaps through the release of factors. In this context, NK
activity of lymphocytes derived from lung carcinoma specimens
is enhanced following intralesional injection of BCG (Nitsuma
et al., 1981). Tumour-infiltrating T-cells may also augment
the reactivity of local NK cells and macrophages and this may
account for the lack of tumour-inhibitory responses detected
with normal splenic NK cells. This is emphasized by studies
already referred to (Robins *et al.*, 1979) where tumour infiltrating lymphocytes derived from a transplanted sarcoma Mc7
were highly effective in suppressing tumour growth when transferred to normal rats in admixture with tumour Mc7 cells at
ratios as low as 3:2. In comparison, normal spleen cells
processed as for the tumour infiltrating lymphocytes were
completely ineffective in modifying tumour growth when transferred in admixture with sarcoma Mc7 into normal rats.

A more precise evaluation of the influence of local delayed
type hypersensitivity responses upon host cells mediating
tumour rejection cannot be made until further information is
obtained about the characteristics of the T-cell subsets
involved in the DTH mediated activation. This is being
approached by defining the capacity of subsets of T-cells
derived from BCG-sensitized rats to adoptively transfer tuberculin hypersensitivity and to initiate rejection of the mammary tumours. Lymphocyte populations are being separated in
a fluorescence activated cell sorter on the basis of their
reactivity with monoclonal antibodies which recognize T-cell
antigens (Robins *et al.*, unpublished studies). In addition,
the functional characteristics of T-cell populations are
being defined in relation to their capacity to produce soluble
factors such as T-cell growth factor (Cantrell *et al.*, 1982).

Even though the influence of local tumour-infiltrating T-
cells stimulated to irrelevant antigens such as BCG has not
been defined, the experiments described in this paper indicate that local immunotherapy with BCG can be enhanced by
this additional treatment. This again draws attention to the
view that regional immunotherapy in which host responses are
generated in the environment of tumours is more effective
than systemic immunostimulation (Baldwin and Byers, 1980).
This has been achieved by intralesional injection of various
agents including BCG as well as subcellular preparations such
as cord factor and synthetic analogues. It has to be recognized, however, that intralesional therapy may have limited
application and alternative procedures for targetting agents

to tumours are being sought. One approach developed with muramyl dipeptide has been to administer the agents in liposomes so as to provide localization in pulmonary tissue in the treatment of pulmonary metastases from several rodent tumours (Fidler *et al.*, 1982). More specific targetting may be achieved using monoclonal antibodies directed against tumour associated antigens as carriers for immunomodulating agents (Baldwin *et al.*, 1981). The potential of this approach with rat mammary carcinoma Sp4 is established by related studies showing that a monoclonal antibody directed against the specific neoantigen on this tumour (Gunn *et al.*, 1980) can be used to target cytotoxic drugs *in vivo* to tumour deposits (Pimm *et al.*, 1982). Moreover, interferon conjugates with a monoclonal antibody directed against a human osteogenic sarcoma have been produced for the local activation of NK cells (Baldwin *et al.*, 1982). These studies indicate that it should be possible to use monoclonal antibodies directed against tumour associated antigens for targetting agents to tumours which will then elicit DTH response and so improve local immunotherapy.

IV. ACKNOWLEDGEMENTS

These studies were carried out under a departmental grant from the Cancer Research Campaign. BCG vaccines were supplied by Glaxo Research Ltd., Greenford, Middlesex, UK and *C. parvum* CN6134 was supplied by Wellcome Research Laboratories, Beckenham, UK. Professor S. Landi, Connaught Medical Research Laboratories, Toronto, Canada, is thanked for supplying a standard preparation of PPD.

V. REFERENCES

Baldwin, R.W. (1980). *In* "Cancer Chemotherapy Annual 2" (Ed. H.M. Pinedo), pp. 150-175. Excerpta Medica, Amsterdam.
Baldwin, R.W. (1981). *In* "Cancer Chemotherapy Annual 3" (Ed. H.M. Pinedo), pp. 178-202. Excerpta Medica, Amsterdam.
Baldwin, R.W. (1982). *In* "Cancer Chemotherapy Annual 4" (Ed. H.M. Pinedo), Excerpta Medica, Amsterdam. In press.
Baldwin, R.W. and Embleton, M.J. (1969). *Int. J. Cancer* **4**, 430-439.
Baldwin, R.W. and Pimm, M.V. (1973). *Brit. J. Cancer* **28**, 281-287.
Baldwin, R.W. and Pimm, M.V. (1978). *Adv. Cancer Res.* **28**, 91-147.
Baldwin, R.W. and Byers, V.S. (1980). *In* "Immunostimulation" (Eds L. Chedid, P.A. Miescher and H.J. Muller Eberhard), pp. 73-94, Springer-Verlag, Berlin, Heidelberg, New York.

Baldwin, R.W., Embleton, M.J. and Price, M.R. (1981). *Molec. Aspects Med.* **4**, 329-368.

Baldwin, R.W., Flannery, G.R., Pelham, J.M. and Gray, J.D. (1982). *Proc. Am. Assoc. Cancer Res.* **23**, 254.

Billiau, A. (1981). *Eur. J. Cancer Clin. Oncol.* **17**, 949-967.

Brooks, C.G., Flannery, G.R., Willmott, N., Austin, E.B., Kenwrick, S. and Baldwin, R.W. (1981). *Int. J. Cancer* **28**, 191-198.

Brooks, C.G., Wayner, E.A., Webb, P.J., Gray, D.J., Kenwrick, S. and Baldwin, R.W. (1982). *J. Immunol.* **127**, 2477-2482

Cantrell, D.A., Robins, R.A. and Baldwin, R.W. (1982). *Cell. Immunol.* Submitted.

Eremin, O., Coombs, R.R.A. and Ashby, J. (1981). *Brit. J. Cancer* **44** (2), 166-176.

Fernandez-Cruz, E., Woden, B.A. and Feldman, J.D. (1980). *J. Exp. Med.* **152**, 823-841.

Fidler, I.J., Barnes, Z., Fogler, W.E., Kirsh, R., Bugelski, P. and Poste, G. (1982). *Cancer Res.* **42**, 496-501.

Flannery, G.R., Robins, R.A. and Baldwin, R.W. (1981). *Cell. Immunol.* **61** (1), 1-10.

Goodnight, J.E. and Morton, D.L. (1978). *Ann. Rev. Med.* **29**, 231-235.

Gunn, B., Embleton, M.J., Middle, J.G. and Baldwin, R.W. (1980). *Int. J. Cancer* **26**, 325-330.

Landi, S. and Held, H.R. (1980). *Annali Sclavo* **22**, 899-907.

Milas, L. and Scott, M.T. (1977). *Adv. Cancer Res.* **26**, 257-260.

Moore, M. and Vose, B.M. (1981). *Int. J. Cancer* **27** (3), 265-272.

Nitsuma, M., Golub, S.H., Edelstein, R. and Carmack Holmes, E. (1981). *J. Nat. Cancer Inst.* **67** (5), 997-1003.

Ogura, T., Namba, N., Hirao, F., Yamamura, Y. and Azuma, I. (1979). *Cancer Res.* **39**, 4706-4712.

Parant, M. (1980). *In* "Immunostimulation" (Eds L. Chedid, P.A. Miescher and H.J. Mueller Eberhard), pp. 111-128. Springer-Verlag, Berlin.

Pimm, M.V. and Baldwin, R.W. (1975). *Nature* **245**, 77-79.

Pimm, M.V. and Baldwin, R.W. (1976). *Brit. J. Cancer* **34**, 453-455.

Pimm, M.V. and Baldwin, R.W. (1977). *Int. J. Cancer* **20**, 923-932.

Pimm, M.V., Cook, A.J., Hopper, D.G., Dickinson, A.M. and Baldwin, R.W. (1978a). *Int. J. Cancer* **22**, 426-432.

Pimm, M.V., Hopper, D.G. and Baldwin, R.W. (1978b). *In* "Developments in Biological Standardization". (Eds A.H. Griffith and R.H. Regamey), **38**, pp. 349-354.

Pimm, M.V., Jones, J.A., Price, M.R., Middle, J.G., Embleton, M.J. and Baldwin, R.W. (1982). *Cancer Immunol. Immunother.* **12**, 125-134.

Robins, R.A., Flannery, G.R. and Baldwin, R.W. (1979). *Brit. J. Cancer* **40**, 946-949.

Sone, S. and Fidler, I.J. (1981). *Cell. Immunol.* **57**, 42-50.

Terry, W.D. and Windhorst, D. (1978). *In* "Immunotherapy of Cancer: Present Status of Trials in Man". pp. 669-671, Raven Press, New York.

Willmott, N., Pimm, M.V. and Baldwin, R.W. (1979). *Int. J. Cancer* **24**, 323-328.

CORYNEBACTERIUM PARVUM, ANTIVIRAL PROTECTION, INTERFERON INDUCTION, AND ACTIVATION OF NATURAL KILLER CELLS

H. Kirchner, E. Storch and L. Schindler

Institute of Virus Research,
German Cancer Research Centre,
Heidelberg, Federal Republic of Germany

I. INTRODUCTION

The so-called immunostimulants, such as Bacillus-Calmette-Guérin and *C. parvum* have received great interest because of their potential value in tumour therapy. There have been quite a few data that clearly documented a beneficial effect of such agents in experimental tumour models of laboratory animals (for review see Scott, 1974; Milas and Scott, 1978). There are serious problems when extrapolating from such animal models to the clinical situation because most laboratory tumour models do not reflect closely enough the pathogenesis of human cancer and the situation observed at the onset of clinical therapy, and therefore are of limited value only (Hewitt, 1978; Kirchner, 1982). Not unexpectedly, therefore, the clinical trials with these "immunostimulants" have been disappointing. On the other hand, one cannot help being impressed by the magnitude of the degree of activation of the lymphoreticular system as for example observed in mice after injection of *C. parvum* (Adlam and Scott, 1973).

Since the immune system and the defence systems of the body in general are systems that are regulated in a complex fashion, indiscriminate activation of these systems may paradoxically cause suppression. Thus, it has been shown by several investigators that injection of *C. parvum* in mice causes immunosuppression (e.g. Howard *et al.*, 1973) and in certain experimental situations a role of suppressor cells has been documented (e.g. Scott, 1972; Kirchner *et al.*, 1975a,b).

Thus, the effects of *C. parvum* (or other "immunostimulants") on different compartments of the defence system are

poorly understood. Nonetheless, there is hope that by systemic analysis some clues to the basic mechanisms that are
operative in the antitumour effects in animal models are found
and that on the next higher level of sophistication we find a
better access to the clinical use of immunostimulants.

Studies are hampered additionally by the complex structure
of the bacteria used. Attempts to isolate the active immunostimulatory principle have met with limited success only. It
therefore may be useful to search for defined bacterial structures with immunostimulatory activity. Below, we will point
to the finding that immunostimulatory bacteria are inducers
of interferon, and that at least some of their effects are
interferon effects. It will be useful to systematically analyse chemically defined molecules isolated from bacterial
cell walls for their interferon inducing effects.

II. SPECIFIC IMMUNITY VERSUS NONSPECIFIC DEFENCE

Immune reactions, the hallmark of which is specificity, are
caused by a *secondary* exposure of the organism to a pathogen.
Primary (nonspecific) defence is activated upon *primary* exposure of the organism to the pathogen. Primary defence as
the specific immune system consists of cellular and of humoral
mechanisms. Some of the mechanisms of primary defence are
preformed such as the phagocytic system whereas others have
to be activated, as for example the interferon system.

At present it is not known if in the defence against
tumours primary defence or specific immune responses are prevalent and the pathobiology of the development of spontaneous
tumours is largely unknown. Obviously, the development of
tumours is a multistep event, but it is not known at which
level the transformed cells are recognized by the defence
system, if at all. One may speculate that at least one step
of tumour cell development in some experimental models is
caused by viruses. From this point of view it may be useful
to study primary defence mechanisms against viruses. We have
in our laboratory investigated primary defence in 2 virus
models of the mouse and we have studied the effects of *C.
parvum* (and of other compounds) in these viral models.

A. Antiviral Effects of *C. parvum*

The 2 virus models which we have studied were infections of
mice with herpes simplex virus type 1 (HSV-1), a human herpes
virus, and with mouse hepatitis virus type 3 (MHV3), a genuine
coronavirus.

In the HSV system, antiviral protection was observed when
C. parvum was given several days before the virus (optimal
protection was seen when there was an interval of 7 days).
In contrast, simultaneous injection of *C. parvum* with HSV did
not result in protection (Kirchner *et al.*, 1978).

In the MHV3 system protection was observed when *C. parvum*
was given on the day of virus infection. However, it did not
matter if *C. parvum* was given several hours before or after
infection (Schindler *et al.*, 1981). Similar data to that
obtained in the MHV3 system have been reported in a mouse system of experimental infection with Junin virus (Budzko *et al.*
1978).

B. Adoptive Transfer Protocols

Mice were injected with *C. parvum*, and 7 days later spleen
cells and peritoneal exudate cells (PEC) were collected. In
adoptive transfer protocols it was tested if these cell populations were protective against i.p. infection with HSV.
Transfer of 5×10^7 spleen cells were indeed protective whereas no protection was observed with PEC (Storch, E. and
Kirchner, H., manuscript in preparation). We have as yet not
defined the cell subpopulation in the spleen of *C. parvum*-
treated mice which transferred protection.

III. INTERFERON INDUCTION *IN VIVO* BY *C. PARVUM*

Since interferons are the best characterized mediator molecules of antiviral defence, we were interested to investigate
if interferon was involved in the antiviral effects of *C.
parvum*. Our results have confirmed the data of Yamamura *et
al.* (1978) that serum interferon could be detected 4 to 8 h
after i.p. injection of *C. parvum* in mice. However, when
compared with conventional interferon inducers or with viruses
the titres were very low. In the peritoneal wash-out fluid
no interferon could be recovered between 1 and 48 h after
injection of *C. parvum*. This is in marked contrast to the
situation observed when, for example, viruses or certian synthetic inducers are injected i.p. In our experiments, HSV
served as a positive control since it induced high interferon
titres in the peritoneal wash-out fluid within 2 to 4 h after
i.p. injection (Zawatzky *et al.*, 1982).

One wonders if the low serum interferon titres observed
after injection of *C. parvum* were sufficient to cause the
protection observed when it was given simultaneously with
MHV3 infection. From our data it appeared to be unlikely
since there was a clearcut dissociation in the dose response
curve of protection and of interferon induction. Optimal

interferon induction could be observed with doses of *C. parvum* that were too low to cause protection against MHV3 (Schindler *et al.*, 1981). In the HSV system no protection was observed when *C. parvum* was given on the day of virus infection. Thus, it appears as if in both viral models the low interferon response observed in the serum is of minor importance.

However, this is another aspect which needs to be discussed. Strong protection against HSV infection was observed when *C. parvum* was given 7 days before virus infection. All investigators agree that maximal systemic activation by *C. parvum* in mice takes about 7 days and it is reasonable to assume that this type of general activation is causing the observed antiviral effect. Furthermore, our data have established that spleens taken out from mice 7 days after injection of *C. parvum* produce considerable titres of interferon. Some interferon was also measured in the serum at this time.

Interferon inducers as conventionally known induce interferon very rapidly. For example, this is found with viruses such as Newcastle Disease Virus (NDV), and HSV, or synthetic inducers such as Poly I-C, or 10-Carboxymethyl-9-acridanone (CMA). Maximal serum titres are observed within a few hours and these are not sustained longer than 1—2 days. Other viruses or inducers appear to induce interferon in a somewhat more prolonged fashion but rarely ever significant titres are found later than 3 days.

The interferon production in spleen cell cultures of *C. parvum*-treated mice which we have observed suggest that there are additional ways of "putting an organism under interferon", i.e. by systemic activation of the lymphoreticular system which then produces considerable titres of interferon for a more prolonged period of time. This may be more useful for the defence of the organism than a short-lasting interferon titre which is usually followed by a period of unresponsiveness to interferon inducers.

However, there is yet another way to explain our data of interferon production by spleen cells from *C. parvum* injected mice. As we will describe below, addition of *C. parvum* organisms *in vitro* to spleen cells of untreated mice causes a mitogenic effect on B lymphocytes (Zola, 1975) and interferon production (Hirt *et al.*, 1978a). It is also established that *C. parvum* after injection in mice is stored in various organs including the spleen and can be detected there for prolonged periods of time (Dimitrov *et al.*, 1977). Thus, conceivably, interferon production observed in spleen cell cultures is a strict *in vitro* phenonemon caused by the exposure of the spleen cells to the bacteria that are still present together with the cell suspension.

Above we have referred to our findings that spleen cells
of mice treated 7 days before with *C. parvum* were protective
against HSV in an adoptive transfer protocol, whereas PEC were
not. Since spleen cells are producers of interferon whereas
no interferon is found in the peritoneal wash-out fluid the
possibility exists that the cells active in the transferred
populations are the interferon producing cells.

IV. *IN VITRO* INDUCTION OF INTERFERON BY *C. PARVUM*

In addition to the *in vivo* induction of interferon our previous work has established that *C. parvum* caused interferon production *in vitro* in various types of leukocyte cultures,
including those of murine and human origin (Hirt *et al.*, 1978
a,b).

   C. parvum is a B-cell mitogen in mouse spleen cell cultures (Zola, 1975) and induces interferon in these cultures.
Although the producer cells of this interferon have not been
unequivocally defined our initial studies have suggested that
B-cells were the producers of interferon when spleen cells
were challenged with *C. parvum* (Hirt *et al.*, 1978a). Interestingly, spleen cells from LPS nonresponsive C3H/HeJ mice
produced equal amounts of interferon, as spleen cells of
C3HeB/FeJ mice, when challenged with *C. parvum*, documenting
that the *C. parvum* effect was not due to contamination with
LPS, or alternatively, that LPS is not the structure in *C.
parvum* organisms that is responsible for interferon induction.

   In our experiments, cultures of peritoneal exudate macrophages did not produce interferon when treated with *C. parvum*
(Hirt *et al.*, 1978a). However, Neumann and Sorg (1981) have
reported that bone marrow macrophages were producing interferon upon addition of *C. parvum*. Our own experiments, using
bone marrow macrophages of C57BL/6 mice have failed to document interferon induction by *C. parvum*, although positive
controls in these cultures including both viruses and synthetic inducers have induced high interferon titres in the culture system of bone marrow-derived macrophages (Storch and
Kirchner, 1982).

   More recently, it has been shown by Evans and Johnson
(1981) that several subtypes of interferons are produced by
mouse spleen cells in response to *C. parvum* and it has been
suggested that these are the products of different cell populations.

   C. parvum also represents a potent inducer of interferon
in cultures of human leukocytes (Hirt *et al.*, 1978b; Sugiyama
and Epstein, 1978). The producer cells of this interferon
appear to be non-T, non-B, non-monocytic cells, i.e. so-
called null cells (Kirchner *et al.*, 1979). Contrary to

initial reports the interferon produced in human leukocyte cultures upon addition of *C. parvum* was alpha interferon and not gamma interferon (Vilcek *et al.*, 1980).

V. ACTIVATION OF NK-CELLS BY *C. PARVUM*

Injection of *C. parvum* has been reported to activate NK-cells in the mouse (Herberman *et al.*, 1977). It has been suggested that this activation was caused by interferon induction since both interferon and interferon inducers are well-established to activate NK-cells (Gidlund *et al.*, 1978). Our experiments have confirmed the strong activation of NK cells by *C. parvum* which can be observed both in the PEC population and in the spleen (Storch, E. and Kirchner, H., manuscript in preparation). However, we are uncertain about the interpretation of the data since the levels of interferon that can be measured after injection of *C. parvum* are low when compared for example with Poly-I-C. Nonetheless, the degree of NK-cell activation is at least as high after injection of *C. parvum* as after injection of Poly-I-C. One possible explanation is that it takes minimal amounts of interferon to activate the NK-cell system *in vivo*. Alternatively, however, there may be other mediators or mechanisms that independently of the interferon system, activate NK-cells.

Our data have confirmed reports that activation of NK-cells in the spleen occurs within 2 days and that at 7 days NK-cell activity in the spleen is in fact depressed. In contrast, NK-cell activity in the PE is continuously high for at least 7 days (Ojo *et al.*, 1978).

In view of the adoptive transfer data referred to above it is of interest that spleen cells 7 days after injection of *C. parvum* that show depressed NK-cell activity are protective whereas PEC that show increased NK-cell activity 7 days after injection of *C. parvum* are not protective.

VI. ACTIVATION OF MACROPHAGES BY *C. PARVUM*

The stimulatory effect of *C. parvum* on the macrophage system has been established several years ago (Adlam and Scott, 1973). Activation of macrophages usually was observed to be maximal 7—10 days after injection of the bacteria, i.e. concomitantly with a general maximal activation of the lymphoreticular system. Interestingly however, Otu *et al.* (1977) have observed a biphasic pattern of macrophage activation after injection of *C. parvum*. An early activation as measured by increased rates of clearance of carbon reached maximal levels 2 days after injection. We would like to speculate that this early peak of activation was caused by interferon induction since interferon is known to activate macrophages. This hypothesis should be tested, particularly since Otu *et al.* have described that the early phase of macrophage activation is due to a lipid extract of the bacteria. One wonders if the "late" macrophage activation is also a secondary effect of interferon produced.

VII. ACTIVATION OF SUPPRESSOR CELLS BY *C. PARVUM*

Above, we have referred to the observation that many different subpopulations of the lymphoreticular system are activated by injection of *C. parvum*. Among these are suppressor cells which belong to various subtypes of leukocytes. Thus, suppressor T-cells, suppressor B-cells, and suppressor macrophages have been described. It is not surprising that suppressor cells have been activated by injection of *C. parvum*.

The original studies of Scott (1972), and later data of our laboratory (Kirchner *et al.*, 1975a,b) have established that 7—10 days after injection of *C. parvum*, suppressor cells were found in the spleens of mice. These had the characteristics of macrophages and were found to suppress a variety of lymphocyte functions, including lymphoproliferation and T-cell cytotoxicity (Kirchner *et al.*, 1975a,b).

Later, it has been described that NK-cell activity is also suppressed in spleen cell cultures obtained from mice 7 days after injection of *C. parvum*. By one group this suppression has been attributed to the presence of suppressor cells (Savary and Lotzova, 1978).

We were interested if in spleen cell cultures of *C. parvum*-treated mice the production of gamma interferon in response to mitogens was altered. Our data have shown that there was no impairment of the production of gamma interferon in response to the mitogens phytohaemagglutinin and Concanavalin A despite the fact that the lymphoproliferative responses to these mitogens were markedly inhibited (H. Kirchner, unpublished data). We interpret these data as an indication that the production of gamma interferon (which in fact is proliferation-independent) — in contrast to lymphoproliferation — is not susceptible to the action of suppressor macrophages.

VIII. HYPORESPONSIVENESS TO INTERFERON INDUCERS

The term hyporesponsiveness has been coined for a state of reduced responsiveness to interferon inducers. For example, tumour-bearing mice or mice infected with certain viruses have been shown to respond to Newcastle Disease Virus (NDV) with reduced levels of serum interferon. Similarly, in mice previously exposed to *C. acnes*, an organism closely related

to or identical to *C. parvum*, serum interferon levels induced
by NDV, Chikungunya virus and Poly I-C are suppressed (Farber
and Glasgow, 1972). However, we have failed to observe such
hyporesponsiveness when HSV was injected into mice 7 days
after injection of *C. parvum* (H. Kirchner, unpublished data).
Thus, it appears that hyporesponsiveness after injection of
C. parvum is observed only with some interferon inducers and
not with others.

One wonders what the significance of the hyporesponsiveness
phenomena is. Perhaps the interferon produced by the lymphoreticular system of the *C. parvum*-treated mice causes "blocking" of subsequent interferon induction.

IX. OUTLOOK

Bacteria have been shown previously to induce interferon *in
vivo* and in leukocyte cultures of various origin and cellular
composition. Although LPS is known to induce interferon,
various observations have suggested that LPS is not the only
bacterial component capable of interferon induction. However,
it will require considerable effort in the future to define
the structures on the bacterial surface that are responsible
for interferon induction by bacteria. Perhaps these studies
may lead to the development of clinically useful interferon
inducers.

Our own data have established interferon induction by *C.
parvum*. Interferons are pleiotropic molecules that cause,
for example, activation of macrophages and NK-cells. Since
these effects were observed after injection of *C. parvum* (or
other immunostimulants) it is not unreasonable to speculate
that many of the effects of these substances are effects
secondary to a primary induction of interferon. Perhaps by
careful investigations of the interferon induction by *C. parvum* we may learn how to manipulate this effect *in vivo* and
get a better insight into how to use immunostimulants effectively in the clinic.

X. REFERENCES

Adlam, C. and Scott, M.T. (1973). Lymphoreticular stimula-
    tory properties of *Corynebacterium parvum* and related
    bacteria. *J. Med. Microbiol.* **5**, 261-274.
Budzko, D., Casal, J. and Waksman, B. (1978). Enhanced resis-
    tance against Junin Virus infection induced by *Corynebac-
    terium parvum*. *Infect. Immun.* **19**, 893-897.
Dimitrov, N.V., Greenberg, C.S. and Denny, T. (1977). Organ
    distribution of *Corynebacterium parvum* labelled with
    Iodine-125. *J. Nat. Cancer Inst.* **58**, 287-294.

Evans, S.R. and Johnson, H.M. (1981). The induction of at least 2 distinct types of interferon in mouse spleen cell cultures by *Corynebacterium parvum*. *Cell. Immunol.* **64**, 64-72.

Farber, P.A. and Glasgow, L.A. (1972). Effect of *Corynebacterium acnes* on interferon production in mice. *Infect. Immun.* **6**, 272-276.

Gidlund, M., Orn, A., Wigzell, H., Senik, A. and Gresser, I. (1978). Enhanced NK-cell activity in mice injected with interferon and interferon inducers. *Nature* **273**, 759-761.

Herberman, R.B., Nunn, M.E., Holden, H.T., Steal, S. and Djeu, J.Y. (1977). Augmentation of natural cytotoxic reactivity of mouse lymphoid cells against syngeneic and allogeneic target cells. *Int. J. Cancer* **19**, 555-564.

Hewitt, H.B. (1978). The choice of animal tumours for experimental studies of cancer therapy. *Adv. Cancer Res.* **27**, 149-200.

Hirt, H.M., Becker, H. and Kirchner, H. (1978a). Induction of interferon production in mouse spleen cell cultures by *Corynebacterium parvum*. *Cell. Immunol.* **38**, 168-175.

Hirt, H.M., Schwenteck, M., Becker, H. and Kirchner, H. (1978b). Interferon production and lymphocyte stimulation in human leukocyte cultures stimulated by *Corynebacterium parvum*. *Clin. Exp. Immunol.* **32**, 471-476.

Howard, J.G., Christie, G.H. and Scott, M.T. (1973). Biological effects of *Corynebacterium parvum*. *Cell. Immunol.* **7**, 290-301.

Kirchner, H., Glaser, M. and Herberman, R.B. (1975a). Suppression of cell-mediated tumour immunity by *Corynebacterium parvum*. *Nature* **257**, 396-398.

Kirchner, H., Holden, H.T. and Herberman, R.B. (1975b). Splenic suppressor macrophages induced in mice by injection of *Corynebacterium parvum*. *J. Immunol.* **115**, 1212-1216.

Kirchner, H., Hirt, H.M., Becker, H. and Munk, K. (1977a). Production of an antiviral factor by murine spleen cells after treatment with *Corynebacterium parvum*. *Cell. Immunol.* **31**, 172-176.

Kirchner, H., Hirt, H.M. and Munk, K. (1977b). Protection against herpes simplex virus infection in mice by *Corynebacterium parvum*. *Infect. Immun.* **16**, 9-11.

Kirchner, H., Scott, M.T., Hirt, H.M. and Munk, K. (1978). Protection of mice against viral infection by *Corynebacterium parvum* and *Bordetella pertussis*. *J. Gen. Virol.* **41**, 97-104.

Kirchner, H., Peter, H.H., Hirt, H.M., Zawatzky, R. and Bradstreet, P. (1979). Studies of the producer cell of interferon in human lymphocyte cultures. *Immunobiology* **156**, 65-75.

Kirchner, H. (1982). Logistics of tumour immunology. *Klin. Wschr*. **60**, 37-47.

Milas, L. and Scott, M.T. (1978). Antitumour activity of *Corynebacterium parvum*. *Adv. Cancer Res*. **26**, 257-306.

Neumann, C. and Sorg, C. (1981). Heterogeneity of murine macrophages in response to interferon inducers. *Immunobiology* **158**, 320-329.

Ojo, E., Haller, O., Kimura, A. and Wigzell, H. (1978). The analysis of the conditions allowing *Corynebacterium parvum* to cause either augmentation or inhibition of natural killer cell activity against tumour cells in mice. *Int. J. Cancer* **21**, 444-452.

Otu, A.A., Russell, R.J. and White, R.G. (1972). Biphasic pattern of activation of the reticuloendothelial system by anaerobic coryneforms in mice. *Immunology* **32**, 255-264.

Savary, C.A. and Lotzova, E. (1978). Suppression of natural killer cell cytotoxicity by splenocytes from *Corynebacterium parvum*-injected, bone marrow-tolerant, and infant mice. *J. Immunol*. **120**, 239-243.

Schindler, L., Streissle, G. and Kirchner, H. (1981). Protection of mice against mouse hepatitis virus by *Corynebacterium parvum*. *Infect. Immun*. **32**, 1128-1131.

Scott, M.T. (1972). Biological effect of the adjuvant *C. parvum*. II. Evidence for macrophage—T-cell interaction. *Cell. Immunol*. **5**, 469-479.

Scott, M.T. (1974). *Corynebacterium parvum* as an immunotherapeutic anticancer agent. *Sem. Oncol*. **1**, 367-378.

Storch, E. and Kirchner, H. (1982). Induction of interferon in murine bone marrow-derived macrophage cultures by 10-Carboxymethyl-9-Acridanone. Eur. J. Immunol. (In press).

Sugiyama, M. and Epstein, L.B. (1978). Effect of *Corynebacterium parvum* on human T-lymphocyte interferon production and T-lymphocyte proliferation *in vitro*. *Cancer Res*. **38**, 4467-4473.

Vilcek, J., Sulea, I.T., Volvovitz, F. and Yip, Y.K. (1980). Characteristics of interferons produced in cultures of human lymphocytes by stimulation with *Corynebacterium parvum* and phytohemagglutinin. *In* "Biochemical Characterization of Lymphokines". Academic Press, New York,

Yamaura, K., Kato, H., Hanazawa, S. and Yamaguchi, Y. (1978). Induction of interferon in mice injected with heat-killed *Corynebacterium anaerobium*. *Japan. J. Exp. Med*. **48**, 69-70.

Zawatzky, R., Engler, H. and Kirchner, H. (1982). Experimental infection of inbred mice with herpes simplex virus.

III. Comparison between newborn and adult C57BL/6 mice. *J. Gen. Virol.* **60**, 25–29.

Zola, H. (1975). Mitogenicity of *Corynebacterium parvum* for mouse lymphocytes. *Clin. Exp. Immunol.* **22**, 514–521.

ONCOLYTIC CLOSTRIDIA

Stanislaw Szmigielski[1], Bozena Dworecka[2],
Slawomir Lipski[1], Janusz Jeljaszewicz[2] and Gerhard Pulverer[3]

[1]*Centre for Radiobiology and Radioprotection,
Warsaw, Poland;*

[2]*National Institute of Hygiene, Warsaw, Poland;*

[3]*Institute of Hygiene, University of Cologne, FRG.*

The possibility of application of anaerobic clostridia in the treatment of neoplastic diseases has long since been investigated (Table 1), but until now unequivocal evaluation of their usefulness has not been established. This is due to the fact that a relatively small number of experimental studies has been undertaken by only a few groups of investigators.

Spores of various pathogenic and nonpathogenic clostridia (*Clostridium tetani, Clostridium histolyticum*) were used in early trials. In 1959 a strain of *Clostridium butyricum* (denominated M55) was isolated and characterized (Möse, 1960) and afterwards classified as *Clostridium oncolyticum* M55 (COM-55). This nonpathogenic strain seems to possess a property of selective germination and growth in necrotic tumour tissue (Möse, 1960; Möse *et al.*, 1970; Möse and Möse, 1959, 1964). In defined cases of cancer, particularly in combination with other forms of antitumour therapy, COM-55 can cause total regression of experimental tumours (Rüster, 1980; Fredette and Plante, 1970). There are also individual cases of COM-55 application in humans with advanced neoplastic disease (Möse *et al.*, 1967; Heppner and Möse, 1966; Kretschmer, 1972). These trials seem to be totally empirical without sufficient experimental support.

In the present paper, biological properties of COM-55 are described, considering their potential application in cancer therapy. Possible mechanisms of their antitumour effects are also discussed, and results of our own studies on the influence of COM-55 on the haemopoietic system and its postirradiation recovery are presented.

TABLE 1

Development of investigations on oncolytic clostridia

1.	Germination of *Cl. tetani* spores selectively in sites of necrosis or injury of tissues.	Villard and Rouget, 1892
2.	Low redox potential needed in tissues for germination of *Cl. tetani* spores.	Fildes, 1929
3.	Clinical improvement in several cases of advanced cancer following injection of sterile filtrates of *Cl. histolyticum.*	Conell, 1935
4.	Prolonged survival of mice with transplantable tumours injected with *Cl. histolyticum* spores.	Parker *et al.*, 1947
5.	Germination of *Cl. tetani* spores after i.v. injection selectively in tumourous tissues.	Malmgren and Flanigan, 1955
6.	Lysis of Ehrlich ascites and solid tumour cells after treatment with *Cl. butyricum* M 55.	Möse and Möse, 1959
7.	Clinical application of *Cl. butyricum s. oncolyticum* M 55.	Möse and Möse, 1964
8.	Role of bacterial kininases in oncolysis by *Cl. butyricum* M 55.	Möse *et al.*, 1972a,b,c
9.	Erradication of experimental tumours after combined therapy with X-rays, microwave hyperthermia and *Cl. butyricum* M 55.	Dietzel *et al.*, 1978; Gericke *et al.*, 1979; Rüster, 1980
10.	Involvement of the immune system in oncolysis caused by *Cl. butyricum* M 55.	Brantner and Schwager, 1980
11.	Stimulation of granulopoiesis and accelerated postirradiation recovery of haemopoiesis in mice treated with *Cl. butyricum* M 55.	Dworecka *et al.*, 1982

I. MICROBIOLOGICAL CHARACTERIZATION OF
   CLOSTRIDIUM ONCOLYTICUM M55

Clostridia are Gram-positive bacilli. They are relatively
large microorganisms with a cylindrical shape and an ability
to form endospores. These bacteria are common in soil, water
and plants, and also are indigenous to the animal and human
digestive tract. According to Bergey *et al.* (1974) these
bacteria form one family — *Bacillacaeae*. It is divided into
5 genera and one of them contains anaerobic clostridia. The
genus *Clostridium* includes about 60 species. All bacteria
belonging to this genus can grow in strictly anaerobic conditions. Their enzymatic system does not contain cytochromes
and cytochrome oxidase, hence lack of ability of atmospheric
oxygen utilization. However, Möse *et al.* (1970) suggested
that COM-55 can grow in microaerophilic conditions. Clostridia also do not demonstrate catalase and peroxidase activity.
The diameter of spores produced by clostridia is about the
width of the bacterial cell.
    Clostridia are divided into 5 groups (Bergey *et al.*, 1974)
on the basis of such criteria as terminal or subterminal
localization of spores, the ability to degrade gelatin and
defined growth conditions. *Clostridium oncolyticum* M55 belongs to group I (spores located subterminally and no gelatin
digestion).
    Several extracellular products have been found and charac-
terized in supernatants of COM-55 cultures (Table 2). At
least some of them may play a significant role in the antitumour effects of these bacteria (Brantner and Schwager,
1980; Fischer *et al.*, 1975; Möse *et al.*, 1972a,b,c). A kininase and an uncharacterized "cytotoxic factor" seem to be of
particular importance for the above effect (Brantner and
Schwager, 1980; Schlechte *et al.*, 1981). It is suggested
that "cytotoxic factor" has the property of selective damage
of tumour cells (Fredette and McSween, 1974). All these
extracellular products of COM-55 are not well characterized
and further studies seem to be necessary to understand the
antitumour effect of COM-55.

II. EXPERIMENTAL ANTITUMOUR ACTIVITY OF
   CLOSTRIDIUM ONCOLYTICUM M55

Some investigators claim that antitumour effect of COM-55 can
be shown not only *in vivo* but also *in vitro* (Rousseau *et al.*,
1970; Schlechte *et al.*, 1981). Administration of viable
spores of COM-55 or sterile supernatants of vegetative forms
results in selective damage of tumour cells with significantly
weaker effect on diploidal cells (Fredette and McSween, 1974).

TABLE 2

Extracellular products of Clostridium butyricum s. oncolyticum *M 55*

Product	Properties	References
Specific nuclease	Highly polymerized DNA	Haller and Brantner, 1979
Protease	Wide spectrum activity; acting on human albumins, casein, collagen and other high molecular proteins and small peptides	Brantner and Schwager, 1979; Haller and Brantner, 1979
Membrane phospholipase A	Injury of granulocytes in combination with protease	Schwager, 1976
Alpha-haemolysin	Activity connected with phospholipase A and/or protease	Brantner and Schwager, 1980
Kininases	Active against bradykinin and other kinins	Brantner and Fischer, 1973
Cytotoxic factor	Injury of heteroploid but not diploid cell cultures *in vitro*	Rousseau *et al.*, 1970; Schlechte *et al.*, 1981

It is assumed that this phenomenon is caused by a not yet
characterized "cytotoxic" factor produced by vegetative forms
of the bacteria. Recently, the effect of COM-55 was examined
on various lines of normal and tumour cells (Schlechte *et
al.*, 1981). In this case germination and growth of spores
was observed in the presence of an oxygen concentration up to
1.5%, whereas in the cultures of normal cells spores did not
germinate in oxygen concentration ranging from 0.3 to 1.5%.
Influence of COM-55 spores and of extracellular products on
cell cultures is not clear.

The majority of published studies concerns the animal
models with transplantable tumours treated intravenously or
intratumourally with COM-55 (for review, see Rüster, 1980).
The administration of even larger numbers of COM-55 does not
result in any significant clinical symptoms and in particular
does not cause generalized bacterial infection (Thiele *et
al.*, 1964b). Our studies demonstrated no elevation of body
temperature after intravenous injection of 4 x 10^7 of COM-55
spores, and also no vegetative forms of bacteria in liver or
spleens were found 12–96 h after administration (Dworecka,
1981). We could not establish the LD_{50} of COM-55 spores as
all the animals survived high doses of 10^8–10^{11} spores per
mouse. The latter dose is approximately 1000 times higher
than that used for antitumour therapy (Möse, 1960).

Spores of COM-55 labelled with ^{51}Cr and injected intravenously were mainly localized in the liver (Fig. 1) where
40–60% of the radioactivity was found (Dworecka, 1981).
Lungs and spleen trapped not more than 10% of the radioactivity, whereas tumour tissue showed only vestigial accumulation
of labelled spores (Fig. 1). It seems to confirm the observations of Möse (1960) and Gericke and Engelbart (1964)
demonstrating that accumulation of spores does not play a
significant role in the mechanism of antitumour effect, but
rather suitable conditions for their germination in tumour
tissue are of importance. Accumulation of the majority of
spores in the liver suggests that they are trapped by the
cells of reticuloendothelial system. The phagocytized spores
can remain for a few weeks in the macrophages without resulting in their damage (Thiele *et al.*, 1964). Slow elimination
of spores by phagocytes is postulated to enhance antitumour
properties of COM-55 (Brantner and Schwager, 1980). However,
interaction between the spores and monocyte–macrophage system
in vivo and *in vitro* is not clear and further studies are
again required.

Möse (1960) listed several features of COM-55 which suggest their possible usefulness in an antitumour therapy
(Table 3). The author points out such properties as lack of
pathogenicity and toxicity (which was confirmed by our

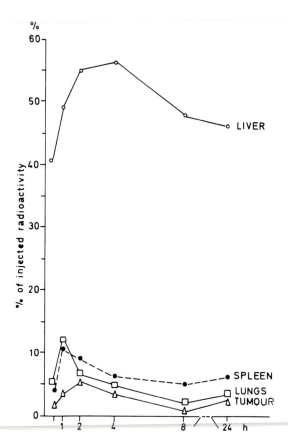

FIG. 1 *Distribution of* ^{51}Cr*-labelled spores of* Cl. oncolyticum *in mice (*10^8 *spores injected i.v.) (Dworecka, 1981).*

studies), low toxicity of COM-55 metabolites and their intracellular products, slow elimination of spores by the reticuloendothelial system, and the ability to germinate only in
tissues with low oxygen partial pressure. Intravenous or
intratumoural administration of COM-55 spores into mice with
a transplantable tumour results frequently in oncolysis and
necrosis of neoplastic lesions. This was observed in various
experimental models (Fig. 2, Table 4). Möse and Möse (1964)
described the histopathological features of tumours after
injection of spores. Softening of tumourous tissue could be
observed several days after the administration of spores. A
large cavity develops in the place of the solubilized tumour.
Usually animals do not survive this process as the open
wound is too large, often penetrating muscles as far as the

TABLE 3

Selected properties of Cl. butyricum s. oncolyticum *M 55*
suggesting its use in therapy of neoplasms (Möse, 1960)

Properties	Observed effects
1. Lack of pathogenicity	No germination in normal tissues, no clinical symptoms after injection of spores in normal mice.
2. Lack of toxicity	LD_{50} in mice and rats undeterminable.
3. No pyrogenic effect	Lack of pyrogenic effect in mice and rats, possible slight fever in rabbits.
4. No toxic metabolites and toxic extracellular products	No clinical symptoms in mice injected with sterile filtrates of *Cl. butyricum.*
5. Slow elimination of spores by phagocytes	Spores injected i.v. are detectable in kidneys after 35 days and in liver after 50 days.
6. Selective germination of spores in tumourous tissues with low partial pressure of oxygen.	PO_2 below 300 Pa needed for germination of *Cl. butyricum* spores.

peritoneum. Although macroscopically the tumour looks completely solubilized, some live tumour cells can be found in microscopic examination. Very often these cells line the inner walls of the cavity and if the animal survives, the tumour recurs at this place. Rarely a complete cure can be observed after a single administration of COM-55. In the case of recurrence, the animals can be again effectively treated with the spores. Spontaneous necrosis of part of the tumourous tissue is observed sometimes in control tumour-bearing animals, but this process has a different character. The lysis of tumourous tissue is never so extensive, and usually just a small central part of the tumour is necrotized.

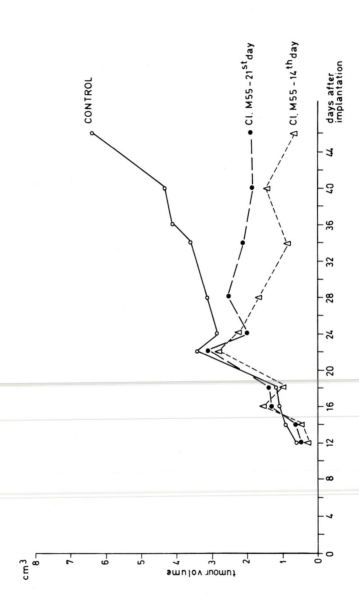

FIG. 2 *Volume of transplantable Madison lung tumour in mice treated with Cl. butyricum M 55 spores (10^8 per mouse) injected intratumourally on days 14 or 21 after implantation of neoplasms (Dworecka, 1981).*

TABLE 4

Sensitivity of transplantable tumours to oncolysis
by Cl. butyricum *M 55*

Species	Tumour type	Sensitivity	References
Mouse	Ehrlich solid tumour	+++	Möse and Möse, 1959
	Ehrlich exudative tumour	++	Möse, 1960
	Harding-Passey melanoma	++	Rüster, 1980
	Spontaneous breast adeno-carcinoma		Möse, 1960
	Methylcholanthrene-induced fibrosarcoma	–	Gericke and Engelbart, 1964
	Transplantable methyl-cholanthrene sarcoma 2677	+/–	Gericke and Engelbart, 1964
	Lymphosarcoma W 946	+/–	Gericke and Engelbart, 1964
Rat	Guerin epithelioma T-8	++	Möse, 1960
	Walker T sarcoma	+/–	Möse, 1960
	Yoshida T sarcoma	+/–	Möse, 1960
	Urine epithelioma	–	Gericke and Engelbart, 1964
	Oberling myeloma	–	Gericke and Engelbart, 1964
	Lymphatic leukaemia	–	Gericke and Engelbart, 1964
Rabbit	Brown-Pearce tumour	+/–	Möse, 1960
Hamster	Amelanotic melanoma	++	Engelbart and Gericke, 1964
	Renal adenocarcinoma	+/–	Engelbart and Gericke, 1964
	Liver cystadenocarcinoma	++	Engelbart and Gericke, 1964
	Adrenal cortical carcinoma	+/–	Engelbart and Gericke, 1964

According to the majority of investigators, the beginning
of the logarithmic phase of tumour growth is optimal for
administration of COM-55 spores (Gericke and Engelbart, 1964).
This is due to the fact that at this time the tumour shows a
tendency to partial spontaneous necrosis, providing better
conditions for the germination of larger number of spores.
It has been also observed that the exudative form of tumours
(e.g. intraperitoneally growing Ehrlich tumour) are also susceptible to therapy with COM-55 (Möse and Möse, 1959; Gericke
and Engelbart, 1964). In this case spores do not grow intra-
cellularly, but they find suitable metabolic conditions in
peritoneal exudate and germinate outside the tumour cells
(Gericke and Engelbart, 1964). This may explain the phenomenon of spore germination and growth in microtumours which do
not exhibit symptoms of necrosis (Urban *et al.*, 1974).

There are considerable differences in sensitivity of
various transplantable tumours to treatment with COM-55
spores (Table 4). Generally, murine tumours are susceptible
to this form of therapy whereas many types of rat and rabbit
tumours are resistant. As was mentioned above, the optimal
time for the beginning of treatment is the logarithmic phase
of tumour growth. In the case of a fibrosarcoma, the optimal
effect is observed when the tumour size is around 12 mm
(Gericke and Engelbart, 1964). It is postulated that early
administration of COM-55 does not cause direct oncolytic
effect. The spores are trapped by the reticuloendothelial
system and germinate in tumourous tissue only after several
days, when the necrosis can be noticed. Similar observations
have been made in our studies (Dworecka, 1981). Balb/c mice
with transplantable Madison lung cancer were locally treated
with 4×10^8 spores at different times after tumour implantation (Fig. 2). Tumour lysis was observed in all examined
groups of animals. It began to appear at approximately the
same time (22—24 days after tumour implantation), independently from the beginning of treatment. Best results of
therapy were obtained when spores were administered on the
14th or 21st day after tumour implantation (Fig. 2). Single
injection of COM-55 spores does not lead to complete lysis
of tumours and to prolongation of survival time of animals
(Dietzel and Gericke, 1977). It is due to the fact that
large and rapid necrosis occurs which provides conditions for
infection with pathogenic bacteria and leads to intoxication
by products of necrotic tissue with subsequent kidney function impairment (Brantner and Schwager, 1980). This, of
course, seriously restricts the possibility of practical
application of this form of therapy in clinical oncology.

III. MECHANISMS OF ONCOLYSIS CAUSED BY *CLOSTRIDIUM* *ONCOLYTICUM* M55

Mechanisms of tumour oncolysis in animals treated with COM-55
are not clear, despite many studies. The whole series of
postulated mechanisms are more hypothetical than experimentally verified facts. Undoubtedly, germination of spores
occurs only at a low partial pressure of oxygen. Such conditions can be found in necrotizing tissue and probably, therefore, a large number of vegetative forms of bacteria are
accumulating there (Möse and Möse, 1964). It has also been
postulated that hydrogen superoxide is produced in the peripheral part of tumours and may lead to neoplastic tissue
damage (Kayser, 1963, 1967). This has not been confirmed by
other studies. Importance has been attached to the extracellular factors produced by vegetative forms of COM-55 such
as kininase (Brantner and Fischer, 1973) or the "cytotoxic
factor" mentioned before (Chagnon *et al.*, 1972; Fredette and
McSween, 1974). But all these substances, including their
biological activity, are not sufficiently characterized to
define their role in the mechanisms of oncolysis. Recent
studies (Brantner and Schwager, 1980; Rüster, 1980) point to
a participation of immune mechanisms in the antitumour
effects of COM-55. Phagocytosis of spores by cells of the
monocyte—macrophage system may result in activation of these
cells with all well known consequences (Currie, 1980), such
as enhancement of nonspecific cytotoxicity of macrophages
directed against tumour cells. Figure 3 summarizes all postulated mechanisms of oncolysis. Most of these may require
confirmation by further studies, but we assume that the above
scheme is useful as a starting point for planning new experiments. It seems undoubtful that germination of spores and
appearance of vegetative forms of bacteria is essential for
initiation of the oncolytic process. Experimental data exist
suggesting that germination of spores and bacterial growth
can be stimulated by both products of anaerobic glycolysis
and those of necrotic tissue degradation. Whole series of
events are triggered by appearance of vegetative forms of
clostridia, which may directly or indirectly lead to neoplastic tissue destruction. Extracellular products of COM-55,
such as kininases, may attack and impair the vascular system
of the tumour (Brantner and Fischer, 1973) and the tumour
cells. Impairment of tumour vessels results in a less efficient local blood supply and consequently leads to further
lowering of tissue oxygen concentration. Significance of
proteases is still unclear. They are extracellular factors
(Table 2) which are thought to enhance inflammatory reactions
in the tumourous tissue. This also concerns the role of the

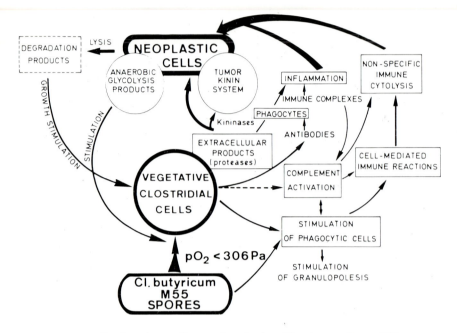

FIG. 3 *Mechanism of oncolysis by* Cl. butyricum *M55*

immune system in the process of oncolysis induced by COM-55
spores. It is known that they stimulate phagocytes (Brantner
and Schwager, 1980). Vegetative forms of bacteria induce
specific antibody formation which results in appearance of
immune complexes (Möse, 1960). Significance of complement
system is also unknown. Experimental studies have not provided evidence of complement system activation by the vegetative forms of COM-55 (Brantner and Schwager, 1980). However,
it is known that both activated macrophages and immune complexes may activate this system (Currie, 1980). The above
reactions may trigger cellular mechanisms leading to nonspecific lysis of tumour cells. Participation of these immune
reactions in the oncolysis produced by spores of COM-55 is
highly probable, although still no direct evidence exists.

IV. COMBINATION OF *CLOSTRIDIUM ONCOLYTICUM* M55
WITH OTHER FORMS OF EXPERIMENTAL TUMOUR THERAPY

As discussed above, administration of COM-55 to tumour-bearing
animals results in rapid and extensive necrosis of neoplastic
tissue but does not lead to complete cure and significant prolongation of survival time. This latter phenomenon is related
to various complications caused by massive necrosis, such as

TABLE 5

Effects of combination of Cl. butyricum M55 and cytostatics in treatment of experimental neoplasms

Neoplasm	Doses of Clostridial spores	Type of additional treatment	Effect	References
Ehrlich solid tumour (mice)	10^7	Sterile extract of tumour cells	Faster oncolysis	Möse and Möse, 1964
Sarcoma 180 (rats)	10^8	Different extracts of clostridial cultures	No enhancement of lysis	Engelbart and Gericke, 1964
Amelanotic melanoma (hamsters)	2×10^8	Vitamin C, Antihistaminica	No enhancement of lysis	Engelbart and Gericke, 1964
Sarcoma 180 (mice and hamsters)	10^8	Bacterial lipopolysaccharides; Mercaptopurine; Thioguanidine; 4-Aminopyrazole; Pirymidine; 5-Fluorouracil; Azaserine; Actinomycin D; Rutine; 5-Fluoro-deoxyuridine (FUDR); Tetramin	No enhancement of lysis; reduction of tumour mass; enhancement of antineoplastic effect	Thiele *et al.*, 1964a
Sarcoma 180 (mice and rats)	10^8	Trenimon	Enhancement of onco-lysis with lowered mortality	Thiele *et al.*, 1964a
Jensen sarcoma (rats)	10^8	Cyclophosphamide	Prolonged survival with retardation of tumour growth	Kretschmer *et al.*, 1974

bacterial infections or intoxication. This fact strongly
restricts the usefulness of COM-55 in cancer therapy (Dietzel
et al., 1978). It is proposed to combine the application of
spores with other forms of antitumour therapy and to establish appropriate schedules of treatment, considering timing
and sequence of the individual elements of the therapy
(Rüster, 1980). For a long time trials have been done on
combination of COM-55 with chemotherapy (Table 5), however,
the results are not encouraging. Usually, neither enhanced
oncolysis nor prolonged survival was noticed. The only exceptions were combination therapy with 5-fluorodeoxyuridine or
with trenimon in the treatment of sarcoma in mice and hamsters (Thiele *et al.*, 1964) and with cyclophosphamide in the
treatment of Jensen sarcoma in rats (Kretschmer *et al.*, 1974).
These observations have not been confirmed in other studies.
Numerous failures in treatment by combination of COM-55 with
cytostatics suggest that this form of therapy will be of
little use. Much more promising results were obtained by
administration of COM-55 with radiotherapy and/or local microwave hyperthermia of transplantable tumours in mice (Dietzel
et al., 1978; Gericke *et al.*, 1979). Recently, local hyperthermia has aroused interest as an adjuvant therapy (Cheung
and Samaras, 1981; Streffer, 1978). Due to the high thermosensitivity of neoplastic cells, local overheating of the
tumour to 43—44°C leads to damage or thermic death of the
majority of tumour cells and even to the complete cure of
animals (Streffer, 1978). Dietzel *et al.* (1978) have found
that intravenous injection of COM-55 spores combined with
local microwave hyperthermia (461 MHz) results in doubly prolonged survival time of tumour-bearing animals. These investigators took advantage of the fact that hyperthermia
increases the number of hypoxic tumour cells which provides
better conditions for germination and growth of the bacilli.
Further progress has been achieved by combination of COM-55
with hyperthermia and local X-ray irradiation of tumours
(Dietzel *et al.*, 1978; Gericke *et al.*, 1979; Rüster, 1980).
In this case, mice with transplantable Harding—Passey melanoma were used as an experimental model. The following
sequence of treatment was applied: 2000 rads locally, 3 min
of local microwave hyperthermia (43°C) and after 12 h administration of COM-55 spores in doses of 10^8 per mouse. This
therapy resulted in significant prolongation of survival time
and even in 18% of the material in a complete cure of animals
after a single course of treatment. Other animals died due
to recurrences of tumours. The second course of treatment
was highly efficient as a prophylaxis of recurrences and led
to a higher percentage of cured animals. Combination therapy

applied twice resulted in 40% of animals in total regression
(Rüster, 1980) with accompanying prolongation of mean survival
time to 69 days, as compared to 27 days in untreated controls.
Repeatedly used courses of therapy resulted in prolongation
of survival time even to 112 days and increased percentage of
completely cured animals to 80% (Rüster, 1980).

V. THE EFFECTS OF *CLOSTRIDIUM ONCOLYTICUM* M55 ON
HAEMOPOIETIC SYSTEM AND ITS POSTIRRADIATION RECOVERY

The complicated mechanisms of oncolysis induced by COM-55
attract attention to the possibility of direct or indirect
stimulation of the haemopoietic and immune systems (Brantner
and Schwager, 1980). There is no evidence of immunomodulatory action of COM-55, similar to that exerted by typical
bacterial immunostimulators such as *Propionibacterium* or BCG.
Injection of COM-55 spores into normal, healthy animals does
not lead to spleno- or hepatomegaly, lymphocytosis or activation of macrophages, or to elevation of body temperature
(Dworecka, 1981). On the other hand, administration of COM-
55 spores produces rapid and pronounced granulocytosis (Fig.
4). The number of granulocytes in peripheral blood reaches
$40 - 50 \times 10^3/mm^3$ in 2 h after injection of spores, without
any other accompanying symptoms. This granulocytosis lasts
for several hours and is slowly decreasing to $10 \times 10^3/mm^3$
after 8 h. On the 3rd to 4th day after single injection of
spores, second rise of granulocytosis reaches $30 \times 10^3/mm^3$
(Fig. 4). Normalization of the granulocyte number in peripheral blood occurs at 6—7 days after spore administration.
As there are no symptoms of functional stimulation of differentiated cells in peripheral blood (not increased reduction
of nitro-blue tetrazolium), it can be assumed that the observed granulocytosis results from massive mobilization of the
reserve pool of the bone marrow (Dworecka *et al.*, 1982).
Secondary stimulation of kinetics of granulocyte proliferation
as well as the increased percentage of undifferentiated
granulocytes was noticed in bone marrow (Fig. 5).

This increased percentage of precursors and differentiated
forms of neutrophils in bone marrow was most pronounced on
the 2nd to 3rd day after single administration of COM-55
spores (Fig. 5). These findings confirm the acceleration of
recovery of this system after mobilization and release of the
reserve pool of granulocytes from bone marrow (Dworecka,
1981). Increased numbers of peroxidase-positive cells were
also observed in spleen on the 2nd to 3rd day after COM-55
spores injection. It was accompanied by slightly increased
splenic index (from 4.3 mg/g of body weight to 7.1 mg/g of
body weight). It suggests that COM-55 spores may also

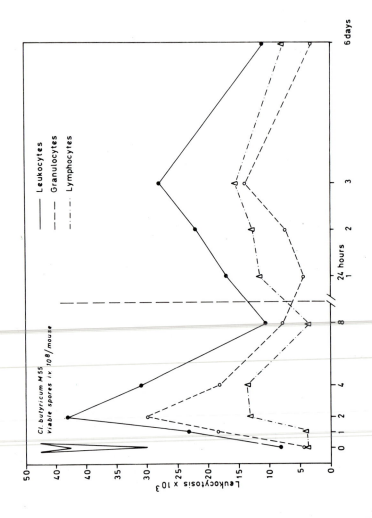

FIG. 4 *Number of leukocytes in peripheral blood of mice injected intravenously with a simple dose of 10^8 Cl. butyricum M 55 spores (Dworecka et al., 1982).*

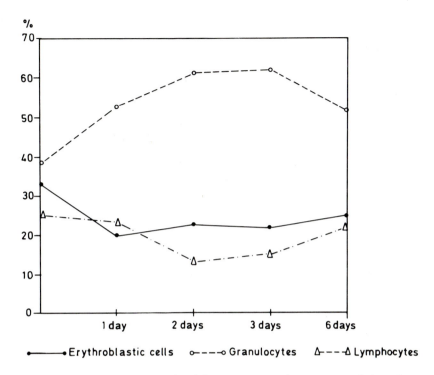

FIG. 5 *Per cent of erythroblasts, granulocytes and lympho-cytes in bone marrow of mice injected i.v. with 10^8 of Cl. butyricum M 55 spores (Dworecka* et al., *1982).*

stimulate other systems of phagocytic cells (macrophages) but
to a lesser extent than granulocytes. Analogous stimulation
of granulocyte system was noticed after the administration of
sterile supernatants of COM-55 cultures (two wave granulocytosis). This was not observed after the injection of killed
and washed COM-55 spores (Dworecka *et al.*, 1982). It points
to the possible significance of COM-55 extracellular products
in the stimulation of the granulocyte system. Potent and
selective stimulation of this system by COM-55 spores or bacterial extracellular products attracted our attention to the
possibility of applying COM-55 as a factor enhancing recovery
of impaired granulopoiesis. It is known that this impairment
is the most common and dangerous side-effect of radio- and
chemotherapy of cancer. Intravenous administration of COM-55
spores into mice sublethally irradiated with X-rays (250 or
400 R) significantly accelerated spontaneous recovery of
haemopoiesis (Dworecka *et al.*, 1982). Moreover, the postirradiation drop in the number of proliferating cells of

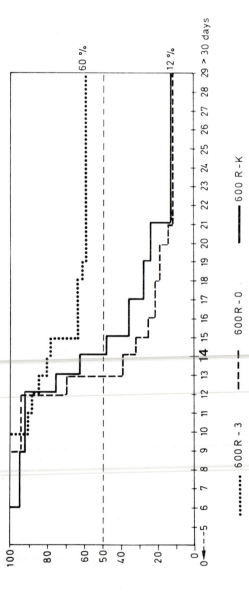

FIG. 6 Survival of mice irradiated with 600 R of X-rays and treated 3 days before or on the day of irradiation (O) with 10^8 viable spores of Cl. butyricum M 55 (Dworecka et al., 1982).

granulocyte system was significantly less pronounced if spores were given simultaneously with, or 3 days before irradiation. A prolongation of mean survival time and the increased number of animals surviving longer than 30 days was observed in the groups of mice irradiated with 500 — 750 R and receiving a single administration of COM-55 spores 3 days before X-ray exposure (Fig. 6). More detailed analysis of obtained results of various schedules of irradiation and spore administration demonstrated that the main mechanism of enhancement of haemopoiesis recovery caused by spores is the above mentioned acceleration of granulopoiesis (Dworecka *et al.*, 1982).

VI. SUMMARY

Undoubtedly, administration of viable spores of strain M 55 of *Clostridium butyricum* results in rapid and massive necrosis of experimental tumours, but it does not lead to a complete cure. Germination of spores seems to be essential in the mechanism of the above phenomenon. Germination can occur only in conditions of low oxygen concentration. The mechanism of oncolytic effect of COM-55 is still unclear. Bacterial extracellular products as well as participation of immunological mechanisms are suggested to play a significant role in this effect. Both these hypotheses have to be confirmed by further experimental studies.

Independently of the oncolytic effect, the bacteria stimulate the granulocyte system and accelerate its postirradiation recovery. It points to beneficial effects of experimental treatment with COM-55 spores in cases of impairment due to side-effects of radio- and chemotherapy.

At present, evaluation of the usefulness of COM-55 in cancer therapy is not equivocal, as many basic experimental studies have not yet been performed. Low toxicity of COM-55 seems to be of particular importance. On the other hand, massive necrosis may lead to serious complications. Moreover, this form of therapy can be dangerous in such cases when the localization of the tumour does not allow evacuation of the necrotic tissue. The therapy with COM-55 spores can be considered only as a potential adjuvant therapy in combination with other forms of treatment. Recently performed studies on the combination of COM-55 with radiotherapy and local hyperthermia look particularly promising. Only further studies will enable the evaluation of real usefullness of this type of combined therapy in clinical oncology.

VII. REFERENCES

Bergey, D.H., Buchanan, R.E. and Gibbsons, N.E. (1974). "Bergey's Manual of Determinative Bacteriology". 8th Edition. pp. 208-211. Williams and Wilkins, Baltimore.

Brantner, H. and Fischer, G. (1973). Untersuchung der Kininasen des onkolytisch wirksamen *Clostridium* Stammes M 55 ATCC 13732 mittels Chelatbildner. *Path. Microbiol.* **39**, 99-106.

Brantner, H. and Schwager, J. (1979). Enzymatische Mechanismen der Onkolyse durch *Clostridium* M 55 ATCC 13732. *Zbl. Bakt. Hyg. I Abt. Orig. A.* **243**, 113-118.

Brantner, H. and Schwager, J. (1980). A theoretical model of oncolysis by *Clostridium oncolyticum* M 55 ATCC 13722. *Arch. Geschwulstforsch.* **50**, 601-612.

Chagnon, A., Hudson, C., McSween, G., Vinet, G. and Fredette, V. (1972). Cytotoxicity and reduction of animal cell growth by *Clostridium* M 55 spores and their extracts. *Cancer* **29**, 431-434.

Cheung, A. and Samaras, G. (1981). Hyperthermia for cancer therapy. *J. Microwave Power* **16**, 85-238.

Currie, C. (1980). "Cancer and the Immune Response". pp. 1-129. E. Arnold, London.

Dietzel, F. and Gericke, D. (1977). Verstärkung der Clostridienonkolyse durch Hochfrequenzhyperthermie im Tierversuch. *Strahlentherapie* **153**, 263-264.

Dietzel, F., Gericke, D., Schumacher, L. and Linhart, G. (1978). Combination of radiotherapy, microwave hyperthermia and Clostridial oncolysis on experimental mouse tumours. *In* "Cancer Therapy by Hyperthermia and Radiation". (Ed. C. Streffer), pp. 233-235. Urban and Schwarzenburg, Baltimore — Munich.

Dworecka, B. (1981). Immunostimulative and antineoplastic properties of *Clostridium butyricum* s. *oncolyticum* M 55 in normal and tumour-bearing experimental animals. Ph.D. Thesis, National Institute of Hygiene, Warsaw.

Dworecka, B., Lipski, S. Szmigielski, S., Jeljaszewicz, J., and Pulverer, G. (1982). Effect of *Clostridium butyricum* s. *oncolyticum* M 55 on haemopoiesis and its recovery from postirradiation injury. *Exp. Pathologie*, In press.

Engelbart, K. and Gericke, D. (1964). Oncolysis by Clostridia. V. Transplanted tumours of the hamster. *Cancer Res.* **24**, 239-246.

Fischer, G., Brantner, H. and Platzer, P. (1975). The kininase activity of Ehrlich ascites solid tumour after treatment with oncolytic Clostridia. *Ztschr. Krebsforsch.* **84**, 203-206.

Fredette, V. and Plante, C. (1970). Oncolytic activity of Clostridium M 55 spores. *Canad. J. Microbiol.* **16**, 249-252.

Fredette, V. and McSween, G. (1974). Complete regression of Ehrlich's solid tumour in the mouse after intravenous injection of soluble products from *Clostridium oncolyticum* M 55. *Zbl. Bakt. Hyg., I. Abt. Orig. A* **240**, 257-263.

Gericke, D. and Engelbart, K. (1964). Oncolysis by Clostridia. II. Experiments on a tumour spectrum with a variety of Clostridia in combination with heavy metals. *Cancer Res.* **24**, 217-222.

Gericke, D., Dietzel, F., König, W., Rüster, I. and Schumacher, L. (1979). Further progress with oncolysis due to apathogenic clostridia. *Zbl. Bakt. Hyg., I. Abt. Orig. A* **243**, 102-112.

Haller, E.M. and Brantner, H. (1979). Enzymologische Untersuchungen am *Clostridium oncolyticum* M 55 ATCC 13732. *Zbl. Bakt. Hyg., I. Abt. Orig. A* **243**, 522-527.

Heppner, F. and Möse, J.R. (1966). Onkolyse maligner Gliome durch apathogene Clostridien Stamm M 55. *Zbl. Neurochirurgie* **37**, 11-16.

Kayser, D. (1963). Uber den Mechanismus der Abtötung von Ascites-Krebszellen durch *Clostridium butyricum*. *Z. Naturforsch.* **18**, 748-752.

Kayser, D. (1967). The use of spores in anaerobic bacteria for special investigations in cancer research. *In* "The Anaerobic Bacteria". pp. 68-72. International Workshop, University of Montreal, Laval-des Rapides.

Kretschmer, H. (1972). Die Behandlung maligner Geschwülste mit Sporen von *Clostridium butyricum*. I. Grundlagen und Entwicklung der Beeinflussung maligner Tumoren durch Mikroorganismen. *Arch. Geschulstforsch.* **39**, 185-194.

Kretschmer, H., Ludewig, R. and Hambsch, K. (1974). Therapeutic experiments with *Clostridium butyricum* spores and cyclophosphamide in carcinosarcoma in rats. *Arch. Geschwulstforsch.* **44**, 65-69.

Malmgren, R.A. and Flanigan, C.C. (1955). Localization of the vegetative forms of *Cl. tetani* in mouse tumours following intravenous spore administration. *Cancer Res.* **15**, 473-478.

Möse, J.R. (1960). Zur Beeinflussbarkeit verschiedener Tiertumoren durch einen apathogenen Clostridienstamm. *Ztschr. Krebsforsch.* **63**, 447-455.

Möse, J.R. and Möse, G. (1959). Onkolyseveruche mit apathogenen Sporenbildnern am Ehrlich Tumor der Maus. *Ztschr. Krebsforsch.* **63**, 63-74.

Möse, J.R. and Möse, G. (1964). Oncolysis by Clostridia. I. Activity of *Clostridium butyricum* M 55 and other nonpathogenic clostridia against the Ehrlich carcinoma. *Cancer Res.* **24**, 212-216.

Möse, J.R., Möse, G., Propst, A. and Fleppner, F. (1967).
Onkolyse maligner Tumoren durch den Clostridienstamm M 55
Med. Klinik. **62**, 220-225.

Möse, J.R., Brantner, H. and Reichenfelser, U. (1970). Untersuchungen über die Züchtung von *Clostridium butyricum*
M 55. *Zbl. Bakt. Hyg. I. Abt. Orig.* A **215**, 360-365. ..

Möse, J.R., Fischer, G. and Mobascherie, T.B. (1972a). Über
Bakterienkininase und deren physiologische Bedeutung.
Zbl. Bakt. Hyg. I. Abt. Orig. A **219**, 465-472.

Möse, J.R., Fischer, G. and Briefs, C. (1972b). Die Wirkung
von *Clostridium butyricum* (Stamm M 55) auf menschliches
Kininogen und ihre Bedeutung für den Onkolyseprozess.
Zbl. Bakt. Hyg. I. Abt. Orig. A **221**, 474-491.

Möse, J.R., Fischer, G. and Mobascherie, T.B. (1972c). Über
Bakterienkininasen und deren physiologische Bedeutung. I.
Untersuchungen an Clostridienstämmen. *Zbl. Bakt. Hyg. I.
Abt. Orig.* A **219**, 530-541.

Parker, R.C., Plummer, H.I., Siebenmann, C.O. and Chapmann,
M.G. (1947). Effect of histolyticus infection and toxin
on transplantable mouse tumours. *Proc. Soc. Exp. Biol.
Med.* **66**, 461-467.

Rousseau, P., Chagnon, A. and Fredette, V. (1970). Effect of
oncolytic anaerobic spores on animal cell cultures. *Cancer
Res.* **30**, 849-854.

Rüster, I. (1980). Heilversuche am Harding-Passey-Melanom
der Maus mit *Clostridium oncolyticum* sive butyricum (M 55)
Hochfrequenzhyperthermie und Röntgenstrahlen. Ein- und
mehrmalige Applikation. pp. 1-104. Ph.D. Thesis, J.W.
Goethe-Universität, Frankfurt/Main.

Schlechte, H., Baumbach, L. and Elbe, B. (1981). Cocultiva-
tion of *Clostridium oncolyticum* with normal and tumour
cell lines. *Arch. Geschwulstforsch.* **51**, 51-57.

Schwager, J. (1976). Vergleichende Untersuchungen an onkolytischen und nicht onkolytischen Clostridienstämmen.
pp. 1-123. Ph.D. Thesis, University of Graz.

Streffer, C. (Ed.) (1978). Cancer Therapy by Hyperthermia
and Radiation. pp. 1-328. Urban and Schwarzenberg,
Baltimore — Munich.

Thiele, E.H., Arison, R.N. and Boxer, G.E. (1964a). Oncolysis by Clostridia. III. Effects of Clostridia and
chemiotherapeutic agents on rodent tumours. *Cancer Res.*
24, 222-233.

Thiele, E.H., Arison, R.N. and Boxer, G.E. (1964b). Oncolysis
by Clostridia. IV. Effect of nonpathogenic Clostridial
spores in normal and pathological tissues. *Cancer Res.*
24, 234-238.

Urban, S., Wildner, G.P., Schneeweiss, U. and Fabricius, E.M. (1974). Das Verhalten von Clostridium butyricum in frühen Ehrlich Karzinoma. Experimentelle Untersuchungen zum Tumor-Clostridien-Phänomen. *Arch. Geschwulstforschung* **43**, 8-14.

Dr SEDLACEK: Dr Schwab, you have shown that bacterial substances may influence various parameters of the immune system.
Could you also show that those substances are immunosuppressive on the final response of the immune system, that means
whether you get reduction of antibody titre or increase or
suppression of cellular immune response or a reduced resistance to infection or any imbalance of the immune system?

Dr SCHWAB: Suppression of antibody response to unrelated
antigens, such as SRBC, by the cytoplasmic membrane preparation has been shown; and this is accompanied by an enhancement of DTH reaction to the same antigen (SRBC). Peptidoglycan-polysaccharide complexes of the cell wall can depress
CMI functions such as response to T-cell mitogens. The
important point is that agents such as whole bacterial cells
contain a variety of components which can modulate the immune
response. Thus, in the animal injected with something like
propionibacterial vaccines, either suppression or enhancement
is possible and this accounts for some of the disappointing
results in immunotherapy.

Dr CHEDID: Since natural antibodies against peptidoglycan
have been described in normal serum, have you determined
whether your animals did have such natural antibodies in view
of your complement activation experiments?

Dr SCHWAB: All immunologically mature animals have measurable
antibody to peptidoglycan in serum and the amount of antibody
is age-related. In our *in vivo* studies of changes in complement levels, alterations of total haemolytic complement were
paralleled by changes in ACP, C3 and factor D. Thus, while
some of the changes in serum complement may be due to immune
complex activation of the classical pathway, a part of the
change is due to activation of the alternate pathway by C3b
binding to peptidoglycan.

Dr KONDO: I agree with you that most bacterial preparations
activate both the alternative and partly the classical path-

way. But by repeated injections you cannot get a decrease of
serum complement level. In contrast, serum complement remains
at a high level. This indicates that the use of bacterial
products results in immunopotentiation rather than immunosuppression, since an initial drop of complement level is
limited to a short period.

Dr SCHWAB: As a generalization, immunosuppressive agents can
function also as immunoenhancers as circumstances such as
dose or timing of injection relative to antigen are changed.
However, injection of LPS 24 h before antigen introduction
will be suppressive and this may involve depression of C3
levels.

Dr PRAGER: In examining the suppression of lymphocyte proliferation in peptidoglycan treated animals, have you
examined techniques other than thymidine incorporation?
Since the peptidoglycan activates macrophages, they might
secrete thymidine which could give a false impression of
depressed DNA synthesis.

Dr SCHWAB: We have also demonstrated suppression of DTH skin
reaction in rats injected with peptidoglycan-polysaccharide
fragments.

Dr GOLDIN: Dr Baldwin, could you indicate the possible mechanism for BCG stimulation of tumour growth?

Dr BALDWIN: BCG exerts multiple effects upon host cells.
This includes the generation of activated macrophages which
may, under certain conditions, function to suppress NK and
T-cells. This has been shown for example in tests where
local BCG stimulation of a tumour draining lymph node can
allow the escape of tumour cells and so lead to the development of metastases.

Dr CHEDID: Dr Baldwin, must the antigen (PPD) be given *in
situ* to elicit the antitumour effect mediated by delayed
hypersensitivity? By what procedure was the monoclonal cell
coupled to FN? And was the interferon activity of the conjugate measured *in vitro*?

Dr BALDWIN: The tumour response initiated by PPD in BCG-
sensitized rats requires that PPD is localized at the tumour
site. This is viewed as sensitized T-cells responding to
PPD producing "factors" which lead to NK and macrophage mediated effects. Interferon is one of the agents we are studying for use with monoclonal antibodies as carriers. The
synthesis of IFN-antibody conjugates is carried out using the
SPDP reagent. Details have recently been presented (Baldwin
et al., 1982).

Dr MITCHELL: Dr Baldwin, why was your monoclonal antibody targeted to the tumour rather than the NK-cell if you want the IFN to be delivered to the latter?

Dr BALDWIN: The objective of the treatment was to target interferon to the tumour for the activation of host cells at the tumour site and to cause further extravasation to the tumour. In our experience, peripheral activation of NK-cells and/or macrophages without a method of attracting cells to the tumour has had limited therapeutic effect.

Dr MITCHELL: There is circumstantial histological evidence in allografts that T-cells act mainly through macrophages, rather than as cytotoxic T-cells. While I agree with your concept that recruitment by T-helper cells is most important in tumour rejection, I might caution that adoptive transfer experiments preferentially identify T-helper cells. These cells recruit other cells in the recipient and their presence would therefore be more easily detected. T-killer cells may require much larger numbers to be present in the injected population for their presence to be detected since they do not recruit additional cells very well.

Dr SEDLACEK: You showed that nonspecific immunostimulation might work in your experimental tumour system but you could not find any indication that CTL may play a role there. The conclusion would be that unspecific immunostimulation might be the candidate for doing the job. However, the vast majority of clinical studies show that BCG or *C. parvum* are not effective, or if at all, the difference to control is rather minor. Facing this discrepancy, don't you think that the tumour models we are using and you have used are not relevant for the clinical situation?

Dr BALDWIN: Our recent experience indicates that activation of host cells in tumour deposits by intralesional injection of agents such as BCG, a subcellular preparation including cell wall, cord factor, etc. leads to tumour rejection. This response involving macrophages and NK-cells can function without T-lymphocytes. However, the response is usually more effective if the tumour is immunogenic, so inducing a T-cell response. In considering these findings, our experience over more than 25 years is that the majority of naturally arising rat tumours do not elicit tumour rejection responses, i.e. these tumours do not express neoantigens — which function to induce T-cell responses. These findings have to be viewed in the context of human tumours where the evidence for tumour associated antigens which elicit T-cell response is equivocal. On this basis, an alternative approach has been developed in which T-cell responses can be generated within tumour deposits

to irrelevant antigens. This is exemplified by the rejection
of mammary carcinoma Sp15 in contact with PPD in BCG-sensitized rats.

The interpretatation of these data is that the local T-
cell response generated by PPD-induced delayed type hypersensitivity reaction leads to localization of cells (NK or
macrophages) which effect tumour rejection. These concepts
have application to clinical immunotherapy since it is also
feasible to elicit local responses in sensitized patients.
This is illustrated in a recent study where injection of BCG
into lung tumours prior to rejection led to an increase of
in situ NK-cell activity.

Dr KIRCHNER: I was impressed by Dr Baldwin's data indicating
the superior effectiveness of BCG in immune animals as compared to nonimmune animals. Do you consider it a possibility
that lymphokines (for example γ-IFN) play a major role?
Also your point that not the killer T-cells but the helper
T-cells play a prevalent role (even in allorecognition) is
extremely exciting. Perhaps you can give references.

Dr BALDWIN: Conceptually, we consider the effects mediated by T-lymphocytes sensitized to irrelevant antigens such
as those generated by PPD injected in admixture with tumour
cells in BCG-sensitized rats, function through the secretion
of lymphokines. This may include γ-IFN but there are as yet
no data on this point. The role of killer T-cells in tumour
rejection has been brought into question in our own studies
showing the lack of correlation between cells adoptively
transferring immunity and their *in vitro* cytotoxicity. This
is becoming clearer from the analysis of the role of T-cell
subsets isolated according to markers such as Ly 1,2 in the
mouse. Specific references to this point are summarized in
our recent review on immunotherapy (Baldwin, *Cancer Chemotherapy Annual* **4**, Excerpta Medica, Amsterdam, in press).

Dr GOLDIN: Dr Kirchner, could you comment on the cross reactivity of the various types of interferon including also
human and animal? If resistance occurs to one type of interferon, is there cross resistance to other types?

Dr KIRCHNER: Thus far, it appears that all known effects of
interferons are shared by all the known interferons. However
the receptor for interferon gamma appears to be different to
the common receptor for interferon alpha and beta. Perhaps
in the future we will learn that certain interferon subtypes
differ in certain effects from other interferon subtypes.
Recently, the rare occurrence of anti-interferon has been
shown (Vallbracht *et al.*, Nature). The one patient of
Vallbracht who developed antibodies against α-IFN was

subsequently successfully treated by interferon beta. It
appears that antibodies are more frequent in patients treated
now with recombinant bacterial interferon.

Dr EGGERS: What is the mechanism of the differential effect
of interferon(s) on tumour cells?

Dr KIRCHNER: It is far from being proven that interferon
acts selectively on tumour cells. In fact, interferon inhibits the proliferation of normal T-lymphocytes. Perhaps in
the future we will learn that interferon inhibits the proliferation of many different cell types and that the problem
will be similar to the one of classical chemotherapy. However, the clinical side-effects of even very high doses of
bacterial recombinant IFN appear to be moderate.

Dr SINKOVICS: I would like to bring up the question of
interferon resistance. This phenomenon was observed long ago
in France by Chany. We also observed it and reported on it
in "Experientia" in the 1960s. It apparently occurs as well
in patients treated with interferon. About one-third of the
patients show a tumour response at least temporarily; then
they relapse with growing tumours while they are still
receiving interferon. If the mechanism of interferon effect
is a direct antiproliferative effect, then loss of interferon
receptors, i.e. non-expression of these receptors, on the
surface of target cells would result in interferon resistance.
But even in this case, Dr Baldwin's monoclonal antibody-
coupled interferon would continue to work, because the binding to the target cell is accomplished by the monoclonal
antibody and not by the interferon. Therefore, activation of
antiproliferative mechanism and NK-cells can take place. But
if the mechanism of interferon resistance is acquisition by
the target cell of resistance to cell mediated cytotoxicity,
then even monoclonal antibody-delivered interferon would fail
to induce an antitumour effect. There is evidence that target cells treated with interferon may become resistant to
NK-cell mediated cytotoxicity. In our experience with osteogenic sarcoma 791-T the dose of interferon needed to activate
NK-cells (10^1–10^3) was lower than that reported to inhibit
target cell NK susceptibility. Nevertheless, the point
raised is important and must be borne in mind when evaluating
the influences of targeting IFN to tumour using monoclonal
antibody as a carrier.

Dr SIECK: Dr Kirchner, you mentioned that IFNs induced certain enzyme cascades. Which kind of enzymes are involved
and are there differences between the different types of IFN?
Are products of the enzyme reactions related to the different
antitumour actions?

Dr KIRCHNER: The enzyme cascade induced by interferon treatment in cells are well studied and have been reviewed. There
are ways to explain the antiviral effects of IFN in at least
some viral systems by the enzyme induction cascade. As far
as the antiproliferative effect of IFN is concerned, its cell-
biological basis is poorly characterized and it is not known
if the enzyme cascade is in part responsible for the antiproliferative effect.

Dr MITCHELL: Human alveolar macrophages from normal individuals do not kill recently explanted tumour cells and cannot
be stimulated to do so by extracted alpha-IFN *in vitro*.
Alveolar macrophages from cancer patients kill spontaneously
and can often be potentiated by IFN. Is there evidence for
the necessity of 2 signals in order for IFN to activate
macrophages?

Dr KIRCHNER: It is important to stress that interferons
activate not only NK-cells but also macrophages. Future
experiments will require pure interferons and highly purified
populations of NK-cells or macrophages to see if, perhaps,
there are differences in cell subpopulations or IFN subpopulations.

Dr KIRN: We obtained similar results to those reported by
Dr Mitchell. Until now we have been unable to "activate"
Kupffer cells with interferon *in vitro*.

Dr KIRCHNER: There is certainly a good possibility that
there are subpopulations of macrophages that are nonresponsive to interferon. Of course, the "natural" IFN (alpha)
contains all alpha-IFN subspecies and perhaps there are more
than additive effects when several subspecies are added
together. However, in order to learn about the exact mechanism, first of all we have to test the subspecies of IFN
alone on pure populations of cells. Later, we can speak
about the effects of combinations of different IFNs.

Dr KIRN: Dr Kirchner, I agree with you that LPS is a poor
interferon inducer; however, it depends what kind of cells
are challenged. For example, Kupffer cells react very poorly
to endotoxin, whereas endothelial liver cells produce high
amounts of interferon with endotoxins. This may be explained
by the fact that endotoxin is detoxified *in vivo* by Kupffer
cells, so that cells should also be able to detoxify it *in
vitro*, whereas endothelial cells do not. Why is there such
a lag phase between the treatment of the animals with *C. parvum* and the efference of interferon in the spleen cell culture? How do you explain that *C. parvum* may induce interferon in leukocytes and not in macrophages?

Dr KIRCHNER: The activation of the lymphoreticular system
by *C. parvum* takes several days and in most investigations it
is observed between 7 and 14 days after injection. Our data
seem to suggest that B-lymphocytes (that are also activated
mitogenically by *C. parvum*) are the producers of interferon
and not the macrophages. True, it is very important to study
the effects of LPS on other cell types besides leukocytes.
As far as leukocytes are concerned, LPS (even if highly purified) is a poor inducer of interferon.

Dr BAŠIĆ: It has been shown recently that pyrimidinone compounds have a strong antitumour effect in mice and that they
induce high levels of interferon in lymphoid organs, lungs
and plasma of treated animals. Does interferon, which is
produced *in vivo* by these compounds or by others such as CP,
directly inhibit tumour cell proliferation or does it work
through the stimulation of antitumour response of the host?

Dr KIRCHNER: The compounds you are referring to are relatively weak inducers of interferon when compared to CMA, for
example. In regard to the general question, how does interferon work against tumours *in vivo*, there are at least 3
known effects of interferon which may contribute:

a) the direct antiproliferative effect on tumour cells;
b) the activation of certain effector cells of the defence
   system;
c) very recently it has been shown that interferon-treatment
   of tumour cells causes a partial reversion of the malignant phenotype.

It will take considerable further work to sort out the relative contribution of these different effects.

Dr PULVERER: Dr Kirchner, acridin orange (AO) is a potent
stimulator of lysogeny at least in staphylococci. Furthermore, staphylococcal phages show differing resistance to this
substance.

Dr KIRCHNER: This is a very interesting comment. I was not
aware of this finding. However, there are numerous effects
of AO, many of which are probably caused by the interaction
of AO with macromolecules of the cell. Why certain AO derivatives induce interferon (by which mechanisms) is essentially unknown.

Dr HILL: Dr Jeljaszewicz, although Clostridial oncolysis may
have no place in curative treatment of cancer might it have
a role in, for example, paliative treatment of large obstructing colorectal tumours that are inoperable? Alternatives
are chemotherapy or radiotherapy, both of which are extremely

debilitating. Perhaps oncolysis might be a more acceptable
treatment to the patient and perhaps this is worth investi-
gating and perhaps oncolysis should not be abandoned completely yet.

Dr JELJASZEWICZ: Too little is known about what oncolytic
Clostridia do in the animal organism, when destructing part
of the tumour mass. Of course, application in such cases as
colorectal carcinoma, when other treatment has failed, may be
of interest. However, animal experiments are not conclusive
enough, and application of live bacteria to humans seems now
much too risky. As for application of Clostridia for paliative therapy in cases not reacting to all other types of
therapy, other and much better means are available, as for
instance, intratumoural injection of propionibacteria in
colorectal carcinoma when radio- and chemotherapy and surgery
are of no further help.

Dr BAŠIĆ: You have pointed out that the mechanism of antitumour activity of *Cl. butyricum* spores is mainly due to the
necrosis of tumour tissue. Since the amount of spores in
tumour tissue was pretty low as shown by radiolabelled spores
it is likely that other mechanisms could also be involved.
Has anybody tried to isolate an active antitumour compound
from those spores? Are disintegrated spores or killed ones
as active as those which give rise to live bacteria?

Dr JELJASZEWICZ: No active compound from spores, vegetative
cells or an extracellular product from Clostridia has been
defined according to contemporary criteria of biochemistry
as an antitumour agent of this origin. Especially, the
spores were not tested. Extracellular products have been
only partially purified or even not at all. One cannot suggest at this moment a substance from Clostridia of this
nature. Extensive work on protease from Clostridium M 55 is
actually being done now.

TUMOUR IMMUNOTHERAPY USING
VIBRIO CHOLERAE NEURAMINIDASE (VCN)

H.H. Sedlacek, E. Weidmann and F.R. Seiler

Research Laboratories of Behringwerke AG,
3550 Marburg/Lahn, FRG

I. INTRODUCTION

During the last 2 decades, immunotherapy has been the slogan
of experimental as well as clinical pioneers in the field of
tumour therapy due to the following reasons: limitations in
the success achieved with the conventional weapons against
cancer, i.e. surgery, radiotherapy and chemotherapy. With
these procedures, the so-called minimal residual disease —
that is the growth of metastases from single tumour cells
spread by the primary tumour — could only be impaired very
insignificantly in the vast majority of tumour diseases.
Thus, research was directed towards the possibilities of
killing specifically those single tumour cells. Based on the
hypothesis that tumour cells quantitatively and/or qualita-
tively differ from normal cells in their cell surface antigenicity, the immune system was investigated for its potential to control tumour growth and to kill tumour cells.
Indeed, in a large number of experimental tumour systems and
even in some selected clinical tumour diseases, manipulation
of the immune system could be shown to have some influence on
the subsequent tumour growth. Even tumour immunotherapy
proved to be at least experimentally successful. However,
clinical studies revealed that the tumour immunotherapy
results gained in experimental tumour systems could not be
transferred to human diseases. Actually, in a vast number
of randomized clinical studies, unspecific or specific (i.e.
immunomodulators in a mixture with tumour antigens or tumour
cells) immunostimulation with bacterial substances — like
BCG, *C. parvum*, MER from BCG or with other compounds such as
levamisole — failed to show any success on the growth of
tumours.

Discussions arose whether the experimental tumour systems, chosen to prove the tumour pharmacologic effect of the respective compound preclinically, were relevant to human tumour diseases. Most of the immunomodulators exhibited tremendous side-effects on the patients. Consequently, the difficulty in proving clinical effectivity of the various immunomodulators and the risk of harming the patient led to the general feeling that immunotherapy as a whole is the wrong way to improve tumour therapy. This attitude, however, tempts one to overlook and to disregard those very compounds and principles which indeed have shown to be effective both preclinically and clinically and which might give the basis for further work to improve tumour immunotherapy and to elaborate its limitations.

The aim of this article is to review and to summarize the preclinical and the clinical data of the specific tumour immunotherapy using *Vibrio cholerae neuraminidase* (VCN). It will be shown that this principle is effective in transplantable tumour systems, in autochthonous spontaneous animal tumours, and in selected human tumour diseases according to randomized prospective studies which have been performed so far. Moreover, a certain immunization procedure called chessboard vaccination, which minimizes the risk of specific tumour immunotherapy to induce tumour enhancement and which therefore can be recommended as an alternative for the injection of VCN-treated cells, will be discussed.

II. EFFECT OF VCN ON THE IMMUNE SYSTEM

The basic hypothesis of the effect of VCN on the immune system was deduced from the finding that VCN treatment of normal and tumour cells could increase their antigenicity *in vitro* and their immunogenicity *in vivo*. It could be shown in some cases that this increase was induced by the potentiation of the cell-specific antigens or cell-specific receptors or by the occurrence of new antigenic determinants (see Table 1). The Thomsen-Friedenreich antigen (Thomsen, 1927) (the so-called T-antigen) belongs to these new antigenic determinants. The relevance of it as a tumour-associated antigen is now being much discussed. On the other hand, VCN itself could be shown to adhere very strongly to the cell membrane (Sedlacek and Seiler, 1974a) and experimental data exist that this sticking of the exogeneous neuraminidase to the cell membrane might indeed also give new antigenic determinants to the cells (Johannsen and Sedlacek, 1974). As cell-membrane-attached VCN also proved to be enzymatically active, the speculation was whether VCN is able to influence the immune system itself. As outlined in Table 1, enzymatically active

VCN is actually able to increase the antigen- or mitogen-
induced lymphocytic transformation rate, to stimulate autologous or allogeneic MLC, to increase macrophage phagocytosis,
macrophage-lymphocyte interaction and sensitivity of lymphocytes to TAF (T-cell activating factor) activity.

As a result of those immunological activities, VCN functions as an adjuvant, which — in a mixture with an antigen —
enhances the humoral and cellular immune response against
this antigen in animal and man. This adjuvant activity of
VCN depends on the dose, on the enzymatic activity and on the
condition that VCN is injected into a mixture with the respective antigen. Under these prerequisites, VCN even enhances
the immune response against BCG (Seiler and Sedlacek, 1980)
or breaks tolerance to albumin. However, it could also be
elaborated that VCN is not active as an adjuvant in mice
which are unspecifically immunostimulated for instance by BCG
(Sedlacek and Seiler, 1978a; Seiler and Sedlacek, 1980).

III. EXPERIMENTAL TUMOUR IMMUNOTHERAPY WITH VCN

In consequence of the finding that the antigenicity and
immunogenicity of tumour cells can be enhanced by neuraminidase, experiments were performed to treat established transplantable experimental tumour diseases of various types by
the injection of tumour cells treated with VCN. In order to
avoid artificial metastases by these injected tumour cells,
their proliferation was stopped by adequate *in vitro* pretreatment with irradiation or cytostatics. Indeed, significant therapeutical effects could be found in experimental
leukaemia as well as in solid tumours (see Table 2). When
using the same or additional tumour models, however, other
groups also found no effect or even an enhancing effect on
tumour growth in this form of therapy (see Table 3). In
subsequent experiments in an autochthonous spontaneous mammary tumour in dogs, it could be shown that the application
of VCN-treated autologous tumour cells was therapeutically
effective even in this tumour system. However, no effect or
stimulation of tumour growth could also be found. The outcome
of this form of therapy strictly depended on the number of
VCN-treated tumour cells injected. For instance, the injection
of 2×10^7 VCN-treated tumour cells revealed to be effective
whereas 1×10^8 tumour cells caused tumour enhancement
(Sedlacek *et al.*, 1975, 1979).

An overall analysis (Sedlacek and Seiler, 1978b) of all
the experiments related to tumour immunotherapy using VCN-
treated tumour cells revealed the following conditions for
success:

TABLE 1

Effect of VCN on the immune system

Increase of cell antigenicity (in vitro)

Leukaemia	man	Kassulke *et al.*, 1971
HLA	man	Grothaus *et al.*, 1971
AML	man	Reed *et al.*, 1974
Bronchus carcinoma	man	Watkins *et al.*, 1971, 1974
Skin melanoma	man	
Hypernephroma	man	
Adenocarcinoma of breast, lung, colon, rectum, gastric carcinoma	man	Akiyoshi *et al.*, 1980
Lewis lung adenocarcinoma	mouse	Alley and Snodgrass, 1977
MC squameous cell carcinoma	mouse	

Occurrence of new antigenic structures

T-antigen	guinea pig	Sanford and Codington, 1971a
	mouse	Sanford and Codington, 1971b
	man	Springer *et al.*, 1974, 1975;
		Springer and Desai, 1975a,b,c
	mouse	Parish *et al.*, 1981
Ia-antigens of foreign haplotype	man	Johannsen and Sedlacek, 1974
Neuraminidase		

Increase of cell immunogenicity (in vivo)

Ehrlich ascites	mouse	Lindenmann and Klein, 1967;
		Sur and Roy, 1979
TA₃ ascites adenocarcinoma	mouse	Sanford, 1967
Landschütz ascites	mouse	Currie, 1967
		Brazil and McLaughlin, 1978

TABLE 1 Contd.

Effect of VCN on the immune system

L1210 leukaemia	mouse	Bagshawe and Currie, 1968; Bekesi and Holland, 1971
AKR leukaemia	mouse	Bekesi and Holland, 1975
MC-42, -43, -10	mouse	Simmons and Rios, 1971, 1974; Faraci *et al.*, 1975
Mast cell tumour	mouse	Sedlacek and Seiler, 1974b
Ependymoblastoma (GL-26)	mouse	Albright *et al.*, 1975
DMBDN fibrosarcoma	mouse	Ray and Sundaram, 1974; Ray *et al.*, 1975, 1976
B16 melanoma	mouse	Jamieson, 1974
Ni3S2-fibrosarcoma	rat	Abandowitz, 1978
E2G leukaemia	mouse	Bekesi and Holland, 1975
Lymphocytes (transplantation antigens)	mouse	Simmons *et al.*, 1971b
Increase of IgG-Fc receptors		
Lymphocytes	man	Seiler and Sedlacek, 1972, 1974
T-cells	man	Schulof *et al.*, 1980; Itoh and Kumagai, 1980
Increase of binding affinity to heterologous and autologous red blood cells of lymphocytes		
Lymphocytes	man	Seiler and Sedlack, 1974
Increase of antigen induced lymphocytic transformation rate	man	Han, 1973, 1975; Johannsen *et al.*, 1978; Pauley *et al.*, 1978

TABLE 1 Contd.

Effect of VCN on the immune system

Increase of mitogen (Con A, PWM) induced lymphocytic transformation rate	man	Han, 1973, 1975, 1978 Johannsen et al., 1978 Pauley et al., 1978
Stimulation of autologous and allogeneic MLC after treatment of stimulator and/or responder cell	rat man	Fidler et al., 1973 Flye et al., 1973; Han, 1972; Johannsen et al., 1975, 1976 Lundren and Simmons, 1971
	mouse	Pauley et al., 1978 Lundgren et al., 1971
	rat	Beyer and Bowers, 1978
Increase of lymphocyte cytotoxicity (liver)	mice	Kolb-Backofen and Kolb, 1979
Increase of differentiation of memory T-cells into cytotoxic effectors (together with galactose oxydase)	mice	Kuppers and Henny, 1977
Increase of complement dependent cytolysis of target cells	mouse man	Ray et al., 1971 Ray and Simmons, 1972 Tompkins et al., 1979
Increase of macrophage phagocytosis	mouse	Lee, 1968
Passive	man	Weiss et al., 1966
Active	mouse	Knop et al., 1978a

TABLE 1 Contd.

Effect of VCN on the immune system

Increase of macrophage cytotoxicity (passive)	mouse	Sethi and Brandis, 1973a,b
Increase of cell adhesion	amoeba	Ray and Challerjee, 1975
	man } mouse}	Weiss *et al.*, 1966
Increase of macrophage—lymphocyte cooperation	mouse	Knop, 1980a
Increase of sensitivity of lymphocytes to TAF activity	mouse	Knop, 1980b
Increase of secretion of MIF and LIF by lymphocytes (together with galactose oxidase)	man	Greineder *et al.*, 1979
Increase of lymphocyte growth factor secretion (together with galactose oxidase)	man	Novogrodsky *et al.*, 1980
Adjuvans activity for antigens (BSA, Rubella, Virus, SRBC, *E. coli, Vibrio cholerae*)	mouse	Knop *et al.*, 1978b
Salmonella typhimurium	mouse	Johannsen and Seiler, 1980
Increase of DTH response		
SRBC	mouse	Sedlacek and Seiler, 1978a
BCG	mouse	Seiler and Sedlacek, 1980
	guinea pig	Ronneberger and Johannsen, 1979
various recall antigens	man	Han, 1974

TABLE 2

Effective tumour immunotherapy with VCN-treated cells

Tumour type	Histology	Animal species	References
Chemically induced not metastasizing	Fibrosarcoma (various types)	Mouse	Simmons et al., 1971a,b,c,d; Rios and Simmons, 1972,1973,1974; Ray et al., 1976; Ray and Seshadri, 1979; Wilson et al., 1974 Song and Levitt, 1975;
	Yoshida sarcoma	Rat	Ray and Seshadri, 1979
	Squamous cell carcinoma	Mouse	Alley and Snodgrass, 1977
Transplantable, metastasizing, unknown origin	B16 melanoma	Mouse	Rios and Simmons, 1973
	Lewis lung	Mouse	Alley and Snodgrass, 1977; Sedlacek et al., 1980*
	Dunn osteosarcoma	Mouse	Miller et al., 1976
	Melanoma	Guinea pig	Egeberg and Jensen, 1974
Spontaneous leukaemia, transplantable	Lymphatic (L1210)	Mouse	Bekesi et al., 1971; Lefever et al., 1976; Killion, 1977
Virus-induced	Mammary adenocarcinoma	Mouse	Simmons and Rios, 1973, 1976
Virus-induced leukaemia	AKR leukaemia	Mouse	Bekesi and Holland, 1975
Spontaneous, not transplantable, partially metastasizing	Mammary tumour (various histological types)	Dog	Sedlacek et al., 1975, 1979, 1981*

*Chessboard vaccination.

TABLE 3

Ineffective tumour immunotherapy with VCN-treated cells

Tumour type	Histology	Animal species	References
Chemically induced, not metastasizing	Fibrosarcoma (various types)	Mouse	Spence et al., 1978; Wilson et al., 1974*
	Fibrosarcoma hepatoma mammary carcinoma	Rat	Pimm et al., 1978
	Glioma	Mouse	Albright et al., 1975
	E14 lymphoma	Mouse	Ghose et al., 1977
Unknown origin, transplantable partially metastasizing	B16 melanoma	Mouse	Jamieson, 1974; Froese et al., 1974*
	Lewis lung adenocarcinoma	Mouse	Sedlacek et al., 1980
	Epithelioma, fibrosarcoma	Rat	Pimm et al., 1978
Virus-induced	Polyoma virus-induced adeno-carcinoma	Mouse	Porwit-Bobr et al., 1974
Spontaneous leukaemia, transplantable	Lymphatic (L1210)	Mouse	Killion, 1977*
Virus-induced leukaemia	Rauscher leukaemia	Mouse	Barinskii and Kobrinskii, 1977
	AKR leukaemia	Mouse	Mathé et al., 1973; Doré et al., 1973*
Spontaneous, not transplantable, partially metastasizing	Mammary tumour (various histological types)	Dog	Sedlacek et al., 1975*

*Enhancement of tumour growth after injection of VCN-treated cells.

1) The injected cells have to be identical in type with the tumour (Simmons and Rios, 1971, 1972, 1974) or with common (for instance viral-induced) membrane antigens (Bekesi and Holland, 1975, 1977).

2) Enzymatically active VCN (Bekesi *et al.*, 1971; Simmons *et al.*, 1971a,c; Simmons and Rios, 1971) has to be used.

3) This VCN has to be applied in an optimal concentration (Bekesi and Holland, 1975; Sethi and Brandis, 1973a,b; Simmons *et al.*, 1971a,b,c).

4) The tumour mass of the tumour bearer has to be as small as possible (Rios and Simmons, 1972, 1973, 1974; Porwit-Bobr *et al.*, 1974).

5) The tumour bearer has to be immunocompetent.

6) Tumour immunotherapy has to be done with an optimal cell dose (Wilson *et al.*, 1974; Sedlacek *et al.*, 1975).

7) Chemotherapy in addition to immunotherapy might improve the results in case the type of drugs, the dose and the treatment intervals are optimized in so far as the chemotherapy is effective on the growth of the tumour without destroying immunological reactivity of the body (Bekesi *et al.*, 1971; Sedlacek *et al.*, 1980).

It is evident that the results of this analysis indeed gives the basis for successful experimental design of specific tumour immunotherapy with VCN-treated tumour cells in the future, but it does not help to avoid the risk of inducing tumour enhancement by the injection of an inadequate number of tumour cells. It is necessary to establish a method to pre-evaluate just this number which induces tumour regression or tumour enhancement and which might depend on type and stage of tumour disease, on the general condition of the tumour bearer and on any additional therapy. Up to now there is no way to define just this cell dose by adequate *in vitro* test. However, an immunization procedure has been developed which seems to minimize the danger of tumour enhancement. This immunization procedure is called chessboard vaccination and has been developed (Seiler and Sedlacek, 1978) on the basis of the experiments in which VCN has been proven to work as a repository adjuvant for the immune response against various antigens. The procedure was stimulated by the design of Bekesi and Holland (1975, 1978) to use just this number of VCN-treated cells in AML patients for immunotherapy, which induces a maximal DTH response in the skin. The chessboard vaccination consists of the intradermal injection of various mixtures of different numbers of tumour cells

(10^5 — 10^8 mitomycin-treated tumour cells) combined with different amounts of VCN (0, 0.65, 6.5, 65 mU). All the different mixtures are simultaneously injected separately from each other in one tumour bearer.

The results in the Lewis lung adenocarcinoma tumour system in mice (Sedlacek *et al.*, 1980) and in spontaneous mammary tumours in dogs (Sedlacek *et al.*, 1981) reveal that the chessboard vaccination is therapeutically effective, that this effectivity is tumour specific and dependent on enzymatically active VCN and that the effectivity seems to be superior to conventional immunotherapy with VCN-treated tumour cells. Important seems to be the risk of inducing tumour enhancement minimized by using chessboard vaccination, but it is not completely abrogated — at least under special experimental conditions in the rat (Pimm and Baldwin, private communication).

IV. CLINICAL STUDIES

First clinical trials to induce regression of solid tumours were performed by the groups of Seigler *et al.* (1972) and of Rosato *et al.* (1974, 1976). Seigler *et al.* treated 7 patients with melanoma who had recurrent disease and multiple metastases. Autologous VCN-treated tumour cells were subcutaneously injected simultaneously with BCG. In 6 of those 7 patients a complete regression of the tumour mass could be found. However, this result could only partially be attributed to the injection of VCN-treated cells since these patients had also been pretreated with various doses of BCG either alone or in combination with autologous lymphocytes, sensitized *in vitro* with autologous melanoma cells. Rosato *et al.* (1974) injected 5 x 10^6 VCN-treated irradiated autologous tumour cells intramuscularly every month for 6 months into 25 patients with various tumours (breast, gastrointestinal, uterus, kidney), who mostly were admitted to the study because of the advanced nature of the disease and the unresponsiveness to other forms of treatment. As no adverse clinical effect and no obvious toxicity but, however, an *in vitro* enhanced cell-mediated cytotoxicity against target cells could be found, Rosato *et al.* (1974) extended their study to 45 patients suffering from various solid tumours. Again they found *in vitro* as well as clinical indications for effectivity of this form of immunotherapy.

Parallel to those very early investigations, prospective and controlled and randomized studies with VCN-treated tumour cells were performed by Holland and Bekesi (1978) in AML and by Simmons *et al.* (1978) in melanoma. Experiences gained in the experimental work of the respective groups were

TABLE 4

Clinical trials with Neuraminidase (VCN)

Tumour (Stage)	No. of patients	Study design	Chemotherapy	Immunotherapy	Results Interim report	Results Final report	Author
AML (No.1)	14	rando. 2 groups	Ara-C and DNR; Ara-C and TG /CP/DNR	I) – II) 10^{10} VCN-blasts (allog.) 48 injection sites; every mth.		Increase in survival and re-mission duration	Bekesi and Holland, 1978 (Buffalo)
AML (No.2)	53	rando. 3 groups	Ara-C and DNR; Ara-C and TG /CP/DNR/CCNU	I) – II) VCN-blasts (allog.) III) VCN-blasts and MER (1 mg)		Increase in survival and remission duration (II) MER (III) abrogates the therapeutic effect of (II)	Bekesi and Holland, 1978 (New York)
AML (No.3)	84	rando. 2 groups	Ara-C and DNR; Ara-C/TG/VC /DM/DNR	I) – II) VCN-blasts (allog.)	Increase in survival and remission duration (II)		Bekesi and Holland, 1978 (New York)
AML	61	rando. 2 groups	Ara-C and DNR (ADM); Ara-C and TG	I) – II) Splenectomy III) 10^{10} VCN-blasts (allog.) 48 injection sites; every mth.	Increase in survival and remission duration (III)		Wiernik *et al.*, (Baltimore) (personal communication)
AML	14	Pilot	Ara-C and ADM and TG/CP/CCNU	Chessboard (2×10^9 blasts (allog.) 5-50 mU VCN)		Improvement compared to historical control possible	Lutz *et al.*, 1981 (Vienna)

TABLE 4 Contd.

Clinical trials with Neuraminidase (VCN)

Tumour (Stage)	No. of patients	Study design	Chemotherapy	Immunotherapy	Interim report	Results Final report	Author
AML	26	rando. 2 groups	Ara-C and DNR; COA/TRA/POM	I) – II) 5×10^8 VCN-blasts (allog.) every 4 wks.		No difference to controls	Rühl *et al.*, 1981 (Berlin)
AML		rando. 2 groups	Ara-C and DNR and TG; Ara-C and DNR /TG/CP	I) – II) 10^{10} VCN-blasts (allog.) 48 injection sites; every mth.	Started 1982		Büchner *et al.*, (Münster) (personal communication)
AML	29	rando. 2 groups	Ara-C and ADM; VCR Ara-C and TG	I) – II) 10^9 blasts (allog.) and 80 mU VCN	No difference to control group		Rosenfeld (EORTC) (Paris) (personal communication)
Melanoma out-treated	7	Pilot	different	VCN-cells (autolog.) and BCG		Tumour regression in 6 Pat.	Seigler *et al.*, 1972
Melanoma Stage II	22	rando. 2 groups	no	I) – II) 2×10^8 VCN-cells (autolog.) and BCG (3×10^8)		No difference	Simmons *et al.*, 1978 (Minneapolis)
Melanoma Stage III	36	rando. 2 groups	DTIC	I) – II) 2×10^8 VCN-cells (autolog.) and BCG (3×10^8)		No difference	Simmons *et al.*, 1978 (Minneapolis)

TABLE 4 Contd.

Clinical trials with Neuraminidase (VCN)

Tumour (Stage)	No. of patients	Study design	Chemotherapy	Immunotherapy	Results Interim report	Results Final report	Author
Melanoma Stages I and II	174	rando. 4 groups	Me.- CCNU (group IV)	I) - II) BCG III) BCG and VCN-cells (allog. cultured cells) IV) Me.- CCNU		No difference	Terry *et al.*, 1980 (Bethesda)
Various solid tumours out-treated	45	Pilot	different	1x10^8 VCN-cells (autolog.) monthly for 6 mth.		Retardation of tumour growth, increase of Ly.- cytotoxicity	Rosato *et al.*, 1974; 1976 (Philadelphia)
Colon-Ca (Duke B, C)	104	rando. 3 groups		I) - II) MMC and 5 FU and Ara-C III) Chessboard (3x) (3-100 mU VCN; 10^5-10^8 cultured cells	Tendency in favour of III not significant		Rainer *et al.*, 1981 (Vienna)
Colon	240	rando. 3 groups		I) - VCN treated autolog. cells II) VCN treated autolog. cells and BCG		Started in 1980; no difference to controls	Gray *et al.*, (Melbourne) (personal communication)

TABLE 4 Contd.

Clinical trials with Neuraminidase (VCN)

Tumour (Stage)	No. of patients	Study design	Chemotherapy	Immunotherapy	Interim report	Results Final report	Author
Gastric Cancer -resect- able		rando. 2 groups	No	I) -- II) Chessboard with VCN and autolog. cells	Started in 1981		Gürsel *et al.*, (Istanbul) (personal communication)
-resect- able and metastases		rando. 2 groups	No	I) -- II) Chessboard with VCN and autolog. cells	Started in 1981		Gürsel *et al.* (Istanbul) (personal communication)
Breast cancer		rando. 2 groups	CFVMP	I) -- II) Chessboard with VCN and allog. cultured cells	Started in 1981		Margreiter *et al.*, (Innsbruck) (personal communication)
Prostata carcinoma out-treated	34	Pilot	different	Multiple chess-board with VCN and autolog. cells		Significant de-crease of Prosta-taphosphatase CEA tumour growth retardation?	Rothauge *et al.*, 1981 (Giessen)
Stage III		rando. 2 groups	--	I) -- II) Chessboard (2x) with VCN and autolog. cells	Started in 1981		Rothauge *et al.*, 1981 (Giessen)

transferred directly to the clinical protocol. Thus, Bekesi
et al. chose such a type, dose and interval of the maintenance chemotherapy in their AML patients that according to DTH
testing and *in vitro* parameters, the immune system of the
patients remained reactive. Immunotherapy consisting of
intradermal injection of 10^{10} VCN-treated allogeneic blast
cells at 48 injection sites on day 15 of the maintenance
course was repeated every month. Simmons *et al.* administered
2×10^8 VCN-treated autologous tumour cells in combination
with BCG (applied via scarification over the inoculation
site) and in addition to surgery and — in the case of advanced
disease — to chemotherapy. The combination with BCG was performed because the preclinical data elaborated by this group
revealed a synergistic effect of specific immunotherapy with
VCN-treated tumour cells and unspecific immunotherapy with
BCG (Simmons and Rios, 1972). While Simmons *et al.* failed
to show any significant therapeutical effect of the combination immunotherapy on survival of disease-free interval in
melanoma patients, Bekesi *et al.* succeeded in prolonging
remission duration and survival time of AML patients.

Subsequently, pilot studies and controlled randomized
clinical trials in AML patients as well as in some selected
solid tumour diseases (melanoma, colon carcinoma, gastric
cancer, breast cancer and prostate cancer) were started.
Interim reports or even final reports for some of those
studies are already available (Table 4). They allow the
following conclusions.

1) The therapeutical success Bekesi *et al.* have found in AML
   with VCN-treated allogeneic blasts could be confirmed in
   subsequent studies by them, and in the meantime also pre-
   sumptively by others (Wiernik *et al.*, submitted for publication). Thus, this therapeutical regimen really seems
   to be of advantage for the patients. However, the success
   is strictly dependent on the number of VCN-treated blasts
   injected after each maintenance chemotherapy. Reduction
   of this number, as done by Rühl *et al.* (1981) results in
   no effectivity.

2) The success of specific tumour immunotherapy with VCN-
   treated cells can be completely deteriorated by combination with unspecific immunotherapy, for instance with MER
   or BCG as shown by Bekesi and Holland (1978). This may
   be in agreement with the adjuvant experiments (Sedlacek
   and Seiler, 1978a, Seiler and Sedlacek, 1980). The sub-
   tractive effect of unspecific immunotherapy on the specific one may at least partially be the reason for the
   negative outcome of the studies by Simmons *et al.* (1978)
   in melanoma patients.

3) Up to now there is no clear proof in AML that the chessboard vaccination may be an effective alternative to the
   cell-spending immunization procedure of Bekesi and Holland
   (1978). This question has to be answered in an adequate,
   controlled study. The lack of difference to historical
   control, found in a pilot study (Lutz, 1981) with altogether 14 patients who underwent chessboard vaccination,
   does not answer this question.

4) To date no final report exists of a randomized controlled
   study in solid tumours, which proves the effectivity of
   specific tumour immunotherapy with VCN. This is true of
   the conventional injection of VCN-treated cells as well as
   of the chessboard vaccination. In melanoma patients, cell
   studies showed lack of effectivity of this treatment,
   partially and possibly due to combination with BCG. In
   carcinoma of the stomach, colon, breast and prostate, the
   various studies with the chessboard vaccination have just
   been started. Interim reports on colon cancer exhibit a
   slight tendency in favour of the chessboard vaccination
   (Rainer *et al.*, 1981) compared to the various control
   groups, but this difference is not yet statistically significant (see Table 5). Moreover, the pilot study in out-
   treated patients with prostate cancer (Gutschank *et al.*,
   1981; Rothauge *et al.*, 1981) also seems to show a tumour
   therapeutical effect of the chessboard vaccination (see
   Table 6). However, this tendency has to be proven by an
   adequate controlled study. Nevertheless, it can already
   be stated that side-effects induced by the injection of
   VCN-treated cells or by the chessboard vaccination are
   minimal and can be neglected. Any indication pointing to
   tumour enhancement as a result of this form of therapy
   could not yet be found. Thus, it seems to be justified
   to prosecute the clinical studies in order to evaluate the
   potential of the specific tumour immunotherapy using VCN.
   Any clear-cut result in those clinical studies would help
   us to improve tumour therapy today and to find our way for
   further research on specific and active tumour immunotherapy.

V. SUMMARY

Immunological as well as tumour immunological activities of
Vibrio cholerae neuraminidase (VCN) and VCN-treated cells
are reviewed. The following conditions for a successful
tumour immunotherapy could be shown: identity of injected
cells or common membrane antigens with the tumour cells, use
of optimal concentrations of enzymatically active VCN, use

TABLE 5

Colon carcinoma stage Duke B and C

Treatment	Patients	Relapse	Death
Control (I)	33	8	5
Chemotherapy with MMC, 5-FK, Ara-C (II)	36	5	2
Immunotherapy with VCN-treated cells (III)	35	4	2

Differences between all groups n.s.
Rainer *et al.*, Clinic for Chemotherapy, Vienna, 1981.

TABLE 6

Chessboard vaccination with autologous tumour cells

	Patients	Pre-treatment	Post-treatment
Prostataphosphatase k/l	34	2.5	1.05
Acid serumphosphatase k/l	34	9.4	5.2
Prostataphosphatase RIA ng/ml	34	23.1	10.7

Mean of all patients: $p < 0.02$; $p < 0.01$
cf. Rothauge *et al.*, Diagnostik and Intensivtherapie **1**, 1-5 (1981).

of an optimal cell number, immunocompetence of the tumour host and minimal tumour burden at the time of treatment. A beneficial effect could also be shown when chemotherapy was used with immunotherapy.

Tumour enhancement by VCN-treated cells, shown in various animal models, could be avoided by using the chessboard vaccination procedure.

Clinical trials are on their way in different tumours, e.g. AML, cancer of the stomach, prostate, colon and breast.

Interim results are not yet available for clinical use of the chessboard vaccination. However, in the treatment of colon cancer, there is a trend in favour of the immunotherapy group. The simultaneous use of specific and unspecific immunotherapy, especially the use of BCG or MER, abrogates the beneficial effects of VCN-treated cells as shown in preclinical as well as clinical studies.

VI. REFERENCES

Abandowitz, H.M. (1978). Neuraminidase effect on the growth of a transplantable nickel sulfide-induced rat tumour. *Japan J. Med. Sci. Biol.* **31**, 421.

Akiyoshi, T., Kawaguchi, M., Miyazaki, S. and Tsuji, H. (1980). Lymphocyte blastogenic response to autologous tumour cells pretreated with neuraminidase in patients with gastric carcinoma. *Oncology* **37**, 309.

Albright, L., Madigan, J.C., Gaston, M.R. and Houchens, D.P. (1975). Therapy in an intracerebral murine glioma model, using bacillus Calmette-Guerin, neuraminidase-treated tumour cells, and 1-(2-chloroethyl)-3-cyclohexyl-1-nitrosourea. *Cancer Res.* **35**, 658.

Alley, C.D. and Snodgrass, M.J. (1977). Effectiveness of neuraminidase in experimental immunotherapy of two murine pulmonary carcinomas. *Cancer Res.* **37**, 95.

Bagshawe, K.D. and Currie, G.A. (1968). Immunogenicity of L1210 murine leukaemia cells after treatment with neuraminidase. *Nature (Lond.)* **218**, 1254.

Barinskii, I.F. and Kobrinskii, G.D. (1977). Inhibitory action of neuraminidase of *Vibrio cholerae* in Rauscher mouse leukaemia. *Bull. Exp. Biol. Med.* **82**/11, 1697.

Bekesi, J.G. and Holland, J.F. (1975). Chemoimmunotherapy of leukaemia in man and experimental animals. Int. Conf. on Immunotherapy of Cancer, Nov. 5-7, New York.

Bekesi, J.G. and Holland, J.F. (1977). Active immunotherapy in leukaemia with neuraminidase-modified leukaemic cells. *Recent Results Cancer Res.* **62**, 78.

Bekesi, J.G. and Holland, J.F. (1978). Immunotherapy of acute myelocytic leukaemia with neuraminidase-treated myeloblast and MER. Proc. Symp. "Immunotherapy of Malignant Diseases", p.375, Schattauer Verlag, Stuttgart, New York.

Bekesi, J.G., Arneault, St. and Holland, J.F. (1971). Increase of leukaemia L1210 immunogenicity by *Vibrio cholerae* neuraminidase treatment. *Cancer Res.* **31**, 2130.

Beyer, C.F. and Bowers, W.E. (1978). Lymphocyte transformation induced by chemical modification of membrane components. II. Effect of neuraminidase treatment of responder cells on proliferation and cytotoxicity in indirect stimulation. *J. Immunol.* **121**, No. 5, 1790.

Brazil, J. and McLaughlin, H. (1978). The modification of the growth characteristics of the Landschütz ascites tumour by neuraminidase. *Europ. J. Cancer* **14**, 757.

Currie, G.A. (1967). Masking of antigens on the Landschütz ascites tumour. *Lancet* II, 1336.

Doré, J.F., Hadjiyannakis, M.J., Guibout, C., Coudert, A., Marholev, L. and Imai, K. (1973). Use of enzyme-treated cells in immunotherapy of a murine leukaemia. *Lancet* VIII, 600.

Egeberg, J. and Jensen, O.A. (1974). The effect of neuraminidase-treated tumour cells on the growth of transplantable malignant melanoma of the Syrian golden hamster (mesocricetus auratus). *Int. Rev. Cytol. Suppl.* **2**, 1573.

Faraci, R.P., Marrone, A.C. and Ketcham, A.S. (1975). Antitumour immune response following injection of neuraminidase treated sarcoma cells. *Ann. Surgery* **181**, 359.

Fidler, I.J., Kahn, J.M. and Montgomergy, P.C. (1973). Effect of neuraminidase on the rat one-way mixed lymphocyte interaction. *Immunol. Commun.* **2**, 573.

Flye, M.W., Reisner, E.G. and Amos, D.B. (1973). The *in vitro* effect of neuraminidase on human lymphocytes. *J. Surg. Res.* **15**, 214.

Froese, G., Berczi, I. and Sehon, A.H. (1974). Brief communication: Neuraminidase enhancement of tumour growth in mice. *J. Natl. Cancer Inst.* **52**, 1905.

Ghose, T., Guclu, A., Tai, J., Mammen, M. and Norvell, S.T. (1977). Immunoprophylaxis and immunotherapy of EL4 lymphoma. *Eur. J. Cancer* **13**, 925.

Greineder, D.K., Rocklin, R.E. and David, J.R. (1979). Production of human lymphocyte mediators after activation with periodate or neuraminidase and galactose oxidase. *J. Immunol.* **123**, No. 6, 2804.

Grothaus, E.A., Flye, M.W., Yunis, E. and Amos, B.D. (1971). Human lymphocyte antigen reactivity modified by neuraminidase. *Science* **173**, 54.

Gutschank, S., Rothauge, C.F., Kraushaar, J., Gutschank, W. and Sedlacek, H.H. (1981). Das entgleiste metastasierende Prostatakarzinom. *Munch. Med. Wschr.* **4**, 133.

Han, T. (1972). Enhancement of mixed lymphocyte reactivity by neuraminidase. *Transplantation* **14**, 515.

Han, T. (1973). Enhancement of *in vitro* lymphocyte response by neuraminidase. *Clin. Exp. Immunol.* **13**, 165.

Han, T. (1974). Enhancement of delayed skin hypersensitivity by neuraminidase in cancer patients. *Clin. Exp. Immunol.* **18**, 95.

Han, T. (1975). Specific effect of neuraminidase on blastogenic response of sensitized lymphocytes. *Immunology* **28**,
283.

Han, T. (1978). Neuraminidase-mediated enhancement of lymphocyte response to pokeweed mitogen: Exclusive action of
neuraminidase on responding T-lymphocytes. *IRCS Med. Sci.*
6, 152.

Holland, J.F. and Bekesi, J.G. (1978). Comparison of chemotherapy with chemotherapy plus VCN-treated cells in acute
myelocytic leukaemia. *In* "Immunotherapy of Cancer: Present Status of Trials in Man". (Eds W.D. Terry and D.
Windhorst), p. 347. Raven Press, New York.

Itoh, K. and Kumagai, K. (1980). Effect of tunicamycin and
neuraminidase on the expression of Fc-IgM and -IgG receptors on human lymphocytes. *J. Immunol.* **124**, 4, 1830.

Jamieson, C.W. (1974). Enhancement of antigenicity of syngeneic murine melanoma by neuraminidase. Int. Cancer
Congress, Florence, (Workshop 1, Abstracts, p. 233).

Johannsen, R. and Sedlacek, H.H. (1974). Specificity of
cytotoxic antibodies to autologous human lymphocytes
treated with neuraminidase from *Vibrio cholerae*. *Behring
Inst. Mitt.* **55**, 209.

Johannsen, R. and Seiler, F.R. (1980). Effective immunostimulation in experimental *S. typhimurium* infection in
the mouse. 4th Int. Congress of Immunology, Paris,
July 21-26, 1980 (Abstract No. 17.3.23).

Johannsen, R., Carlsson, A.B. and Sedlacek, H.H. (1975). *In
vitro* transformation of human lymphocytes by neuraminidase
from *Vibrio cholerae* (VCN). 6th Workshop on Leukocyte
Cultures. Basel, March 17-19, 1975. (Abstracts, p. 86).

Johannsen, R., Knop, J., Sedlacek, H.H. and Seiler, F.R.
(1976). Augmentation of *in vitro* lymphocyte response by
neuraminidase. *Z. Immun. Forsch.* **152**, 90.

Johannsen, R., Sedlacek, H.H. and Seiler, F.R. (1978).
Adjuvant effect of *Vibrio cholerae* neuraminidase on the
in vitro and *in vivo* immune response. *In* Proc. Symp.
"Immunotherapy of Malignant Diseases". p. 244, Schattauer
Verlag, Stuttgart, New York.

Kassulke, J.T., Stutman, O. and Yunis, E.J. (1971). Blood-
group isoantigens in leukaemic cells. Reversibility of
isoantigenic changes by neuraminidase. *J. Natl. Cancer
Inst.* **46**, 1201.

Killion, G.J. (1977). The immunotherapeutic value of a
L1210 tumou- cell vaccine depends upon the expression of
cell-surface carbohydrates. *Cancer Immunol. Immunother.*
3, 87.

Knop, J. (1980a). Effect of *Vibrio cholerae* neuraminidase on the mitogen response of T-lymphocytes. I. Enhancement of macrophage T-lymphocyte cooperation in Concanavalin A-induced lymphocyte activation. *Immunobiology* **157**, 474.

Knop, J. (1980b). Effect of *Vibrio cholerae* neuraminidase on the mitogen response of T-lymphocytes. II. Modulation of the lymphocyte response to macrophage released factors by neuraminidase. *Immunobiology* **157**, 486.

Knop, J., Ax, W., Sedlacek, H.H. and Seiler, F.R. (1978a). Effect of *Vibrio cholerae* neuraminidase on the phagocytosis of *E. coli* by macrophages *in vivo* and *in vitro*. *Immunology* **34**, 555.

Knop, J., Sedlacek, H.H. and Seiler, F.R. (1978b). Stimulatory effect of *Vibrio cholerae* neuraminidase on the antibody response against various antigens. *Immunology* **34**, 181.

Kolb-Bachofen, V. and Kolb, H. (1979). Autoimmune reactions against liver cells by syngeneic neuraminidase-treated lymphocytes. *J. Immunol.* **123**, 6, 2830.

Kuppers, R.C. and Henney, C.S. (1977). The effects of neuraminidase and galactose oxidase on murine lymphocytes. *J. Immunol.* **119**, 6, 2163.

Lee, A. (1968). Effect of neuraminidase on the phagocytosis of heterologous red cells by mouse peripheral macrophages. *Proc. Soc. Exp. Biol. Med.* **128**, 891.

Lefever, A.V., Killion, J.J. and Kollmorgen, G.M. (1976). Active immunotherapy of L1210 leukaemia with neuraminidase-treated drug-resistant L1210 sublines. *Cancer Immunol. Immunother.* **1**, 211.

Lindenmann, J. and Klein, P.A. (1967). Immunological aspects of viral oncolysis. *Recent Results Cancer Res.* **9**, 66.

Lundgren, G. and Simmons, R.L. (1971). Effect of neuraminidase on the stimulatory capacity of cells in human mixed lymphocyte cultures. *Clin. Exp. Immunol.* **9**, 6, 915.

Lundgren, G., Jeitz, L., Lundin, L. and Simmons, R.L. (1971). Increased stimulation by neuraminidase-treated cells in mixed lymphocyte cultures. *Fed. Proc.* **30**, 395.

Lutz, D. (1981). Klinik und Immunologie bei akuten myeloischen Leukamien unter einer Chemo-Immunotherapie. *Onkologie* **4**, 202.

Mathé, G., Halle-Pannenko, O. and Bourut, C. (1973). Active immunotherapy of AKR mice with spontaneous leukaemia. *Rev. Eur. Etudes Clin. Biol.* **17**, 997.

Miller, C.W., DeBlasi, R.F. and Fisher, S.J. (1976). Immunological studies in murine osteosarcoma. *J. Bone Joint Surg. (Am.)* **58**, 312.

Novogrodsky, A., Suthanthiran, M., Saltz, B., Newman, D., Rubin, A.L. and Stenzel, K.H. (1980). Generation of a lymphocyte growth factor by treatment of human cells with neuraminidase and galactose oxidase. *J. Exp. Med.* **151**, 755.

Parish, C.R., Higgins, T.J. and McKenzie, I.F.C. (1981). Lymphocytes express Ia antigens of foreign haplotype following treatment with neuraminidase. *Immunogenetics* **12**, 1.

Pauly, J.L., Germain, M.J. and Han, T. (1978). Neuraminidase alteration of human lymphocyte reactivity to mitogens, antigens and allogenic lymphocytes. *J. Medicine* **9**, 3, 223.

Pimm, M.V., Cook, A.J. and Baldwin, R.W. (1978). Failure of neuraminidase treatment to influence tumourigenicity of syngeneically transplanted rat tumour cells. *Eur. J. Cancer* **14**, 869.

Porwit-Bobr, Z., Slowik, M. and Tomecki, J. (1974). Effect of neuraminidase-treated and mitomycin C-treated polyoma tumour cells on the established tumour growth in CBA mice. I. An attempt at evaluation of polyoma tumour destruction using the distribution of lissamine green method. *Folia Histochem. Cytochem. (Krakow)* **12**, 315.

Rainer, H., Kovats, E., Lehmann, H.G., Micksche, M., Rauhs, R., Sedlacek, H.H., Seidl, W., Schemper, M., Schiessl, R., Schweiger, B. and Wunderlich, M. (1981). Effectiveness of postoperative adjuvant therapy with cytotoxic chemotherapy (cytosine arabinoside, mitomycin C, 5-fluorouracil) or immunotherapy (neuraminidase-modified allogeneic cells) in the prevention of recurrence of Duke's B and C colon cancer. *Recent Results Cancer Res.* **79**, 41.

Ray, P.K. and Simmons, R.L. (1971). Failure of neuraminidase to unmask allogeneic antigens on cell surface. *Proc. Soc. Exp. Biol. Med.* **138**, 600.

Ray, P.K. and Simmons, R.L. (1972). Comparative effect of viral and bacterial neuraminidase on the complement sensitivity of lymphoid cells. *Clin. Exp. Immunol.* **10**, 139.

Ray, P.K. and Sundaram, K. (1974). Neuraminidase induced immunotherapy of cancer. Int. Cancer Congress Florence, 1974.

Ray, P.K. and Challerjee, S. (1975). Neuraminidase treatment enhances the lysolecithin induced intercellular adhesion of Amoeba proteus. *Z. Naturforsch.* **30C**, 551.

Ray, P.K. and Seshadri, M. (1979). Inhibition of growth of rat Yoshida sarcoma using a neuraminidase treated tumour vaccine. *Indian J. Exp. Biol.* **17**, 36.

Ray, P.K., Gewurz, H. and Simmons, R.L. (1971). Complement sensitivity of neuraminidase treated lymphoid cells. *Transplantation* **12**, 327.

Ray, P.K., Thankur, V.C. and Sundaram, K. (1975). Antitumour immunity. I. Differential response of neuraminidase-treated and X-irradiated tumour vaccine. *Eur. J. Cancer* **11**, 1.

Ray, P.K., Thakur, V.S. and Sundaram, K. (1976). Antitumour immunity. II. Viability, tumourgenicity and immunogenicity of neuraminidase-treated tumour cells: effective immunization of animals with a tumour vaccine. *J. Natl. Cancer Inst.* **56**, 83.

Reed, R.C., Guttermann, J.U., Mavligit, G.M. and Hersh, E.M. (1974). Sialic acid on leukaemia cells: Relation to morphology and tumour immunity (37896). *Proc. Soc. Exp. Biol. Med.* **145**, 790.

Rios, A. and Simmons, R.L. (1972). Comparative effect of *mycobacterium bovis-* and neuraminidase-treated cells on the growth of established methylcholanthrene fibrosarcomas in syngeneic mice. *Cancer Res.* **3**, 16.

Rios, A. and Simmons, R.L. (1973). Immunospecific regression of various syngeneic mouse tumours in response to neuraminidase-treated tumour cells. *J. Natl. Cancer Inst.* **51**, 637.

Rios, A. and Simmons, R.L. (1974). Active specific immunotherapy of minimal residual tumour: Excision plus neuraminidase-treated tumour cells. *Int. J. Cancer* **13**, 71.

Ronneberger, H. and Johannsen, R. (1979). BCG sensitisation of guinea pigs. An animal model for demonstrating the immunostimulating effects of various substances. *Int. J. Immunopharmacol.* **1**, 113-118.

Rosato, F.E., Brown, A.S., Miller, E.E., Rosato, E.F., Mullis, W.F., Johnson, J. and Moskovitz, A. (1974). Neuraminidase immunotherapy of tumours in man. *Surg. Gynaec. Obstet.* **139**, 675.

Rosato, R.E., Miller, E., Rosato, E., Brown, A., Wallack, M., Johnson, J. and Moskovitz, A. (1976). Active specific immunotherapy of human solid tumours. *Ann. N.Y. Acad. Sci.* **277**, 332.

Rothauge, C.F., Kraushaar, J. and Gutschank, S. (1981). Spezifische immunologische Intensivtherapie des entgleisten metastasierenden Prostata-Karzinoms. *Diagnostik & Intensivtherapie* **1**, 1-5.

Rühl, H., Fülle, H.H., Kopeen, K.M. and Schwerdtfeger, R. (1981). Adjuvant specific immunotherapy in maintenance treatment of adult acute non-lymphocytic leukaemia. *Klin. Wochenschr.* **59**, 1189.

Sanford, B.H. (1967). An alteration in tumour host compatibility induced by neuraminidase. *Transplantation* **5**, 1273.

Sanford, B.H. and Codington, J.F. (1971a). Alteration of the tumour cell surface by neuraminidase. *Transplant. Proc.* **3**, 1155.

Sanford, B.H. and Codington, J.F. (1971b). Further studies on the effect of neuraminidase on tumour cell transplantability. *Tissue Antigens* **1**, 153.

Schulof, R.S., Fernandes, G., Good, R.A. and Gupta, S. (1980). Neuraminidase treatment of human T-lymphocytes: Effect on Fc-receptor phenotype and function. *Clin. Exp. Immunol.* **40**, 611.

Sedlacek, H.H. and Seiler, F.R. (1974a). Demonstration of *Vibrio cholerae* neuraminidase (VCN) on the surface of VCN-treated cells. *Behring Inst. Mitt.* **55**, 254.

Sedlacek, H.H. and Seiler, F.R. (1974b). Dose dependency of the effect of syngeneic, *Vibrio cholerae* neuraminidase treated mastocytoma cells on the life span of DBA f/2 mice. *Behring Inst. Mitt.* **55**, 343.

Sedlacek, H. and Seiler, F.R. (1978a). Effect of *Vibrio cholerae* neuraminidase on the cellular immune response *in vivo*. In Proc. Symp. "Immunotherapy of Malignant Diseases" p. 268, Schattauer-Verlag, Stuttgart, New York.

Sedlacek, H.H. and Seiler, F.R. (1978b). Immunotherapy of neoplastic diseases with neuraminidase: contradictions, new aspects, and revised concepts. *Cancer Immunol. Immunother.* **5**, 153.

Sedlacek, H.H., Meesmann, H. and Seiler, F.R. (1975). Regression of spontaneous mammary tumours in dogs after injection of neuraminidase-treated tumour cells. *Int. J. Cancer* **15**, 409.

Sedlacek, H.H., Weise, M., Lemmer, A. and Seiler, F.R. (1979). Immunotherapy of spontaneous mammary tumours in mongrel dogs with autologous tumour cells and neuraminidase. *Cancer Immunol. Immunother.* **6**, 47.

Sedlacek, H.H., Bengelsdorff, H.J. and Seiler, F.R. (1980). Minimal residual disease may be treated by chessboard vaccination with *Vibrio cholerae* neuraminidase (VCN) and tumour cells. *In* "Metastasis Clinical and Experimental Aspects" (Eds K. Hellmann, P. Hilgard and S. Eccles), p. 310. Martinus Nijhoff Publ., The Hague, Boston, London.

Sedlacek, H.H., Hagmeyer, G. and Seiler, F.R. (1981). Tumour immunotherapy using the adjuvant neuraminidase. *In* Proc. "Int. Symp. Immunomodulation by Microbial Products and Synthetic Compounds", Osaka, Japan, July 27-29, 1981, Excerpta Medica, In press.

Seigler, H.F., Shingleton, W.W., Metzgar, R.S., Buckley, C.E., Bergoc, P.M., Miller, D.S., Fetter, B.F. and Phaup, M.B. (1972). Nonspecific and specific immunotherapy in patients with melanoma. *Surgery* **72**, 1, 162.

Seiler, F.R. and Sedlacek, H.H. (1972). Über die Brauchbarkeit immunologischer Nachweismethoden zur Differenzierung
funktionell verschiedener Lymphozyten: Spontanrosetten,
Komplement-rezeptor-Rosetten, und Immunglobulinrezeptoren.
Behring Inst. Mitt. **52**, 26.

Seiler, F.R. and Sedlacek, H.H. (1974). Alterations of
immunological phenomena by neuraminidase: Marked rise in
the number of lymphocytes forming rosettes or bearing
immunoglobulin receptors. *Behring Inst. Mitt.* **55**, 258.

Seiler, F.R. and Sedlacek, H.H. (1978). Chessboard vaccination: A pertinent approach to immunotherapy of cancer
with neuraminidase and tumour cells? *In* Proc. Symp.
"Immunotherapy of Malignant Diseases", p. 479, Schattauer-
Verlag, Stuttgart, New York.

Seiler, F.R. and Sedlacek, H.H. (1980). BCG versus VCN:
The antigenicity and the adjuvant effect of both compounds.
Recent Results in Cancer Research **75** (Eds G. Mathe and
F.M. Muggia), Springer-Verlag, Berlin, Heidelberg.

Seiler, F.R., Sedlacek, H.H., Kanzy, E.J. and Lang, W. (1972).
Über die Brauchbarkeit immunologischer Nachweismethoden
zur Differenzierung funktionell verschiedener Lymphozyten:
Spontanrosetten Komplementrezeptorrosetten in Immunglobulinrezeptoren. *Behring Inst. Mitt.* **52**, 26.

Sethi, K.K. and Brandis, H. (1973a). Synergistic cytotoxic
effect of macrophages and normal mouse serum on neuraminidase-treated murine leukaemia cells. *Eur. J. Cancer* **9**,
809.

Sethi, K.K. and Brandis, H. (1973b). Neuraminidase-induced
loss in the transplantability of murine leukaemia L1210,
induction of immunoprotection and the transfer of induced
immunity to normal DBA/2 mice by serum and peritoneal
cells. *Br. J. Cancer* **27**, 106.

Simmons, R.L. and Rios, A. (1971). Immunotherapy of cancer:
Immunospecific rejection of tumours in recipients of
neuraminidase-treated tumour cells plus BCG. *Science* **174**,
591.

Simmons, R.L. and Rios, A. (1972). Immunospecific regression
of methylcholanthrene fibrosarcoma using neuraminidase.
III. Synergistic effect of BCG and neuraminidase-treated
tumour cells. *Ann. Surg.* **176**, 188.

Simmons, R.L. and Rios, A. (1973). Differential effect of
neuraminidase on the immunogenicity of viral associated
and private antigens of mammary carcinomas. *J. Immunol.*
111, 1820.

Simmons, R.L. and Rios, A. (1974). Immunospecific regression
of methylcholanthrene fibrosarcoma with the use of neuraminidase. V. Quantitative aspects of the experimental
immunotherapeutic model. *Isr. J. Med. Sci.* **10**, 925.

Simmons, R.L., Rios, A., Lundgren, G., Ray, P.K., McKhann, C.F. and Haywood, G. (1971a). Immunospecific regression of methylcholanthrene fibrosarcoma using neuraminidase. *Surgery* **70**, 38.

Simmons, R.L., Rios, A. and Ray, P.K. (1971b). Immunogenicity and antigenicity of lymphoid cells treated with neuraminidase. *Nature (Lond.)* **23**, 179.

Simmons, R.L., Rios, A., Ray, P.K. and Lundgren, G. (1971c). Effect of neuraminidase on growth of a 3-methylcholanthrene-induced fibrosarcoma in normal and immunosuppressed syngeneic mice. *J. Natl. Cancer Inst.* **47**, 5, 1087.

Simmons, R.L., Aranha, G.V., Gunnarsson, A., Grage, T.B. and McKhann, C.F. (1978). Active specific immunotherapy for advanced melanoma utilizing neuraminidase-treated autochthonous tumour cells. *In* "Immunotherapy of Cancer: Present Status of Trials in Man". (Eds W.D. Terry and D. Windhorst), p. 123, Raven Press, New York.

Song, C.W. and Levitt, S.H. (1975). Immunotherapy with neuraminidase treated tumour cells after radiotherapy. *Radiat. Res.* **64**, 485.

Spence, R.J., Simon, R.M. and Baker, A.R. (1978). Failure of immunotherapy with neuraminidase-treated tumour cell vaccine in mice bearing established 3-methylcholanthrene-induced sarcomas. *J. Natl. Cancer Inst.* **60**, 451.

Springer, G.F. and Desai, P.R. (1975a). Increase in anti-T titre scores of breast carcinoma patients following mastectomy. *Naturwisschenschaften* **62**, 587.

Springer, G.F. and Desai, P.R. (1975b). Depression of Thomsen-Friedenreich (anti-T) antibody in humans with breast carcinoma. *Naturwisschenschaften* **62**, 302.

Springer, G.F. and Desai, P.R. (1975c). Human blood group MN and precursor specificities: Structural and biological aspects. *Carbohydr. Res.* **40**, 183.

Springer, G.F., Desai, P.R. and Banatwala, I. (1974). Blood group MN-specific substances and precursors in normal and malignant human breast tissues. *Naturwissenschaften* **61**, 457.

Springer, G.F., Desai, P.R. and Banatwala, I. (1975). Brief communication: Blood group MN antigens and precursors in normal and malignant human breast glandular tissue. *J. Natl. Cancer Inst.* **54**, 335.

Sur, P. and Roy, d.K. (1979). Antigenicity of Ehrlich ascites through neuraminidase treatment of cells. *Indian J. Exp. Biol.* **17**, 953.

SEDLACEK *et al.*

Terry, W.D., Hodes, R.J., Rosenberg, S.A., Fisher, R.I., Makuch, R., Gordon, H.G. and Fisher, S.G. (1980). Treatment of stage I and II malignant melanoma with adjuvant immunotherapy of chemotherapy: Preliminary analysis of a prospective, randomized trial. *In* 2nd Int. Conf. "Immunotherapy of Cancer: Present Status of Trials in Man". (Eds W.D. Terry and D. Windhorst), Raven Press, New York.

Thomsen, O. (1927). Ein vermehrungsfähiges Agens als Veränderer des isoagglutinatorischen Verhaltens der roten Blutkörperchen, eine bisher unbekannte Quelle der Fehlbestimmungen. *Z. Immunitaetsforsch.* **52**, 85.

Tompkins, W.A.F., Schmale, J.D., Mock, R., Tick, N.R., Lock, T., Sidell, N. and Palmer, J.L. (1979). Neuraminidase-augmented anticarcinoembryonic antigen (CEA)-complement lysis of human colon tumour cells: Lack of correlation with anti-CEA-mediated K-cell lysis and iodine-125 binding. *J. Natl. Cancer Inst.* **62**, 3, 503.

Watkins, E. Jr. (1974). Neuraminidase accentuation of cancer cell immunogenicity. *Behring Inst. Mitt.* **55**, 355.

Watkins, E. Jr., Ogata, Y., Anderson, L.L., Watkins, E. III. and Waters, M.R. (1971). Activation of host lymphocytes cultured with cancer cells treated with neuraminidase. *Nature New Biol.* **231**, 20, 83.

Watkins, E. Jr., Gray, B.N., Anderson, L.L., Baralt, O.L., Nebril, L.R., Waters, M.F. and Connery, C.K. (1974). Neuraminidase-mediated augmentation of *in vitro* immune response of patients with solid tumours. *Int. J. Cancer* **14**, 799.

Weiss, L., Mayhew, E. and Ulrich, K. (1966). The effect of neuraminidase on the phagocytic process in human monocytes. *Lab. Invest.* **15**, 1304.

Wilson, R.E., Sonis, S.T. and Godrick, E.A. (1974). Neuraminidase as an adjunct in the treatment of residual systemic tumour with specific immune therapy. *Behring Inst. Mitt.* **55**, 334.

ANTITUMOUR AND OTHER EFFECTS OF ASPARAGINASE

Morton D. Prager, F. Samuel Baechtel and Aaron Heifetz

Departments of Surgery and Biochemistry,
University of Texas Health Science Centre at Dallas,
Dallas, Texas 75235, USA

I. INTRODUCTION

Studies of asparaginase as an antitumour agent have been the subject of a number of reviews (Capizzi *et al.*, 1970; Cooney and Handschumacher, 1970; Prager, 1971; Patterson, 1975; Oettgen, 1975). This presentation, therefore, makes no pretence at being exhaustive but rather emphasizes certain issues which have been of particular interest to the authors. The major effort with the enzyme derives from the identification by Broome (1961, 1963a) of asparaginase as the antileukaemic agent in guinea pig serum which had been described by Kidd (1953). This finding gave heightened significance to the observation that certain malignant cells failed to grow in tissue culture medium devoid of asparagine (McCoy *et al.*, 1956, 1959). Broome (1963b) further demonstrated that as the asparaginase sensitive 6C3HED lymphoma underwent a selection process which permitted its growth in asparagine depleted medium, it lost susceptibility *in vivo* to asparaginase administration. Although these findings were of great fundamental importance, they did not reach the point of practical application until the discovery by Mashburn and Wriston (1964) that asparaginase from *E. coli* also possessed antileukaemic activity. The bacterial source provided the potential of a supply adequate for therapeutic trials and has indeed become the principal enzyme used in clinical application.

II. SOURCES OF ASPARAGINASE

Asparaginase is widely distributed in nature, and enzymes from many sources have been tested for antitumour activity (Prager, 1971; Patterson, 1975). Most such trials were

conducted with lymphomas or leukaemias in rodents. Because
a significant number of asparaginases failed to alter the
growth of susceptible tumours, the issue of whether asparaginase activity coincided with antitumour activity had to be
addressed. If an enzyme preparation is to produce a therapeutic response, ability to hydrolyze asparagine (Asn) is a
necessary but insufficient characteristic. Additional advantageous properties are significant catalytic activity and
stability under physiologic conditions, an appreciable biologic half-life, and capacity to effect hydrolysis of Asn at
a significant rate at the concentrations of substrate (40 —
70 μM) found in blood. With regard to the latter point, however, asparaginases with relatively high Km values have been
reported to demonstrate antitumour activity (Jayaram *et al.*,
1968; Peterson and Ciegler, 1969). The mostly widely used
enzyme source is *E. coli* which makes 2 asparaginases, only
one of which has antitumour activity (Roberts *et al.*, 1966;
Campbell *et al.*, 1967). The form producing tumour regression
and commonly referred to as EC-2, has properties consistent
with those just enumerated.

III. ASPARAGINASE SUBSTRATES

For EC-2 the physiologically significant substrates are L-asparagine (Asn) and L-glutamine (Gln), the latter undergoing
cleavage at 2—3% the rate of the former. The evidence
strongly indicates that a single enzyme protein acts on both
substrates (Campbell and Mashburn, 1969). Other substrates
for EC-2 include D-asparagine, β-aspartyl hydroxamate, β-
cyanoalanine, and 5-diazo-4-oxo-norvaline (DONV). In the
presence of 50% dimethyl sulfoxide, DONV binds irreversibly
to EC-2 with a stoichiometry indicating 4 subunits/molecule,
each with one active site (Jackson and Handschumacher, 1970).
Asparaginases from various sources have variable ratios of
activity with Asn and Gln as substrates. Among selected
antitumour asparaginases, which have been or are of current
interest, the guinea pig serum enzyme lacks glutaminase
activity, and the enzyme from *Vibrio succinogenes* has only
0.015% the activity with Gln that it has with Asn (Table 1).
Since significant toxicity associated with asparaginase
therapy appears related to Gln depletion, low glutaminase
activity may be a desirable property. On the other hand,
regression of experimental tumours resistant to EC-2 has been
reported for enzyme preparations with high glutaminase activity (Table 1). Two other substrates reported for EC-2 are
the glycoprotein fetuin (Bosmann and Kessel, 1970) and asparaginyl-tRNA[Asn] (Kessel, 1971), but these activities are of
doubtful significance for therapeutic effectiveness.

TABLE 1

Glutaminase activity of selected asparaginases

Source	$\left(\dfrac{\text{Gln as substrate}}{\text{Asn as substrate}} \times 100 \right)$	Reference
Pseudomonas 7A	200	Roberts *et al.*, 1979
Acinetobacter glutaminasificans	120	Roberts *et al.*, 1979
Erwinia carotovora	9	Howard and Carpenter, 1972
Escherichia coli	2	Campbell and Mashburn, 1969
Vibrio succinogenes	0.015	Distasio *et al.*, 1976
Guinea pig serum	0	Tower *et al.*, 1963

Carbohydrate was released from fetuin at 1000 units/ml, a concentration considerably higher than is achieved *in vivo*, and hydrolysis of $tRNA^{Asn}$ was slower than hydrolysis of Asn by a factor of 2.5×10^7. It seems reasonable to question, especially for the glycoprotein cleavage, whether hydrolysis might have been effected by a minor contaminating enzyme.

IV. DETERMINANTS OF SENSITIVITY

The critical biochemical lesion in leukaemic cells which renders them sensitive to the action of asparaginase has been considered to be inability to synthesize adequate asparagine to meet their requirements. Thus asparagine synthetase (ASase) which catalyses the following reaction,

Aspartate + Glutamine + ATP ———>

Asparagine + Glutamate + AMP + pyrophosphate

becomes the key enzyme determining susceptibility or resistance. A number of studies demonstrated that in sensitive leukaemic cells ASase levels were extremely low or absent while in asparaginase resistant cells, the enzymatic activity was substantially higher (Patterson and Orr, 1967; Prager and Bachynsky, 1968a, 1968b; Horowitz *et al.*, 1968; Broome, 1968a). These observations lead to the conclusion that asparaginase sensitive cells must derive their asparagine from an exogenous supply, and when that supply is grossly diminished by asparaginase administration, the tumour cells fail to survive. In accord with this view is the observation of rapid decrease in circulating Asn during treatment.

Modulation of ASase activity is a factor of importance bearing on sensitivity to asparaginase. Increased asparagine biosynthetic capacity has been observed under conditions which diminish circulating levels of the amino acid. Administration of asparaginase itself leads to enhanced ASase activity, particularly in resistant tumours (Prager and Bachynsky, 1968a, 1968b; Haskell *et al.*, 1970). ASase has also been shown to increase in certain normal tissues during asparaginase administration (Prager and Bachynsky, 1968b), during liver regeneration, in the presence of active tumour growth, or as a result of nutritional deprivation of Asn (Patterson and Orr, 1969). All of these conditions impose metabolic stress characterized by a high Asn requirement. Asparagine concentration in both blood and tissues falls after asparaginase administration (Broome, 1968b; Haskell *et al.*, 1970). Thus as the exogenous asparagine supply diminishes, resistant malignant cells and certain normal cells have the capacity of satisfying their need by an enhanced biosynthetic rate. Since ASase is product inhibited by Asn, part of the increase observed, when using crude cell extracts as a source of enzyme for assay purposes, reflects the decreased Asn content of the preparations. However, Patterson (1971) has presented immunochemical evidence that there is indeed an increase in asparagine synthetase protein following asparaginase administration. This finding suggests that Asn acts not only as a feedback inhibitor of preformed ASase but that it also represses synthesis of the enzyme.

While asparaginase selects for resistant populations, several studies demonstrate that resistant cells pre-exist in a sensitive tumour (Patterson *et al.*, 1969; Morrow, 1971). For the Walker 256 rat carcinosarcoma the rate of mutation from Asn dependence to Asn independence was $1.4 - 3.5 \times 10^{-6}$/cell/generation.

Attention has been given to the possibility that asparaginyl-tRNA may also play a role in regulating ASase and in determining tumour cell sensitivity to asparaginase. This line of investigation evolved from studies with bacteria in which there was a demonstrated correlation between decreased aminoacylation of tRNA and increased expression of biosynthetic enzymes for the cognate amino acid. Asparaginase sensitive L5178Y leukaemia cells have been reported to contain an extra asparaginyl-tRNA not found in resistant L1210 cells, leading to the suggestion that this unique tRNA might act as a repressor or corepressor of the formation of ASase (Gallo *et al.*, 1970). Interesting data relating to this issue have been generated using temperature sensitive mutants of Chinese hamster ovary (CHO) cells with defects in specific aminoacyl-tRNA synthetases. Although there was an inverse relation

between Asn attachment to tRNA^{Asn} and ASase activity, a
similar elevation of ASase was observed when the aminoacylation of several other tRNAs was restricted. Since a specific
relationship was anticipated, the lack of specificity failed
to support the contention that tRNA^{Asn} might be involved in
the regulation of ASase (Andrulius *et al.*, 1979). The
asparaginyl-tRNA synthetase mutants were even more sensitive
to asparaginase *in vitro* than were ASase mutants (Waye and
Stanners, 1981) even though ASase activity in the former
approached that for certain resistant tumours; other cell
lines (B-lymphocyte lines), however, with low asparaginyl-
tRNA synthetase and high ASase were relatively insensitive
to asparaginase *in vitro*. These results do not clearly support a role for tRNA^{Asn} in regulating ASase nor the idea that
a defect in asparaginyl-tRNA synthetase is critical for asparaginase sensitivity. However, the mutants with relatively
high ASase and high asparaginase sensitivity could present a
challenge to the excellent correlation from earlier work
already cited of asparaginase resistance when there is elevated ASase activity.

Two additional pieces of information about these mutants
would be most welcome. Was ASase fully expressed under the
culture conditions when it was measured, or is there an
increase in response to the metabolic stress imposed by
asparaginase in the culture medium, and does the sensitivity
reflect Asn or Gln depletion? Caution in interpreting these
in vitro data should be exercised because cells exhibiting
sensitivity to *E. coli* asparaginase *in vitro* may do so because of glutamine, rather than asparagine, depletion.
Recent studies with a human renal tumour cell line (Caki)
may be cited to support this contention. At 2 units/ml, EC-2
markedly inhibited growth. In other experiments the culture
medium was depleted of Asn and Gln, either by treatment with
asparaginase linked to Sepharose so that the enzyme could be
removed by centrifugation or alternatively by placing EC-2
(2.5 units) inside a dialysis bag in 40 ml of medium and
treating for 48 h. Growth of Caki cells was then determined
in complete medium, treated medium, and the latter repleted
with asparagine (137 μM) and/or glutamine (2 μM) at their
usual concentrations. Treated medium or treated Asn repleted
medium failed to support growth. In contrast, glutamine re-
pletion with or without added asparagine permitted growth
comparable to that in complete medium.

Loss of therapeutic response to asparaginase also results
when the host becomes sensitized to the enzyme. Anti-aspara-
ginase antibodies have been identified in both patients and
experimental animals. Mouse (Baechtel and Prager, 1973) and
human (Peterson *et al.*, 1971) antibody inhibits catalytic

activity about 50%, and increasing the antibody to enzyme
ratio fails to increase inhibition. The diminished catalytic
activity is characterized by a 3-fold reduction in V_{max} with
little change in K_m. The major effect of host sensitization
resulting in loss of therapeutic effect is the marked decrease
in biologic half-life ($t_{\frac{1}{2}}$) of asparaginase. After allowing
for the initial decline as the enzyme is diluted through its
volume of distribution, the $t_{\frac{1}{2}}$ in the blood following i.v.
injection was about 27 h for nonsensitized tumour bearing
mice while in the sensitized tumour-bearer it was less than
2 min. Comparable values for mice without tumour were 7.5
and 0.5 h, respectively (Baechtel and Prager, 1973). The
mean half-life for EC-2 (Merck) for patients with a variety
of tumour types was 22.6 h (Oettgen, 1975), and a rapid fall
in the plasma enzyme level can suggest an impending anaphyl-
actic reaction (Peterson *et al.*, 1971). Antibody to EC-2
does not cross-react with asparaginases of *Erwinia carotovora*,
Vibrio succinogenes, or *Proteus vulgaris*, making it possible
to continue therapy with one of these enzymes after the host
has become sensitized to EC-2.

V. IMMUNOSUPPRESSION

Interest in asparaginase as an immunosuppressive agent followed the observation by Astaldi *et al.* (1969) that EC-2 inhibited phytohemagglutinin (PHA)-stimulated blastogenesis of
lymphocytes. A variety of test systems which give evidence
of immunosuppressive effects of asparaginase include:

1) plaque forming spleen cells;

2) haemagglutinin and haemolysin formation;

3) graft-vs.-host (GVH) reaction;

4) graft rejection;

5) skin hypersensitivity;

6) immunoglobulin formation;

7) production of allergic encephalomyelitis;

8) immune cytolysis;

9) antibody dependent cellular cytotoxicity (ADCC);

10) reconstitution of the immune response.

While *in vitro* demonstration of immunosuppressive effects may
be questioned with regard to pharmacologic or physiologic
significance, there is little doubt concerning the *in vivo*
demonstrations with asparaginase as an immunosuppressive

agent. The latter effects are most readily observed when
enzyme is administered simultaneously or shortly after antigen.

Several issues concerning mechanism of immunosuppressive
action have commanded the attention of investigators. Since
Gln depletion accompanies the decrease in Asn following EC-2
administration, a number of studies have been directed
towards determining the contribution of the glutamine depletion to immunosuppression (Hersh, 1971). Several lines of
evidence suggest that asparaginase without accompanying
glutaminase activity can exert an immunosuppressive effect.
Asparaginases of agouti and guinea pig serum, which lack
glutaminase activity, are potent inhibitors of lymphoblasto-
genesis (Miura *et al.*, 1970; Prager and Mehta, 1973). Additional evidence for this view comes from the observation of
inhibition of blast transformation by the asparagine antago-
nist β-aspartyl hydroxamate (Miura *et al.*, 1970) and the
report of a critical Asn requirement for blasting lymphocytes
(Weiner *et al.*, 1969). Evidence from *in vivo* observations
is more tenuous. Glutamine levels recover more rapidly than
asparagine following EC-2 administration (Miller *et al.*,
1969), and in a report demonstrating that EC-2 inhibits allograft rejection, mention is made of preliminary experiments
in which agouti serum had a comparable effect (Schulten *et*
al., 1969). The glutamine synthetase activity of lymphoid
tissue was found to be 2000 times as high as ASase, suggesting a greater dependence on exogenous Asn (Prager and Derr,
1971). Contrary evidence comes from comparing immunosuppressive activity of EC-2 and *Vibrio succinogenes* asparaginase
in vivo. Durden and Distasio (1980, 1981) found that under
conditions where EC-2 significantly inhibited antibody response, *V. succinogenes* asparaginase had no suppressive activity. They postulated that the difference was the result of
the glutaminase activity of EC-2 (Table 1). This observation
is potentially of considerable importance if immunosuppression and other undesirable side-effects can be minimized by
the use of an enzyme source lacking glutaminase activity.

A number of studies have been concerned with determining
the cell type affected by asparaginase during an immune response. Inhibition of delayed hypersensitivity, allograft
rejection, and the GVH reaction indicate the sensitivity of
T-cell mediated functions to asparaginase suppression. The
great sensitivity of the PHA response is in accord with such
a conclusion. Furthermore, certain T-cell lines derived from
patients with acute lymphocytic leukaemia exhibit 1000-fold
greater sensitivity to asparaginase in tissue culture than
do B-cell or non-T-non-B-cell lines (Ohnuma *et al.*, 1980).
It has also been noted that several times as much asparagin-

ase was required for a given amount of inhibition of the
response of mouse B-splenocytes to LPS as was required to
inhibit a T-cell response to Con A (Gordon *et al.*, 1976).
Friedman (1971) examined reconstitution of the ability of
X-irradiated mice to form antibodies to sheep erythrocytes.
Marrow cells from asparaginase-treated animals given with
normal thymocytes were less able to reconstitute the irradiated mice than were the converse preparations. In this
experiment the primary target would thus appear to be a bone
marrow stem cell required for B-cell differentiation. An
antibody forming cell rather than the effector cell population was implicated in studies with ADCC, suggesting suppressed B-cells in asparaginase treated animals (Durden and
Distasio, 1981). Furthermore, immunoglobulin production by
spleen cells from rats injected earlier with *Mycobacterium
tuberculosis* was significantly reduced by asparaginase (Jasin
and Prager, 1972). It thus appears that both T and B-cell
functions may be compromised by asparaginase with the former
exhibiting somewhat greater sensitivity.

Histologic studies have given variable results with respect to the effect of EC-2 on T-cell- and B-cell-rich areas
of lymphoid organs. Rabbits injected with 1000—3000 units
EC-2/kg exhibited destruction of cortical areas of lymphoid
follicles with an accompanying increase in size of germinal
centres and frequency of plasma cells (Astaldi *et al.*, 1971).
With a dose of 50,000 units/kg to mice there was necrosis of
lymphoid cells in the thymic cortex and the germinal centres
of spleen and lymph nodes. Germinal centres which were
maximally affected in 4—8 h returned to normal by 24 h, but
the thymic cortex atrophied at 24—48 h (Berenbaum and
Bondurant, 1971). In contrast to these acute effects injection of 50 units EC-2/day for 4 days reduced the size of
splenic lymphoid follicles. The reduced germinal centres
also exhibited fewer immunoglobulin containing cells
(Distasio *et al.*, 1982). This finding is in accord with
earlier work from the same laboratory (Durden and Distasio,
1980, 1981) which claimed that B-cells were particularly
affected by asparaginase treatment. In contrast, the asparaginase from *V. succinogenes* had no effect on spleen histology,
correlating with the lack of immunosuppressive activity of
this enzyme.

The attempts to use rescue experiments to relieve the
immunosuppression produced by asparaginase administration
have failed to be persuasive. At times asparagine alone,
glutamine alone, or both in combination, have reversed the
inhibitory effect both *in vivo* and *in vitro*. One problem
with the *in vitro* studies which might account for some of
the variability is failure to take into account the potential

of ammonia toxicity. It has been demonstrated that exposure
of PHA stimulated splenocytes to high NH_4^+ concentration for
the first 24 h of 72 or 96 h cultures inhibits transformation
measured by thymidine incorporation by 50%. Thus, even
though the cells were restored to medium with normal NH_4^+
levels, they failed to transform in a normal fashion
(Baechtel et al., 1976). There is little doubt that NH_4^+
concentration builds up in asparaginase treated cultures, and
the issue of irreversible or only partially reversible damage to the cells should be scrutinized more closely.

VI. BIOCHEMICAL ALTERATIONS

In trying to elucidate the lesion produced in sensitive cells
by asparaginase, Sobin and Kidd (1966) reported that protein
synthesis was inhibited prior to DNA and RNA synthesis.
This is consistent with incorporation into protein being the
only role for Asn in mammalian cells. Whether suppressed
formation of specific proteins is responsible for the rapid
loss of cell viability remains uncertain but of interest.
There are several studies which demonstrate changes in cell
membrane structure. Con A binding to lymphocytes was inhibited following EC-2 treatment (Fidler and Montgomery, 1972),
and this led to a clustering effect on the distribution of
the Con A binding sites (Fidler et al., 1973). The 40%
reduction in electrophoretic mobility of human lymphocytes
as a result of EC-2 treatment is also consistent with membrane alterations (Lajolo et al., 1970). These changes
recorded for lymphocytes may relate significantly to the
immunosuppressive effect of asparaginase. They are also consistent with alteration of plasma membrane glycoproteins, and
several studies have recorded inhibitory effects on glycoprotein synthesis (Bosmann and Kessel, 1970; Wu et al., 1978).
    Recent studies in our laboratory with the Caki human
renal carcinoma cell line have been directed towards this
issue. After a 2 h exposure to 2 units of EC-2, ^{3}H-mannose
incorporation into glycoprotein decreased 40% during a 3 h
pulse while in medium depleted of Asn as described earlier,
the decrease was 25%. Since lengthy incubation with EC-2
also depletes Gln, and cells begin to die, for longer term
observation of glycoprotein synthesis relative to total protein formation, measurements were made in Asn-free medium in
the absence of EC-2 (Table 2). Cell growth and protein synthesis (leucine incorporation) relative to that in complete
medium reached a nadir on day 3 and then caught up on day 4.
This finding suggests increased ASase activity between days
3 and 4, but this remains to be determined. Glycoprotein
synthesis (mannose incorporation) was inhibited to a somewhat

TABLE 2

Protein and glycoprotein synthesis by human renal carcinoma
(Caki) cells in asparagine-free medium

Day	$\dfrac{\text{Asn-free medium}}{\text{Complete medium}} \times 100$		
	Cell count	^{3}H-Mannose incorp.	^{14}C-leucine incorp.
1	113	48	69
2	86	51	75
3	73	–	50
4	107	–	112

greater degree than protein synthesis (50% vs. 28%) on days
1 and 2, but data for mannose incorporation on days 3 and 4
are unfortunately not available. Two principal types of
glycoprotein are characterized structurally by carbohydrate
attachment either by O-glycosidic linkage to serine or thre-
onine residues or alternatively through N-glycosidic bonding
to asparagine residues. It will be of interest to determine
whether formation of the latter type of glycoprotein is
selectively inhibited by Asn deprivation. Ultimately comparisons must be made between cells sensitive or resistant
to asparaginase *in vivo*.

VII. CLINICAL ASPECTS

Perhaps the most extensive review of the clinical evaluation
of L-asparaginase has been that of Oettgen (1975). Patients
with acute lymphocytic leukaemia (ALL) have been by far the
most responsive group, exhibiting a remission rate of
approximately 60%. With the greater sensitivity of lympho-
blasts with T-cell markers to asparaginase in tissue culture
(Ohnuma *et al.*, 1980), it might be anticipated that T-cell
ALL would be the most responsive form of disease. Since
T-cell leukaemia accounts for only about 25% of the ALL
population there must be responders from among patients with
either B-cell or non-T-non-B-cell leukaemia. Application of
the battery of determinations of cell markers which enhance
ability to distinguish lymphocyte sub-populations should be
a fruitful area of investigation for seeking correlation with

patient response. For patients with acute leukaemia of non-lymphoblastic types the remission rate has been about 20%. There have also been reports of response among patients with lymphosarcoma. The drug has not proved useful in the treatment of chronic leukaemias and the results with solid tumours have also been almost uniformly disappointing. Asparaginase continues to be used in combination chemotherapy. Following asparaginase administration leukaemic cells were shown to attain maximum sensitivity to methotrexate after about one week (Capizzi, 1974), and on this regimen acute leukaemia patients who had relapsed after a variety of other treatments have yielded encouraging remission rates (Amadori *et al.*, 1980; Harris *et al.*, 1980).

A variety of side-effects have been reported as an accompaniment of asparaginase therapy (Oettgen, 1975). Those which occur in >50% of the treated patient population include fever, anorexia, nausea, vomiting, weight loss, abnormal liver function tests, elevated BUN, depressed circulating albumin, fibrinogen, and cholesterol. Patients not uncommonly develop hypersensitivity and demonstrable antibodies to EC-2 with a resultant loss of therapeutic effectiveness as described earlier. Certain asparaginases do not cross-react immunologically with EC-2, and it has been possible to continue therapy with the *Erwinia* enzyme after development of hypersensitivity to EC-2. By virtue of the immunosuppressive activity of asparaginase, some success has been recorded in the treatment of autoimmune disorders with the enzyme (Patterson, 1975).

A number of different test procedures have been explored for predicting response to asparaginase therapy. While some success has been recorded with experimental systems, these tests have generally not been adequate for prediction in a clinical setting. Although asparagine synthetase activity in leukaemic cells prior to initiating asparaginase treatment has not been predictive, elevation of ASase during therapy correlates with development of resistance and may be taken as an indication to explore other therapeutic measures.

VIII. CONCLUDING STATEMENT

Asparaginase remains as one of the useful drugs in the treatment of ALL, primarily in combination with other agents. It is disappointing that its therapeutic efficacy has been so narrow and that resistant leukaemic cells can so easily appear during therapy. Applying the specificity of an enzymatic reaction to deplete a nutrient required to a greater extent by malignant cells than normal ones is conceptually appealing, and a considerable research effort on other enzymes followed the discovery of the antitumour effect of asparaginase. Although some of these enzymes have shown promise in isolated experimental systems, none has evolved to the extent that asparaginase has, and this too has been disappointing. Twenty years have elapsed since Broome (1961) identified asparaginase as the antileukaemic agent of guinea pig serum, and if this approach to cancer management is to remain viable, it would seem that it is necessary to produce additional examples of effective enzymatic agents.

REFERENCES

Amadori, S., Tribalto, M., Pacilli, L., DeLaurentis, C., Papa, G. and Mandelli, F. (1980). Sequential combination of methotrexate and L-asparaginase in the treatment of refractory acute leukaemia. *Cancer Treat. Rep.* **64**, 939-942.

Andrulius, I.L., Hatfield, W. and Arfin, S.M. (1979). Asparaginyl tRNA aminoacylation levels and asparagine synthetase expression in cultured Chinese hamster ovary cells. *J. Biol. Chem.* **254**, 10629-10633.

Astaldi, G., Burgio, G.R., Krc, J., Genova, R. and Astaldi, A.A., Jr. (1969). L-asparaginase and blastogenesis. *Lancet* **1**, 423.

Astaldi, G., Micu, D., Astaldi, A., Jr., Burgio, G.R. and Krc, I. (1971). Further Investigation on L-Asparaginase and Immune Reactions. Int. Symp. No. 197 of the Centre National de la Recherche Scientifique, 205-219.

Baechtel, S. and Prager, M.D. (1973). Basis for loss of therapeutic effectiveness of L-asparaginase in sensitized mice. *Cancer Res.* **33**, 1966-1969.

Baechtel, F.S., Gregg, D.E. and Prager, M.D. (1976). The influence of glutamine, its decomposition products and glutaminase on the transformation of human and mouse lymphocytes. *Biochim. Biophys. Acta* **421**, 33-43.

Berenbaum, M.C. and Bondurant, S. (1971). Effect of L-asparaginase on germinal centre haemolysin. *Nature* **231**, 318-319.

Bosmann, H.B. and Kessel, D. (1970). Inhibition of glycoprotein synthesis in L5178Y mouse leukaemic cells by L-asparaginase *in vitro*. *Nature* **226**, 850-851.

Broome, J.D. (1961). Evidence that the L-asparaginase activity of guinea pig serum is responsible for its antilymphoma effects. *Nature* **191**, 1114-1115.

Broome, J.D. (1963a). Evidence that the L-asparaginase of guinea pig serum is responsible for its antilymphoma effects. I. Properties of the L-asparaginase of guinea

pig serum in relation to those of the antilymphoma substance. *J. Exp. Med.* **118**, 99-120.

Broome, J.D. (1963b). Evidence that the L-asparaginase of
guinea pig serum is responsible for its antilymphoma
effects. II. Lymphoma 6C3HED cells cultured in a medium
devoid of L-asparagine lose their susceptibility to the
effects of guinea pig serum *in vivo*. *J. Exp. Med.* **118**,
121-148.

Broome, J.D. (1968a). Evolution of a new tumour inhibitory
agent. *Trans. N.Y. Acad. Sci.* **30**, 690-704.

Broome, J.D. (1968b). Studies on the mechanism of tumour
inhibition by L-asparaginase. *J. Exp. Med.* **127**, 1055-
1072.

Campbell, H.A. and Mashburn, L.T. (1969). L-asparaginase
from *Escherichia coli*. Some substrate specificity characteristics. *Biochemistry* **8**, 3768-3775.

Campbell, H.A., Mashburn, L.T., Boyse, E.A. and Old, L.J.
(1967). Two L-asparaginases from *Escherichia coli* B.
Their separation, purification and antitumour activity.
Biochemistry **6**, 721-730.

Capizzi, R.L. (1974). Schedule-dependent synergism and
antagonism between methotrexate and asparaginase. *Biochem. Pharmacol.* **23** (suppl. 2), 151-161.

Capizzi, R.L., Bertino, J.R. and Handschumacher, R.E. (1970).
L-asparaginase. *Ann. Rev. Med.* **21**, 433-444.

Cooney, D.A. and Handschumacher, R.E. (1970). L-asparaginase
and L-asparagine metabolism. *Ann. Rev. Pharmacol.* **10**,
421-440.

Distasio, J.A., Niederman, R.A., Kafkewitz, D. and Goodman,
D. (1976). Purification and characterization of L-aspara-
ginase with antilymphoma activity from *Vibrio succinogenes*.
J. Biol. Chem. **251**, 6929-6933.

Distasio, J.A., Durden, D.L., Paul, R.D. and Nadji, M. (1982).
Alteration in spleen lymphoid populations associated with
specific amino acid depletion during L-asparaginase treatment. *Cancer Res.* **42**, 252-258.

Durden, D.L. and Distasio, J.A. (1980). Comparison of the
immunosuppressive effects of asparaginases from *Escherichia coli* and *Vibrio succinogenes*. *Cancer Res.* **40**, 1125-
1129.

Durden, D.L. and Distasio, J.A. (1981). Characterization of
the effects of asparaginase from *Escherichia coli* and a
glutaminase-free asparaginase from *Vibrio succinogenes* on
specific cell-mediated cytotoxicity. *Int. J. Cancer* **27**,
59-65.

Fidler, I.J. and Montgomery, P.C. (1972). Effects of L-asparaginase on lymphocyte surface and blastogenesis.
Cancer Res. **32**, 2400-2406.

Fidler, I.J., Montgomery, P.C. and Cesarini, J.P. (1973). Modification of surface topography of lymphocytes by L-asparaginase. *Cancer Res*. **33**, 3176-3180.

Friedman, H. (1971). L-asparaginase induced immunosuppression: Inhibition of bone marrow derived antibody precursor cells. *Science* **174**, 139-141.

Gallo, R.C., Longmore, J.L. and Adamson, R.H. (1970). Asparaginyl-tRNA and resistance of murine leukaemias to L-asparaginase. *Nature* **227**, 1134-1136.

Gordon, W.C., Mandy, W.J. and Prager, M.D. (1976). Differential effect of asparaginase on lymphocyte subpopulations and alleviation of the immunosuppression by lipopolysaccharide. *Cancer Immunol. Immunother*. **1**, 205-209.

Harris, R.E., McCallister, J.A., Provisor, D.S., Weetman, R.M. and Baehner, R.L. (1980). Methotrexate/L-asparaginase combination chemotherapy for patients with acute leukaemia in relapse. *Cancer* **46**, 2004-2008.

Haskell, C.M., Canellos, G.P., Cooney, D.A. and Hansen, H.H. (1970). Biochemical and pharmacologic effects of L-asparaginase in man. *J. Lab. Clin. Med*. **75**, 763-770.

Hersh, E.M. (1971). L-glutaminase: suppression of lymphocyte blastogenic responses *in vitro*. *Science* **172**, 736-738.

Horowitz, B., Madras, B.K., Meister, A., Old, L.J., Boyse, E.A. and Stockert, E. (1968). Asparagine synthetase activity of mouse leukaemias. *Science* **160**, 533-535.

Howard, J.B. and Carpenter, F.H. (1972). L-asparaginase from *Erwinia carotovora*. Substrate specificity and enzymatic properties. *J. Biol. Chem*. **247**, 1020-1030.

Jackson, R.C. and Handschumacher, R.E. (1970). *Escherichia coli* L-asparaginase. Catalytic activity and subunit nature. *Biochemistry* **9**, 3585-3590.

Jasin, H.E. and Prager, M.D. (1972). Effects of L-asparaginase on *in vitro* immunoglobulin synthesis by rat spleen cells. *Clin. Exp. Immunol*. **10**, 515-523.

Jayaram, H.N., Ramakrishnan, T. and Vaidyanathan, C.S. (1968). L-asparaginases from *Mycobacterium tuberculosis* strains $H_{37}R_v$ and $H_{37}R_a$. *Arch. Biochem. Biophys*. **126**, 165-174.

Kessel, D. (1971). Asparaginyl-transfer RNA: A substrate for L-asparaginase. *Biochim. Biophys. Acta* **240**, 554-557.

Kidd, J.G. (1953). Regression of transplanted lymphomas *in vivo* by means of normal guinea pig serum. I. Course of transplanted cancers of various kinds in mice and rats given guinea pig serum, horse serum or rabbit serum. *J. Exp. Med*. **98**, 565-581.

Lajolo, D., Astaldi, A., Jr., Pecco, P., Bert, G. and Astaldi, G. (1970). Electrophoretical mobility of human lymphocytes in the presence of *E. coli* L-asparaginase. *Exp. Cell Res*. **60**, 458-459.

Mashburn, L.T. and Wriston, J.C., Jr. (1964). Tumour inhibitory effect of L-asparaginase from *Escherichia coli*. *Arch. Biochem. Biophys*. **105**, 450-452.

McCoy, T.A., Maxwell, M. and Neuman, R.E. (1956). The amino acid requirements of the Walker carcinosarcoma 256 *in vitro*. *Cancer Res*. **16**, 979-984.

McCoy, T.A., Maxwell, M. and Kruse, P.F., Jr. (1959). The amino acid requirements of the Jensen sarcoma *in vitro*. *Cancer Res*. **19**, 591-595.

Miller, H.K., Salser, J.S. and Balis, M.E. (1969). Amino acid levels following L-asparagine aminohydrolase (EC.3.5.1.1) therapy. *Cancer Res*. **29**, 183-187.

Miura, M., Hirano, M., Kakizawa, K., Morita, A., Uetani, T. and Yamada, K. (1970). Inhibitory effect of L-asparaginase in lymphocyte transformation induced by phytohemagglutinin. *Cancer Res*. **30**, 768-772.

Morrow, J. (1971). Mutation rate from asparagine requirement to asparagine non-requirement. *J. Cell Physiol*. **77**, 423-426.

Oettgen, H.F. (1975). L-asparaginase: Current status of clinical evaluation. *In* "Antineoplastic and immunosuppressive agents II". (Eds A.C. Sartorelli and D.G. Johns) Handbook Exp. Pharm. **38**, 723-746, Springer-Verlag, New York.

Ohnuma, T., Arkin, H. and Holland, J.F. (1980). Differences in chemotherapeutic susceptibility of human T-, B-, and non-T-non-B-lymphocytes in culture. *Recent Results Cancer Res*. **75**, 61-67.

Patterson, M.K., Jr. (1971). Effects of L-asparaginase on asparagine synthetase levels of normal and malignant tissue. Int. Symp. No. 197 of the Centre National de la Recherche Scientifique, 107-113.

Patterson, M.K., Jr. (1975). L-asparaginase: Basic aspects. *In* "Antineoplastic and Immunosuppressive Agents II". (Eds A.C. Sartorelli and D.G. Johns), Handbook Exp. Pharm. **38**, 695-722, Springer-Verlag, New York.

Patterson, M.K., Jr. and Orr, G. (1967). L-asparagine biosynthesis by nutritional variants of the Jensen sarcoma. *Biochem. Biophys. Res. Commun*. **26**, 228-233.

Patterson, M.K., Jr. and Orr, G.R. (1969). Regeneration, tumour, dietary, and L-asparaginase effects on asparagine biosynthesis in rat liver. *Cancer Res*. **29**, 1179-1183.

Patterson, M.K., Jr., Maxwell, M.D. and Conway, E. (1969). Studies on the asparagine requirement of the Jensen sarcoma and the derivation of its nutritional variant. *Cancer Res*. **29**, 296-300.

Peterson, R.E. and Ciegler, A. (1969). L-asparaginase production by *Erwinia aroideae*. *Appl. Microbiol*. **18**, 64-67.

Peterson, R.G., Handschumacher, R.E. and Mitchell, M.S. (1971). Immunological responses to L-asparaginase. *J. Clin. Invest.* **50**, 1080-1090.

Prager, M.D. (1971). Tumour inhibitory enzymes. *In* "Oncology 1970", Proc. 10th Int. Cancer Congress. (Eds R.L. Clark R.W. Cumley, J.E. McCay and M.M. Copeland), Vol. 11, pp. 237-253, Yearbook Medical Publishers, Chicago.

Prager, M.D. and Bachynsky, N. (1968a). Asparagine synthetase in asparaginase resistant and susceptible mouse lymphomas. *Biochem. Biophys. Res. Commun.* **31**, 43-47.

Prager, M.D. and Bachynsky, N. (1968b). Asparagine synthetase in normal and malignant tissues; correlation with tumour sensitivity to asparaginase. *Arch. Biochem. Biophys.* **127**, 645-654.

Prager, M.D. and Derr, I. (1971). Metabolism of asparagine, aspartate, glutamine, and glutamate in lymphoid tissue: basis for immunosuppression by L-asparaginase. *J. Immunol.* **106**, 975-979.

Prager, M.D. and Mehta, J.M. (1973). Enzymes as immunosuppressants: basic considerations. *Transpl. Proc.* **5**, 1171-1175.

Roberts, J., Prager, M.D. and Bachynsky, N. (1966). The antitumour activity of *Escherichia coli* L-asparaginase. *Cancer Res.* **26**, 2213-2217.

Roberts, J., Schmid, F.A. and Rosenfeld, H.J. (1979). Biologic and antineoplastic effects of enzyme-mediated *in vivo* depletion of L-glutamine, L-tryptophan and L-histidine. *Cancer Treat. Rep.* **63**, 1045-1054.

Schulten, H.K., Giraldo, G., Boyse, E.A. and Old, L.J. (1969). Immunosuppressive action of L-asparaginase. *Lancet* **2**, 644-645.

Sobin, L.H. and Kidd, J.G. (1966). Alterations in protein and nucleic acid metabolism of lymphoma 6C3HED-OG cells in mice given guinea pig serum. *J. Exp. Med.* **123**, 55-74.

Tower, D.B., Peters, E.L. and Curtis, W.C. (1963). Guinea pig serum L-asparaginase. Properties, purification, and application to determination of asparagine in biological samples. *J. Biol. Chem.* **238**, 983-993.

Waye, M.M.Y. and Stanners, C.P. (1981). Role of asparagine synthetase and asparagyl-transfer RNA synthetase in the cell killing activity of asparaginase in Chinese hamster ovary cell mutants. *Cancer Res.* **41**, 3104-3106.

Weiner, M.S., Waithe, W.I. and Hirschhorn, K. (1969). L-asparaginase and blastogenesis. *Lancet* **2**, 748.

Wu, M., Arimura, G.K. and Yunis, A.A. (1978). Mechanism of sensitivity of cultured pancreatic carcinoma to asparaginase. *Int. J. Cancer* **22**, 728-733.

DISCUSSION IV
CHAPTERS 13-14

Dr HILL: Dr Sedlacek, in your study of colon cancer patients were there any side-effects of the "chessboard immunization" suffered by the patients?

Dr SEDLACEK: Of course, patients get the DTH reaction. However, this side-effect is minor and the reaction goes within about 4 to 5 days. We could not find other side-effects. We are facing the fact that neuraminidase produced by bacteria plays a distinct role in the pathogenicity of the haemolytic uremic syndrome, but neither preclinically nor clinically have we got any information that the "chessboard vaccination" has any influence or is causing this disease. Altogether we can say that this treatment causes almost no side-effects.

Dr BALDWIN: What is the status of work on whether VCN exposure of antigens upon tumour cells or VCN attached to tumour cells can promote hapten-like response?

Dr SEDLACEK: According to my information there are a lot of data showing that VCN increases antigenicity and immunogenicity of cells. But with the exception of the Thomsen-Friedenreich antigen the data are contradictory as to whether VCN increases the amount of antigen on the cell surface. Indeed, it might be that VCN attached to the cell surface may function as an additional antigen and the data of Johannsen *et al.* strongly support this assumption. We are restricting our main research in this field to unmarked antigens (including Thomsen-Friedenreich antigen) and to the adjuvant activity of VCN.

Dr GOLDIN: This is an important area and I wonder whether you could indicate your plans for further study?

Dr SEDLACEK: Of course, we are trying to get rid of tumour cells and to isolate the antigen in question and to combine it with an adequate carrier or an adjuvant. The outcome of the clinical studies mentioned in my talk will show us whether this method has any chance of being successful. But we do know that this work is extremely difficult.

Dr KONDO: Dr Sedlacek, in your experiment, could you demonstrate the fixation of the antibody or complement to the
residual tumour cells after treating the animal by mitomycin-
neuraminidase-treated cells?

Dr SEDLACEK: I cannot answer your question correctly because
we have not looked for antibodies binding to the residual
tumour mass.

Helen NAUTS: Have you ever injected neuraminidase directly
into the mammary tumour in your dogs to see if it might be
more effective?

Dr SEDLACEK: Simmons *et al*. have injected neuraminidase into
the mammary tumour of mice and he got secondaries. With
respect to these data and facing the problem of cell dose
dependence shown in our dog experiments, we cannot recommend
the procedure because of the increased risk of inducing
tumour enhancement.

Dr MITCHELL: Dr Prager, why does asparaginase, as an immunosuppressive agent, not suppress the antibody response to
itself?

Dr PRAGER: When asparaginase therapy is continuous, development of hypersensitivity appears minimized. Sensitization
occurs with intermittent treatment. This suggests that
asparaginase may indeed suppress an immune response against
itself, but when therapy is stopped, a rebound occurs.

Dr GOLDIN: For patients in which no response was observed,
is there any difference in asparaginase requirement or status?

Dr PRAGER: Attempts have been made to determine this, but
there appears to be no correlation between circulating asparagine levels and response to asparaginase. Measurement of
asparagine synthesis as an indicator of probable asparaginase
dependence has been useful in only one situation. Asparaginase activity is rather low in most tissues, but significant
increases after asparaginase administration indicates that
the tissue has become independent of exogenous asparagine.
For a tumour, increased asparaginase indicates asparaginase
resistance, but we know this only after initiating drug
treatment.

Dr SINKOVICS: Mouse lymphoma cells can be quite resistant to
L-asparaginase. We have described such a mouse lymphoma. L-
asparaginase was immunosuppressive to the mice without
afflicting damage to the lymphoma cells. An enhancement
growth of the lymphoma resulted.

MICROBIAL PRODUCTS AND CANCER THERAPY

Abraham Goldin[1] and Franco M. Muggia[2]

[1]*Division of Medical Oncology,*
Vincent T. Lombardi Cancer Research Centre,
Georgetown University School of Medicine,
3800 Reservoir Road NW, Washington D.C. 20007, USA

[2]*Director, Division of Oncology,*
School of Medicine,
New York University Medical Centre,
550 First Avenue, New York, NY 10016, USA

I. INTRODUCTION

From a historical point of view the microbial products that
have been discovered have clearly played a significant and
increasingly important role in cancer therapy. They have
emerged among the most highly important drugs constituting
the armamentarium of the clinical oncologist, with significant activity against the refractory solid tumours that
account for the highest incidence of cancer deaths (Carter,
1978). The importance of microbial products is reflected
for example in a listing of the total number of approved
antitumour drugs in the USA. Of the 31 drugs approved
through 1980 (endocrines excluded), 7 are microbial products
including actinomycin D, mithramycin, bleomycin, mitomycin C,
adriamycin, L-asparaginase and daunomycin (Table 1) (De Vita
et al., 1979). It is of interest that although the first
drug to be approved was nitrogen mustard, in the late 1940s,
the initial microbial product to be approved, namely actinomycin D, was not until some 15 years later in the mid 1960s.
The pioneer in the investigation of microbial metabolites
for the treatment of cancer is clearly Hamao Umezawa. Rokujo
et al. (1949) in the screening of compounds against Yoshida
sarcoma of the rat observed that an actinomycin that Umezawa
had found in a Streptomyces isolated from the soil in Japan
was capable of destroying the Yoshida rat sarcoma cells in
in vitro culture (Umezawa, 1978). Hackmann (1952) in

TABLE 1

*Chronology of approval of microbial products
as antitumour drugs in the USA*

Actinomycin D	1964
Mithramycin	1970
Bleomycin	1973
Mitomycin	1974
Adriamycin	1974
L-asparaginase	1978
Daunomycin	1980

(De Vita *et al.*, 1979).

investigations of actinomycin C found that it was capable of
inhibiting the growth of experimental mouse and rat tumours.
Reilly *et al.* (1953) in their investigation on the effect of
antibiotics on the growth of Sarcoma 180 *in vivo* observed
that among the number of antibiotics tested, actinomycin
resulted in significant inhibition of the growth of the
tumour. Since as indicated by Umezawa (1978), in 1951 the
progress in the discovery of antibacterial antibiotics had
been extensive and the significant emergence of resistant
pathogenic organisms had not occurred, Umezawa and his colleagues initiated a programme of investigation of antitumour
antibiotics. They discovered the active compounds caliomycin
(Umezawa and Yamamoto, 1952), substance No. 289 (Umezawa *et
al.*, 1953) and sarkomycin (Umezawa *et al.*, 1954), thereby
emphasizing that it was possible to discover new antitumour
compounds by the investigation of the action of microbial
culture filtrates in inhibiting experimental tumours.

II. MICROBIAL PRODUCTS APPROVED FOR CLINICAL USE
IN THE TREATMENT OF NEOPLASIA IN THE USA

The actinomycins were isolated from actinomycetes by Waksman
and Woodruff (1940). Actinomycin D has a broad spectrum of
activity in experimental animal tumour systems. A listing
of tumour models in which actinomycin D has demonstrated
definitive activity is given in Table 2 (Goldin and Johnson,
1974). At the National Cancer Institute, USA, a new screening programme was initiated in 1975, designed to be subject
to prospective analysis of screening data on new drugs with

TABLE 2

Experimental tumour systems in which actinomycin D demonstrated definitive activity

---

Tumours which could be cured by actinomycin D

        P388 leukaemia
        Ridgeway osteogenic sarcoma

Tumours against which actinomycin D markedly increased life span or inhibited local tumour growth

        B16 melanotic melanoma
        L1210 leukaemia
        Gardner 6 C3HED lymphosarcoma
        Mecca lymphosarcoma
        P288 lymphocytic leukaemia
        P1081 chloroleukaemia
        P329 reticulum cell sarcoma
        Lymphoma 4
        Sarcoma 180
        Adenocarcinoma 755

(Goldin and Johnson, 1974).

respect to prediction of clinical activity. The revised programme incorporated a screening panel of both human tumours growing in xenograft in athymic (nude) animals and corresponding animal tumours of similar histologic type. The animal data for the series of microbial agents that have been approved for clinical use in the USA, including actinomycin D, in the new National Cancer Institute screen, are summarized in Table 3 (Goldin *et al.*, 1981). Actinomycin D was active in most of the animal tumour models of the new screen. Although not active in the mammary (MX-1), lung (LX-1) or colon (CX-1) human tumours xenografted subcutaneously, it was active against the MX-1 and CX-1 human tumours when the tumours were transplanted in the subrenal capsule of the athymic mice.

    Mithramycin was isolated by Rao and his colleagues in 1962 (Calabresi and Parks, 1975) from *Streptomyces tanashiensis*. Structurally, olivomycin A and chromomycin A_3, are related to mithramycin (Gause, 1975). It may be noted that with respect to animal test systems, the leukaemia P388 system is considerably more sensitive to microbial products than is the leukaemia L1210 system, which

TABLE 3

Activity in the NCI screening panel for microbial products with established clinical activity

| | | | | | | | | Xenografts | | | | | |
| | | | | | | | | Subcutaneous | | | Subrenal capsule | | |
	L1210	P388	B16 Melanoma	Lewis Lung	Colon 26	Colon 38	CD8F$_1$	MX-1	LX-1	CX-1	MX-1	LX-1	CX-1
Actinomycin D	+	+	+		+	+	+			+	+		+
Mithramycin		+											
Bleomycin	+	+	+	+		+	+	+					
Mitomycin C	+	+	+	+	+	+	+	+	+				
Daunomycin	+	+	+		+								
Adriamycin	+	+	+	+	+	+	+						
L-aspara- ginase	+												

has been used as a standard screening model for many years. Mithramycin, for example, was not active in the L1210 system and on this basis would not have qualified as an agent of interest, whereas it was highly active in the leukaemia P388 system. A comparison of the relative activity of a series of microbial agents in the leukaemia L1210 and P388 systems is presented in Table 4 (Venditti and Abbott, 1967; Goldin and Kline, 1978).

TABLE 4

Relative activity of microbial products in the L1210 and P388 systems

	Percent increase in life span over controls (ILS %)	
	L1210	P388
Actinomycin D	45	>175
Mithramycin	10	90
Bleomycin	<25	50
Mitomycin C	70	150
Daunomycin	58	127
Adriamycin	64	>200
L-asparaginase	<25	36
Azaserine	50	97
Neocarzinostatin	63	>200
Azotomycin	68	120
Streptonigrin	<25	36
Chromomycin A3	<25	153
Tubercidin	<25	50
Sangivamycin	67	90
Streptovitacin A	55	81

(Venditti and Abbott, 1967; Goldin and Kline, 1978)

Bleomycin is produced by a strain of *Streptomyces verti-cillus*. It is a mixture of antibiotics and was discovered by Umezawa *et al.* in 1966 (Umezawa *et al.*, 1966a). Umezawa *et al.* (1966b) separated the bleomycin mixture into the fractions A_2 which is the main component and B_2, employing Sephadex chromatography. As is the case with actinomycin D, bleomycin has exhibited activity against a wide spectrum of animal tumours (Takeuchi and Yamamoto, 1968; Umezawa *et al.*, 1968; Goldin and Kline, 1978). In the new screen at the National Cancer Institute, USA, it was active against leukaemia P388, B16 melanoma, Lewis lung carcinoma, colon 38 and $CD8F_1$ mammary carcinoma (Table 3) (Goldin *et al.*, 1981). It also evidenced activity against the mammary MX-1 subcutaneous xenograft.

The mitomycins were isolated from fermentation filtrates of *Streptomyces caespitosus*. Mitomycins A and B were isolated by Hata *et al.* (1956) and a series of related fractions including mitomycin C was isolated by Wakaki *et al.* (1958). It should be noted that although mitomycin C was reported on as early as 1958 (Wakaki *et al.*, 1958) the delay in approval for standard clinical practice in the USA was approximately 16 years, to 1974. Mitomycin C has demonstrated broad spectrum activity against mouse tumours, including leukaemia, lymphosarcoma, sarcoma, carcinoma and melanoma and it has also exerted activity against rat and hamster tumours (Sugiura, 1959, 1961; Kojima *et al.*, 1972; Driscoll *et al.*, 1974). In the revised screen of the National Cancer Institute, it was active in leukaemias P388 and L1210, B16 melanoma, Lewis lung carcinoma, colon tumours 26 and 38, $CD8F_1$ mammary tumour and the subcutaneous human tumour xenografts MX-1 and LX-1 (Goldin *et al.*, 1981).

Daunomycin was isolated from *Streptomyces peucetius* culture in 1963 (Di Marco *et al.*, 1963, 1964). It was reported as having relatively broad spectrum activity against a variety of ascitic tumours, including Yoshida 130 hepatoma, Walker carcinoma 256, Sarcoma 180 and Ehrlich carcinoma (Di Marco *et al.*, 1963, 1964; Venditti *et al.*, 1966). It was active in the new screen against the 2 leukaemias L1210 and P388, B16 melanoma and colon 26, but was not tested in the human tumour xenograft systems (Table 3).

Adriamycin was also isolated from cultures of *Streptomyces peucetius* (Di Marco *et al.*, 1963) and this drug, too, exhibited broad spectrum activity against a variety of experimental tumours. Adriamycin differs from daunomycin in the substitution of a hydroxyl group for a hydrogen atom in the acetyl radical of the aglycone moiety (Arcamone *et al.*, 1969). The broad spectrum activity, based on the experience

of a number of investigators, has been delineated (Goldin
and Johnson, 1975a) and the activity in the new screen is
shown in Table 3. In the new screen it was active against
the mouse leukaemias L1210 and P388, B16 melanoma, Lewis lung
carcinoma, colon tumour 26 and CD8F$_1$ mammary tumour (Goldin
et al., 1981). It did not demonstrate activity in the human
tumour xenograft systems.

L-asparaginase, an enzyme which catalyses the hydrolysis
of L-asparagine, was reported on as far back as 1904 (Lang,
1904; Patterson, 1975). Clementi (1922) demonstrated L-asparaginase activity in guinea pig serum and Kidd (1953)
reported that guinea pig serum was capable of inhibition of
growth of transplantable mouse and rat tumours. More
recently, Tsuji (1957) obtained L-asparaginase from *Escherichia coli*, and Mashburn and Wriston (1964) found that an *E.
coli* preparation could inhibit experimental tumour growth.
The therapeutic activity of L-asparaginase has been demonstrated in lymphomas and leukaemias in mice, against primary
lymphosarcoma in dogs and against acute lymphoblastic leukaemia in children (Boyse *et al.*, 1967; Broome, 1963; Dolowy *et
al.*, 1966; Hill *et al.*, 1967; Old *et al.*, 1967; Oettgen *et
al.*, 1967). In the new screen at the National Cancer Institute, L-asparaginase was active only in the P388 system
(Table 3) (Goldin *et al.*, 1981). It was, however, not tested
in the human tumour xenograft systems.

III. MICROBIAL PRODUCTS IN DEVELOPMENTAL INVESTI-
GATION FOR THE TREATMENT OF CLINICAL NEOPLASIA

In addition to the above microbial products that have been
demonstrated to exert antitumour activity, there is a broad
range of microbial products in various stages of development,
which are being subjected to intensive preclinical investigation, and some of the products have received preliminary
clinical trials. A listing of additional microbial substances in various stages of development is given in Table 5.
Here, too, the broad spectrum activity of a number of the
products is evident (Goldin *et al.*, 1981).

Umezawa (1978) pointed out that with respect to antitumour
antibiotics, more than 200 compounds have been discovered
that are of interest. The potential, then, in this area is
quite high with respect to the uncovering of new and more
effective materials for the clinic.

IV. THE TYPES OF MICROBIAL PRODUCTS

Microbial products for cancer treatment may be classified in
a general way as: (a) cytotoxics including actinomycin D,

TABLE 5

Activity in the NCI screening panel for microbial products in various stages of development

| | | | | | | | | Xenografts | | | | | |
| | | | | | | | | Subcutaneous | | | Subrenal capsule | | |
	L1210	P388	B16 Melanoma	Lewis Lung	Colon 26	Colon 38	CD8F$_1$	MX-1	LX-1	CX-1	MX-1	LX-1	CX-1
Neocarzinostatin	+	+	+		+	+							
D-0-Norleucine	+	+			+	+	+	+	+		+		
L-Alanosine	+	+			+	+	+						
AT-125 Acivicin	+	+	+	+	+		+	+	+		+		
Aclacinomycin A	+	+	+				+				+		+
2'-Deoxycoformycin													
AD-32	+	+	+		+	+	+						
Valinomycin	+	+	+		+	+							+
Macromomycin	+	+	+		+	+					+		
Quinomycin A		+	+										

mitomycin C, adriamycin and so forth; (b) enzymes including
L-asparaginase, glutaminase and other amino acid depleting
enzymes, carboxypeptidase and other enzymes that act in novel
fashion; (c) biological response modifiers including immunoaugmenting agents. Immunoaugmenting agents may increase the
defence response of the host to tumour by stimulation of macrophage and reticuloendothelial system functions. Some
examples of microbial products that may serve as immunoaugmenting agents are listed in Table 6. In addition to the
above there could be novel types of approaches, such as selective delivery systems including liposome encapsulation of
cytotoxics, coupling of drugs to diphtheria toxin or antibodies and so forth that could result in therapeutic improvement.

TABLE 6

Microbial products that may act as immunoaugmenting agents

Bacillus Calmette-Guerin (BCG)

Methanol extraction residue (MER) of BCG

Mycobacterial cell wall skeleton

Corynebacterium parvum and extracts

Nocardia rubra cell wall skeleton

Bestatin

Forphenicine

Phage lysate of *Staphylococcus aureus*

Bru-Pell (extract of *Brucella* strain)

V. MICROBIAL PRODUCTS IN COMBINATION
CHEMOTHERAPY AND COMBINED MODALITIES

There is extensive interest in the employment of combinations
of drugs, including antitumour microbial products, as a result of the efficacy of drug combination therapy in the
clinic. Several preclinical investigations indicating therapeutic synergism (therapeutic enhancement) in protocols
involving antitumour microbial products may be cited.

   On weekly administration of actinomycin D plus daily
treatment with methotrexate, the combination resulted in
therapeutic synergism, since by definition, it was more
effective than either agent alone in the treatment of an
advanced stage of subcutaneously inoculated leukaemia L1210

(Goldin and Johnson, 1974). That the schedule of therapy is important in the achievement of therapeutic synergism is indicated by the observation that a sequential schedule in which a single dose of actinomycin D followed by daily methotrexate did not result in a therapeutic response that was more extensive than that obtained with methotrexate alone (Goldin and Johnson, 1974).

The combination of actinomycin D plus L-asparaginase elicited a markedly synergistic effect, resulting in long-term survivors in the treatment of leukaemia L5178Y, a tumour which is L-asparaginase sensitive (Jacobs et al., 1970). Leukaemia L1210 does have the enzyme L-asparagine synthetase and is thereby resistant to the cytotoxic action of L-asparaginase. With this combination, no therapeutic synergism was elicited.

The combination of mitomycin C with drugs such as adriamycin, 5-fluorouracil or cytosine arabinoside has resulted in enhanced therapeutic response (Hoshino et al., 1972). In one of the experiments (Hoshino et al., 1972) the combination of 3 drugs, namely mitomycin C, 5-fluorouracil and cytosine arabinoside resulted in a greater increase in therapeutic effectiveness in the treatment of leukaemia L1210 than that observed with the drugs employed in combinations of 2 or when they were employed individually.

The combination of bleomycin with drugs such as cyclophosphamide, vincristine, CCNU and procarbazine was more effective in each instance than the drugs employed individually in increasing the survival time of mice that had been inoculated intramuscularly or subcutaneously with Lewis lung carcinoma (Goldin and Kline, 1978). The combination of bleomycin plus vinblastin resulted in therapeutic synergism in the treatment of luekaemia P388 inoculated intraperitoneally as well as B16 melanoma inoculated either intraperitoneally or subcutaneously (Goldin and Kline, 1978). In the same series of investigations it was also observed that the combination of bleomycin plus cis-platinum II resulted in therapeutic synergism in the treatment of B16 melanoma inoculated subcutaneously.

Adriamycin elicited broad spectrum therapeutic synergism when it was employed in combination with alkylating agents, antimetabolites and a variety of other types of drugs (Goldin and Johnson, 1975a; Schabel, 1975; Griswold et al., 1973; Goldin and Johnson, 1975b). In the leukaemia L1210 system for example adriamycin exerted therapeutic synergism in combination with cyclophosphamide, melphalan, BCNU, CCNU, methyl CCNU, vincristine, methotrexate, cytosine arabinoside, cyclocytidine, 5-azacytidine, ICRF 159, procarbazine and camptothecin (Goldin and Johnson, 1975a).

In the treatment of leukaemia L1210 with the combination of adriamycin plus ICRF 159, and adriamycin plus cyclophosphamide, therapeutic synergism was observed with maintenance or reduction in the optimal dosage of adriamycin (Goldin and Johnson, 1975b). This is of interest since such reduction in the dosage of an anthracycline may result in the avoidance of cardiac toxicity.

In addition to the interest in preclinical investigations with combinations of drugs there is also a strong interest in combined modalities of therapy. The combination of surgery plus microbial antitumour products may result in an improvement in therapeutic response, and investigations in this area should be encouraged. With the anthracyclines it was demonstrated by Schabel *et al.* (1979) that administration of adriamycin prior to or subsequent to surgical adjuvant chemotherapy of mammary adenocarcinoma implanted subcutaneously was more effective than either surgery alone or adriamycin alone in increasing the life span and the number of long-term survivors. Another study may be cited (Giuliani *et al.*, 1979) in which surgery plus the combination of adriamycin and cyclophosphamide was markedly effective in reducing the lung metastases of the MS-2 sarcoma in BALB/c mice.

Investigations of the combination of microbial products and radiation are important. In one study, as an example, Ridgway osteogenic sarcoma was treated with the combination of ^{60}Co gamma-irradiation plus adriamycin and this resulted in a greater increase in survival time than was observed with radiation alone or drug treatment alone (Goldin *et al.*, 1978).

There have been a number of studies of the utilization of combinations of biologic response modifiers with surgery or in conjunction with the combination of chemotherapy plus surgery. In a study by Karrer *et al.* (1979), treatment of moderately advanced Lewis lung carcinoma with the combination of *C. parvum* and surgery resulted in an increase in the number of survivors. In another study, Schabel *et al.* (1979) observed that administration of *C. parvum* at one day after the inoculation of Lewis lung carcinoma, followed by surgery on the seventh day, resulted in an increase in the number of long-term (90 day) survivors. The potential for microbial products employed alone or in combination with other agents or in combined modalities is great and this area, as part of a total biologic response modifier programme, as recently instituted at the National Cancer Institute, USA, is worthy of considerable attention.

VI. ANALOGUES OF KNOWN MICROBIAL PRODUCTS

The criteria for the selection of analogues of known microbial products that may exert more extensive or more desirable
antitumour activity is a highly important subject and one
review by Goldin (1978) may be cited. Such preclinical criteria for the selection of new analogues may include:

1) Greater activity in an appropriate test system relative
   to the parent compound.

2) Broader spectrum of effectiveness against experimental
   tumours including transplantable, carcinogen induced or
   spontaneous.

3) More desirable or different scheduling and route characteristics.

4) Diminished limiting or undesirable acute, chronic or
   delayed host toxicity.

5) Delay in origin of tumour cell resistance and/or increased
   activity against tumours resistant to the parent drug.

6) More extensive therapeutic synergism than the parent compound in combination with a structurally unrelated agent,
   or therapeutic synergism when employed in combination with
   the parent drug, which would imply different pharmacologic
   or biochemical action.

7) Greater penetration to sequestered tumour sites including
   the intracranial site.

8) Greater effectiveness in preventing metastasis or in
   treating metastatic disease.

9) Greater action against oncogenic virus, virus transformation of cells to malignancy, viral induced tumour or rein-
   duction of tumour.

10) More desirable pharmacologic or biochemical characteris-
    tics, including greater maintenance of blood and tissue
    levels, diminished detoxification or excretion, depot
    form with steady release and so forth.

11) Differing desirable action in relation to tumour and/or
    vital host cell kinetics, i.e. effect at different phase
    of the cell cycle than the parent compound, more exten-
    sive selective tumour cell kill and so forth.

12) Reduced carcinogenicity.

13) More desirable action on the immune status of the host,
    i.e. less immunosuppressive or more effective immunosti-
    mulation, or more effective in altering tumour cell
    antigenicity.

14) Superiority over the parent compound in combined modalities with surgery, radiation or immunotherapy.

With respect to the above it is of interest to mention 3 doxorubicin derivatives of current interest, that were modified in the 4' position of the amino surgar, i.e. 4'-epi doxorubicin, 4'-deoxy doxorubicin and 4'-0-methyl doxorubicin (Arcamone *et al.*, 1975, 1976; Cassinelli *et al.*, 1979; Casazza *et al.*, 1980). As compared with doxorubicin these analogues have evidenced differences in their toxicologic, pharmacologic and therapeutic characteristics in preclinical systems. The acute toxicity in mice (LD_{50}) of 4'-deoxy doxorubicin and 4'-0-methyl doxorubicin were at lower dosages than for doxorubicin (Casazza *et al.*, 1980). The LD_{50} for 4'-epi doxorubicin was moderately higher than that of doxorubicin.

The minimal cumulative cardiotoxic dose levels for 4'-deoxy doxorubicin and 4'-0-methyl doxorubicin were higher than that for doxorubicin in mice treated with the drugs. The ratio of the minimal cumulative cardiotoxic dose to the optimal cumulative antitumour dose was increased for 4'-deoxy doxorubicin and for 4'-0-methyl doxorubicin as compared with doxorubicin, indicative of a greater margin of safety in the employment of these analogues with respect to cardiotoxicity. The ratio of the cardiotoxic dose to the tumour dose was slightly higher for 4'-epi doxorubicin as compared with doxorubicin (Casazza *et al.*, 1980). The cardiac toxicity of 4'-deoxy doxorubicin in rabbits was markedly reduced and that of 4'-epi doxorubicin somewhat reduced as compared with doxorubicin (Casazza *et al.*, 1980).

Notably, although adriamycin was essentially ineffective in a series of colorectal tumours heterotransplanted into athymic (nude) mice, 4'-deoxy doxorubicin and 4'-0-methyl doxorubicin evidenced definitive activity against a number of these tumours (Giuliani and Kaplan, 1980).

Thus, it would appear that there were sufficiently significant differences among the 3 analogues 4'-deoxy doxorubicin, 4'-epi doxorubicin and 4'-0- methyl doxorubicin as compared with doxorubicin to make them worthy of extensive preclinical and clinical investigation. Clearly, the effort to obtain analogues of the known microbial antitumour agents is an approach worthy of extensive effort.

VII. CLINICAL ASPECTS FOR MICROBIAL PRODUCTS

Landmarks in cancer treatment with microbial products are listed in Table 7. The clinical activity of a number of microbial products with definitive activity when employed alone or in combination are listed in Table 8 (Davis *et al.*,

TABLE 7

Landmarks in cancer treatment with microbial products

Disease (year)	Drugs
Choriocarcinoma (1958)	Actinomycin
Wilms' tumour (1960)	Bleomycin
Testicular cancer (1970)	Daunorubicin
Acute myeloid leukaemia (1970)	Adriamycin
Sarcomas, lymphomas (1972)	Streptozotocin
Islet cell tumours (1972)	Actinomycin

TABLE 8

Clinical activity of microbial products
(Definite activity alone or in combination with other agents)

Microbial products	Efficacy
Actinomycin D	Lymphomas, childhood solid tumours[2], bone sarcomas[2], soft tissue sarcomas, melanoma, uterus, ovary[3], testis[3], trophoblastic disease[3], Kaposi's sarcoma.
Mithramycin	Ovary[3], testis[3]
Bleomycin	Lymphomas, Kaposi's sarcoma, squamous head and neck, cervix, oesophagus, ovary, testis, non-small lung.
Mitomycin C	Lymphomas, breast cancer, gastrointestinal adenocarcinoma, kidney, bladder, prostate.
Adriamycin	Acute leukaemia[1], lymphomas, childhood solid tumours[2], bone sarcomas[3], soft tissue sarcomas, breast cancer, squamous head and neck, cervix, oesophagus, gastrointestinal adenocarcinoma, thyroid, kidney, bladder, prostate, ovary, uterus, small cell lung, non-small cell lung.
Daunomycin	Acute leukaemia[1].

[1] Major impact in acute myelogenous leukaemia.
[2] Potentially curative in combined modality treatment.
[3] May be curative alone or combined with other drugs.
(Davis *et al.*, 1978).

TABLE 9

Clinical activity of additional microbial products

Microbial product	Efficacy
Streptozotocin	Islet cell tumours, carcinoid
Streptonigrin	Lymphomas
Porfiromycin	Squamous head and neck, cervix, oesophagus, gastrointestinal adenocarcinoma, ovary, uterus
Azotomycin	Colon(?)
Neocarzinostatin	Acute leukaemia
Pyrazofurin	Leukaemia(?), lymphomas(?)
Tubercidin	Islet cell tumours(?)
5-azacytidine	Acute leukaemia

(Muggia, 1978; Davis *et al.*, 1978).

1978). The "carry over" of the antitumour activity observed in the preclinical experimental models to definitive clinical activity is clear. The broad spectrum antitumour activity observed in the animal tumour systems was also observed in the treatment of various categories of clinical neoplasia. There was, however, not a necessary correspondence of activity for the different histologic types of animal and clinical tumours.

The status in the clinic of a number of additional microbial products is listed in Table 9 (Muggia, 1978; Davis *et al.*, 1978). The full potential of these substances awaits detailed clinical investigation. However, even at this time there is evidence of clinical activity for some of them.

VIII. SUMMARY

The microbial products comprise a significant percentage of the antitumour agents that have been introduced into the clinic. A number of them have established roles in clinical therapy of neoplasia. Others are in earlier stages of clinical investigation and still others are being developed with a view to their introduction into the clinic. The potential for this broad class of substances is great, with respect to the possibility for the discovery of new agents, the preparation of analogues of known active microbial

324   GOLDIN and MUGGIA

products that may have increased antitumour activity and the
optimization of therapy through improved application, when
the microbial products are employed individually or in combination with other drugs or in combined modality approaches.

IX. REFERENCES

Arcamone, F., Cassinelli, G., Fantini, G., Grein, A., Orezzi,
    P., Pol, C. and Spalla, C. (1969). Adriamycin, 14-hydro-
    xydaunomycin, a new antitumour antibiotic from *S. peucet-
    ius var. caesius*. *Biotech. Bioengin.* **11**, 1101.
Arcamone, F., Penco, S., Vigevani, A., Redaelli, S., Franchi,
    G., DiMarco, A., Casazza, A.M., Dasdia, T., Formelli, F.,
    Necco, A. and Soranzo, C. (1975). Synthesis and anti-
    tumour properties of new glycosides of daunomycinone and
    adriamycinone. *J. Med. Chem.* **18**, 703-707.
Arcamone, F., Penco, S., Redaelli, S. and Hanessian, S.
    (1976). Synthesis and antitumour activity of 4'-deoxy
    daunorubicin and 4'-deoxy adriamycin. *J. Med. Chem.* **19**,
    1424-1425.
Boyse, E.A., Old, L.J., Campbell, H.A. and Mashburn, L.T.
    (1967). Suppression of murine leukaemias by L-asparagin-
    ase. Incidence of sensitivity among leukaemias of various
    types: comparative inhibitory activity of guinea pig serum
    L-asparaginase and *Escherichia coli* L-asparaginase. *J.
    Exp. Med.* **125**, 17-31.
Broome, J.D. (1963). Evidence that L-asparaginase of guinea
    pig serum is responsible for its antilymphoma effects.
    I. Properties of the L-asparaginase of guinea pig serum
    in relation to those of the antilymphoma substance. *J.
    Exp. Med.* **118**, 99-120.
Calabresi, P. and Parks, R.E. Jr. (1975). Chemotherapy of
    neoplastic diseases. Alkylating agents, antimetabolites,
    hormones and other antiproliferative agents. *In* "The
    Pharmacological Basis of Therapeutics". (Eds L.S. Goodman
    and A. Gilman), 5th Edtn., pp. 1248-1307, Macmillan Pub-
    lishing Co., New York.
Carter, S. (1978). Antitumour antibiotics — Thoughts for
    the future. *In* "Recent Results in Cancer Research.
    Antitumour Antibiotics". (Eds S.K. Carter, H. Umezawa,
    J. Douros and Y. Sakurai), pp. 298-303, Springer-Verlag,
    New York.
Casazza, A.M., DiMarco, A., Bonadonna, G., Bonfante, V.,
    Bertazzoli, C., Bellini, O., Pratesi, G., Sala, L. and
    Ballerini, L. (1980). Effects of modifications in posi-
    tion 4 of the chromophore or in position 4' of the amino-
    sugar, on the antitumour activity and toxicity of dauno-
    rubicin and doxorubicin. *In* "Anthracyclines: Current

Status and New Developments". (Eds S.T. Crooke and S.D. Reich), pp. 403-430. Academic Press, New York.

Cassinelli, G., Ruggieri, D. and Arcamone, F. (1979). Synthesis and antitumour activity of 4'-0-methyl daunorubicin, 4'0-methyladriamycin and 4'-epianalogues. *J. Med. Chem.* **22**, 121-123.

Clementi, A. (1922). La désamidation enzymatique de l'asparagine chez les différentes éspèces animales et la signification physiologique de sa présence dans l'organisme. *Arch. Int. Physiol.* **19**, 369-398.

Davis, H.L. Jr., Von Hoff, D.D., Henney, J.E. and Rozencweig, M. (1978). Role of antitumour antibiotics in current oncology practice. *In* "Recent Results in Cancer Research: Antitumour Antibiotics". (Eds S.K. Carter, H. Umezawa, J. Douros and Y. Sakurai), pp. 21-29. Springer-Verlag, New York.

De Vita, V.T., Oliverio, V.T., Muggia, F.M., Wiernik, P.W., Ziegler, J., Goldin, A., Rubin, D., Henney, J. and Schepartz, S. (1979). The drug development and clinical trials programme of the Division of Cancer Treatment, National Cancer Institute. *Cancer Clin. Trials* **2**, 195-216.

DiMarco, A., Gaetani, M., Dorigotti, L., Soldati, M. and Bellini, O. (1963). Studi sperimentali sull'attivita antineoplastica del nuovo antibiotica daunomicina. *Tumori* **49**, 203-220.

DiMarco, A., Gaetani, M., Orezzi, P., Scarpinato, B.M., Silvestrini, R., Soldati, M., Dasdia, T. and Valentini, L. (1964). Daunomycin, a new antibiotic of the rhodomycin group. *Nature* **201**, 706-707.

Dolowy, W.C., Henson, D., Cornet, J. and Sellin, H. (1966). Toxic and antineoplastic effects of L-asparaginase. Study of mice with lymphoma and normal monkeys and report on a child with leukaemia. *Cancer* **19**, 1813-1819.

Driscoll, J.S., Hazard, G.F., Wood, H.B. Jr. and Goldin, A. (1974). Structure-antitumour activity relationships among quinone derivatives. *Cancer Chemother. Rep.* **4**, 1-362.

Gause, G.F. (1975). Chromomyin, olivomycin, mithramycin. *In* "Antineoplastic and Immunosuppressive Agents". (Eds A.C. Sartorelli and D.G. Johns), Vol. 2, pp. 615-622, Springer-Verlag, New York.

Giuliani, F.C. and Kaplan, N.O. (1980). New doxorubicin analogues active against doxorubicin-resistant colon tumour xenografts in the nude mouse. *Cancer Res.* **40**, 4682-4687.

Giuliani, F., DiMarco, A., Casazza, A.M. and Savi, G. (1979). Combination chemotherapy and surgical adjuvant chemotherapy on MS-2 sarcoma and lung metastases in mice. *Europ. J. Cancer* **15**, 715-723.

Goldin, A. (1978). Criteria for selection of new analogues of antitumour antibiotics. *In* "Recent Results in Cancer Research. Antitumour Antibiotics". (Eds S.K. Carter, H. Umezawa, J. Douros and Y. Sakurai), pp. 99-112. Springer-Verlag, New York.

Goldin, A. and Johnson, R.K. (1974). Evaluation of actinomycins in experimental systems. *Cancer Chemother. Rep. (1)* **58**, 63-77.

Goldin, A. and Johnson, R.K. (1975a). Experimental tumour activity of adriamycin (NSC-123127). *Cancer Chemother. Rep. (3)* **6**, 137-145.

Goldin, A. and Johnson, R.K. (1975b). Antitumour effects of adriamycin in comparison with related drugs and in combination chemotherapy. *In* "Adriamycin Review". (Eds M. Staquet, H. Tagnon, Y. Kenis *et al.*), pp. 37-54. European Press, Medikon, Ghent, Belgium.

Goldin, A. and Kline, I. (1978). The bleomycins: Experimental tumour activity. *In* "Bleomycin: Current Status and New Developments". (Eds S.K. Carter, S.T. Crooke and H. Umezawa), pp. 91-106. Academic Press, New York.

Goldin, A., Wodinsky, I., Merker, P.C. and Venditti, J.M. (1978). Search for new radiation potentiators. *Int. J. Radiat. Oncol. Biol. Phys.* **4**, 23-35.

Goldin, A., Venditti, J.M., Macdonald, J.S., Muggia, F.M., Henney, J.E. and DeVita, V.T. (1981). Current results of the screening programme at the Division of Cancer Treatment, National Cancer Institute. *Europ. J. Cancer* **17**, 129-142.

Griswold, D.P., Laster, W.R. Jr. and Schabel, F.M. Jr. (1973). Therapeutic potentiation by adriamycin and 5-(3,3-dimethyl-1-triazeno)-imidazole-4-carboxamide against B16 melanoma, C3H breast carcinoma, Lewis lung carcinoma and leukaemia L1210. *Proc. Am. Assoc. Cancer Res.* **14**, 15.

Hackmann, C.H. (1952). Experimentelle untersuchungen uber die wirkung von aktinomycin C bei bösartigen. *Geschwulsten. Z. Krebsforsch.* **58**, 607-613.

Hata, T., Sano, Y., Sugawara, R., Matsumae, A., Kanamori, K. Shima, T. and Hoshi, T. (1956). Mitomycin, a new antibiotic from Streptomyces. *J. Antibiot. (Tokyo)* **9**, 141-146.

Hill, J.M., Roberts, J., Loeb, E., Kahn, A., MacLellan, A. and Hill, R.W. (1967). L-asparaginase therapy for leukaemia and other malignant neoplasms. Remission in human leukaemia. *J. Am. Med. Assoc.* **202**, 882-888.

Hoshino, A., Kato, T., Amo, H. and Ota, K. (1972). Antitumour effects of adriamycin on Yoshida rat sarcoma and

L1210 mouse leukaemia — cross resistance and combination chemotherapy. *In* "Int. Symp. on Adriamycin". (Eds S.K. Carter, A. DiMarco, M. Ghione, I.H. Krakoff and G. Mathé), pp. 75-89. Springer-Verlag, New York.

Jacobs, S.P., Wodinsky, I., Kensler, C.J. and Venditti, J.M. (1970). Combination therapy studies with L-asparaginase (NSC-109229) and other antitumour agents. *Cancer Chemother. Rep.* **54**, 329-335.

Karrer, K., Rella, W. and Goldin, A. (1979). Surgery plus *Corynebacterium parvum* immunotherapy for Lewis lung carcinoma in mice. *Europ. J. Cancer* **15**, 867-873.

Kidd, J.G. (1953). Regression of transplanted lymphomas induced *in vivo* by means of normal guinea pig serum. I. Course of transplanted cancers of various kinds in mice and rats given guinea pig serum, horse serum, or rabbit serum. *J. Exp. Med.* **98**, 565-582.

Kojima, R., Goldin, A. and Mantel, N. (1972). The influence of schedule of administration of mitomycin C (NSC-26980) in the treatment of L1210 leukaemia. *Cancer Chemother. Rep.* **3**, 111-119.

Lang, S. (1904). Über Desamidierung im Tierkörper. *Beitr. Chem. Physiol. Path.* **5**, 321-345.

Mashburn, L.T. and Wriston, J.C. Jr. (1964). Tumour inhibitory effect of L-asparaginase from *Escherichia coli*. *Arch. Biochem. Biophys.* **105**, 450-452.

Muggia, F.M. (1978). Clinical evaluation of new antitumour antibiotics. *In* "Recent Results in Cancer Research. Antitumour Antibiotics". (Eds S.K. Carter, H. Umezawa, J. Douros and Y. Sakurai), pp. 288-297. Springer-Verlag, New York.

Oettgen, H.F., Old, L.J., Boyse, E.A., Campbell, H.A., Philips, F.S., Clarkson, B.D., Tallal, L., Leeper, R.D., Schwartz, M.K. and Kim, J.H. (1967). Inhibition of leukaemias in man by L-asparaginase. *Cancer Res.* **27**, 2619-2631.

Old, L.J., Boyse, E.A., Campbell, H.A., Brodey, R.S., Fidler, J. and Teller, J.D. (1967). Treatment of lymphosarcoma in the dog with L-asparaginase. *Cancer* **20**, 1066-1070.

Patterson, M.K. Jr. (1975). L-asparaginase: Basic aspects. *In* "Antineoplastic and Immunosuppressive Agents". (Eds A.C. Sartorelli and D.G. Johns), Vol. 2, pp. 695-722. Springer-Verlag, New York.

Reilly, H.C., Stock, C.C., Buckley, S.M. and Clark, D.A. (1953). The effect of antibiotics upon the growth of sarcoma 180 *in vivo*. *Cancer Res.* **13**, 684-687.

Rokujo, T., Ukita, T., Tsumita, T. and Kosaka, S. (1949). Screening test of the chemical compounds which inhibit Yoshida sarcoma of rat. *Gann* **40**, 140-142.

Schabel, F.M. Jr. (1975). Animal models as predictive sys-
    tems. *In* "Proc. 19th Annual Clinical Conf. on Cancer
    Chemotherapy — Fundamental Concepts and Recent Advances".
    pp. 323-355. Yearbook Medical Publishers, Chicago.
Schabel, F.M. Jr., Griswold, D.P. Jr., Corbett, T.H., Laster,
    W.R. Jr., Dykes, D.J. and Rose, W.C. (1979). Recent
    studies with surgical adjuvant chemotherapy or immuno-
    therapy of metastatic solid tumours of mice. *In* "Adjuvant
    Therapy of Cancer II". (Eds S.E. Jones and S.E. Salmon),
    pp. 3-17, Grune and Stratton, New York.
Sugiura, K. (1959). Studies in a tumour spectrum. VIII. The
    effect of mitomycin C on the growth of a variety of mouse,
    rat and hamster tumours. *Cancer Res.* **19**, 438-445.
Sugiura, K. (1961). Antitumour activity of mitomycin C.
    Cancer Chemother. Rep. **13**, 51-65.
Takeuchi, M. and Yamamoto, T. (1968). Effects of bleomycin
    on mouse transplantable tumours. *J. Antibiot.* **A21**, 631-
    637.
Tsuji, Y. (1957). Amidase action of bacteria. *Naika Hokan.*
    4, 222-223.
Umezawa, H. (1978). Advances in bioactive microbial second-
    ary metabolites useful in the treatment of cancer. *In*
    "Advances in Cancer Chemotherapy". (Eds S.K. Carter, A.
    Goldin, K. Kuretani, G. Mathé, Y. Sakurai, S. Tsukagoshi
    and H. Umezawa), pp. 27-52. Japan Sci. Soc. Press,
    Tokyo/Univ. Park Press, Baltimore.
Umezawa, H. and Yamamoto, T. (applied in 1952). Process for
    production of caliomycin, a new antitumour antibiotic.
    Japan Patent Sho 30-396.
Umezawa, H., Takeuchi, T., Nitta, K., Maeda, K., Yamamoto, T.
    and Yamaoka, S. (1953). Studies on antitumour substances
    produced by microorganisms. I. On the antitumour sub-
    stance No. 289 resembling to luteomycin. *J. Antibiot.*
    6A, 45-51.
Umezawa, H., Takeuchi, T. and Nitta, K. (1954). Sarkomycin,
    an antitumour substance produced by Streptomyces. *J.
    Antibiot.* **6A**, 101.
Umezawa, H., Maeda, K., Takeuchi, T. and Okami, Y. (1966a).
    New antibiotics, bleomycin A and B. *J. Antibiot.* **A19**,
    200-209.
Umezawa, H., Suhara, Y., Taketa, T. and Maeda, K. (1966b).
    Purification of bleomycins. *J. Antibiot.* **A19**, 210-215.
Umezawa, H., Ishizuka, M., Kimura, K., Iwanaga, J. and
    Takeuchi, T. (1968). Biological studies on individual
    bleomycins. *J. Antibiot.* **A21**, 592-602.
Venditti, J.M. and Abbott, B.J. (1967). Studies on oncolytic
    agents from natural sources. Correlations of activity

against animal tumours and clinical effectiveness. *Lloydia* **30**, 332–348.

Venditti, J.M., Abbott, B.J., DiMarco, A. and Goldin, A. (1966). Effectiveness of daunomycin (NSC-82151) against experimental tumours. *Cancer Chemother. Rep.* **50**, 659–665.

Wakaki, S., Marumo, H., Tomioka, K., Shimizu, G., Kato, E., Kamada, H., Kudo, S. and Fujimoto, Y. (1958). Isolation of new fractions of antitumour mitomycins. *Antibiot. Chemother.* **3**, 228–240.

Waksman, S.A. and Woodruff, H.B. (1940). Bacteriostatic and bactericidal substances produced by a soil actinomyces. *Proc. Soc. Exp. Biol. (NY)* **45**, 609–614.

CLINICAL EXPERIENCE IN TREATMENT OF CANCER BY PROPIONIBACTERIA

K. Roszkowski[1], W. Roszkowski[2], H.L. Ko[3],
S. Szmigielski[4], G. Pulverer[3] and J. Jeljaszewicz[5]

[1]*Postgraduate Medical Centre, Warsaw, Poland;*

[2]*National Institute of Tuberculosis, Warsaw, Poland;*

[3]*Institute of Hygiene, University of Cologne, Cologne, FRG;*

[4]*Centre for Radiobiology and Radioprotection, Warsaw, Poland;*

[5]*National Institute of Hygiene, Warsaw, Poland.*

The interest in anaerobic propionibacteria (formerly classi-
fied as corynebacteria) as microorganisms possessing immunostimulatory and antitumour activity comes from Halpern's *et
al.* (1964) demonstration of a stimulatory effect of these
bacteria on the reticuloendothelial system in mice. These
authors and Woodruff and Boak (1966) also observed inhibition of transplantable tumour growth in experimental animals
treated by *Corynebacterium parvum*. Since then many experimental studies have been made in an attempt to explain the
mechanisms of immunological and antineoplastic effects of
bacteria. At the same time the first clinical trials began
with the aim of establishing the usefulness of propionibacteria in human cancer therapy.

Until now the evaluation of the efficacy of these bacteria in clinical oncology is unequivocal and results of
treatment reported by investigators are often contradictory.
It seems important to mention that in the majority of these
clinical trials nonstandardized criteria of antitumour
effect evaluation have been used. In some cases normalization of haematologic or immunologic parameters has been
recognized as a positive effect of treatment with propionibacteria whereas in others lack of pronounced regression of
the tumour has been considered as a negative result of the
therapy. Various combinations of application of propionibacteria with other methods of antitumour treatment (e.g.

chemotherapy) provide an additional difficulty in evaluation
of immunotherapy efficiency.

Four years ago we started our studies on clinical application and evaluation of antitumour immunotherapy with *Propionibacterium granulosum*-strain KP-45 (P.KP-45). These studies
included 301 patients with lung, breast and digestive tract
cancer and with malignant melanoma. One hundred and fifty
one of these 301 patients were treated with P.KP-45.

We decided on the following basic assumptions:

1) Randomized selection of patients;
2) Uniform combination of immunotherapy with other methods of
   antitumour treatment;
3) Uniform criteria of therapeutic effects evaluation;
4) Analysis of immunotherapy efficiency for individual types
   of tumours.

The study on local immunotherapy was the only exception from
the above-mentioned assumptions as the patients were not
randomly selected and criteria other than Karnofsky's were
used for evaluation of clinical state (Karnofsky and Burchenal, 1949).

I. GENERAL IMMUNOTHERAPY

The studies were undertaken to evaluate the usefulness and
effectiveness of immunotherapy with P.KP-45 in patients
receiving oncostatic treatment. These studies were performed
on patients of both sexes, all aged below 70 years and in
good general clinical state at the beginning of treatment.
Clinical state was evaluated according to Karnofsky's index,
assuming 50 points or more as a criterion qualifying the
patient to the study.

All patients undergoing general immunotherapy received
P.KP-45 in intravenous infusions at weekly intervals in
single doses of 10 mg (dry weight of whole cells, in sterile
suspension).

The following parameters were taken as criteria of evaluation of chemoimmunotherapy or chemotherapy effects:

1) Percentage of patients with observed
   Objective remission (OR) — appearance of complete or partial response;
   Complete response (CR) — total regression of all clinically detectable lesions; in the case of lesions localized
   in bones — their total recalcification;
   Partial response (PR) — decrease of lesion size more than
   50% of initial value and lack of new lesions;

Stable disease (SD) — changes of neoplastic lesion size
lower than 25% of initial value and no appearance of new
lesions;
Progressive disease (PD) — increase of lesion size higher
than 50% of initial value; stabilization of existing
lesions with simultaneous appearance of new ones.

2) Time of remission — time between application of treatment
and first symptoms of progression.

3) Survival time.

A. Chemoimmunotherapy of Lung Cancer

The studies were performed on 79 patients with small cell
undifferentiated lung carcinoma, disqualified from surgical
treatment because of advancement of disease. All patients
received chemotherapy according to regimen VCM (vincristine,
cyclophosphamide, methotrexate) and 38 of them additionally
received intravenous infusions of P.KP-45. Regimen of treatment is presented in Fig. 1.

   In preliminary studies performed on the other group of
patients we used P.KP-45 at longer intervals to observe the
eventual side-effects and to avoid their risk. The results
of these pilot studies (Roszkowski *et al.*, 1981a) showed
that frequency of P.KP-45 treatment can be increased without
serious risk.

   Both groups of patients, undergoing chemo- and chemoimmunotherapy, were comparable when mean age, general clinical
state and location of tumour lesions were analysed (Table 1).
Advancement of the neoplastic process was also taken into
consideration in the analysis of results and both groups of
patients were divided into 2 subgroups, one having tumour
lesions localized only in the chest (limited disease) and
the other one with distant metastases (diffuse disease).
This classification was done on the basis of clinical examination — X-ray, isotope scan and computerized tomography.

   Significant differences were not observed in frequency of
objective remissions between the group of patients receiving
chemotherapy and those treated by chemoimmunotherapy (VCM-
28/41, VCM + P.KP-45 - 29/38) independent of the advancement
of the disease. Treatment with P.KP-45 also did not significantly influence the time of remission of disease except in
patients with distant metastases. In this case, the time of
remission was prolonged from 3.5 months in the group treated
only with oncostatics to 6 months in the group receiving
additionally P.KP-45. The summary of our results is presented in Table 2. Basic immunological and haematological parameters (total number of lymphocytes and their subpopulation

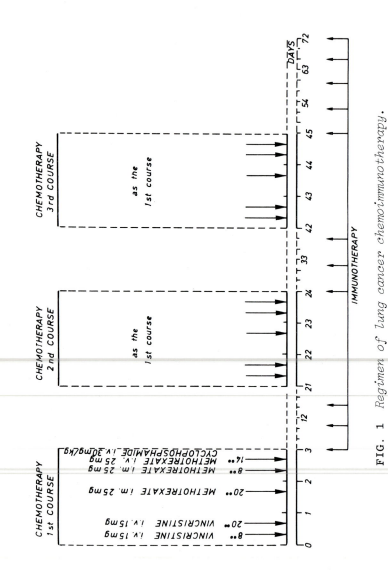

FIG. 1 Regimen of lung cancer chemoimmunotherapy.

TABLE 1

Clinical characteristics of patients with lung cancer

Group	VCM	VCM + P
Number of patients	41	38
Age (years)	49 (29-64)	51 (34-61)
Karnofsky index (points)	70 (50-90)	70 (50-90)
Number of patients with limited intrathoracic tumour	14	12
Distribution of metastases in patients with diffuse disease		
liver	14	9
brain	4	6
bone	8	10
peripheral lymph nodes	17	14

VCM = vincristine, cyclophosphamide, methotrexate;
P   = *Propionibacterium granulosum* KP-45

T and B, levels of IgG, IgM, IgA, fraction C3 of complement, α_2 macroglobulin, delayed skin hypersensitivity to recall antigens, number of erythrocytes, haemoglobin level) were analysed in all patients with lung cancer. It was observed that the drop in number of leukocytes caused by chemotherapy was significantly less pronounced in patients receiving P.KP-45 (Fig. 2). Differences in other parameters were not noticed between both examined groups of patients.

It has to be emphasized that in the group of 41 patients treated only by oncostatic drugs, 6 cases of infective complications were observed (2 oral candidiasis, 2 upper respiratory tract viral infections, 1 pneumonia, 1 sepsis), whereas in the group of 38 patients receiving additionally P.KP-45 only one case of oral candidiasis was noted. The appearance of infective complications caused a temporary discontinuation of chemotherapy which could have been the reason for the observed shortened mean time of remission in the group of patients with distant metastases not receiving immunotherapy.

An additional group of 17 patients with squamous lung cancer was not treated with chemotherapy. Lung lavage was performed in these patients which enabled us to investigate

TABLE 2

*Results of VCM versus VCM + P.KP-45 treatment of patients
with advanced lung cancer*

		No. of patients	CR	PR	SD	PD	Duration of remission median	range (mths)
VCM	Total	41	11	17	5	8	8	1 - 8
	Limited disease	14	9	4	1		7	2 - 10
	Diffuse disease	27	2	13	4	8	3.5	1 - 6
VCM + P.KP-45	Total	38	9	20	4	5	7	2 - 11
	Limited disease	12	7	3	1	1	8	4 - 11
	Diffuse disease	26	2	17	3	4	6	2 - 9

CR = complete response; PR = partial response; SD = stable
disease; PD = progressive disease; VCM = vincristine, cyclophosphamide, methotrexate; P.KP-45 = *Propionibacterium
granulosum* KP-45.

parameters of local resistance in this organ. Ten patients
from this group were treated with 10 mg of P.KP-45 (intravenous infusion) 7 days before lung lavage. A significantly
higher total number of cells (more than 80% were alveolar
macrophages) was observed in lavage obtained from such
treated patients than in untreated controls (Table 3). The
cells were examined for adherent, phagocytic and bacteriolytic activity as well as cytostatic effect against tumour
cells. The results of this study are shown in Table 4 and
Fig. 3. Treatment with P.KP-45 resulted in a pronounced
increase of adherent and phagocytic properties of the cells.
We also observed enhanced bacteriolytic activity of macrophages obtained from lungs of P.KP-45 treated patients. In
addition, the cells showed pronouncedly increased cytotoxic
effect directed against the established line of lung cancer
cells.

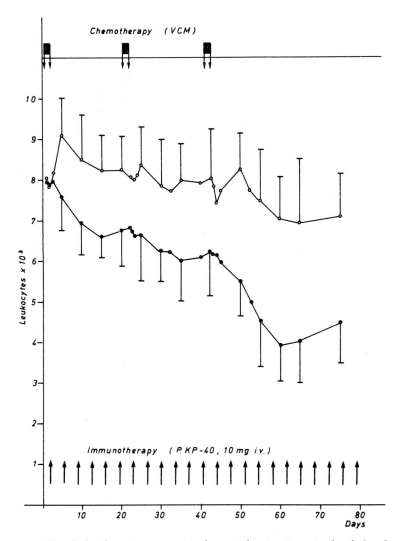

FIG. 2 *Blood leukocyte counts in patients treated with chemo-immunotherapy (O ——— O) and chemotherapy alone (● ——— ●). Results are expressed in mean value ± S.D.*

Stimulation of the cells may the reason for the less fre-quent infective complications of chemotherapy in patients receiving P.KP-45. Increased activity of lung macrophages observed *in vitro* was not confirmed by clinical observations, having no antitumour effect in advanced lung cancer. At least 2 possibilities for this discrepancy can be considered:

TABLE 3

Cellular features of broncho-alveolar lavages in P.KP-45 treated patients with lung cancer

	Total number of cells x 10^6 (mean ± SD)	Macrophages %	Lymphocytes				Granulocytes %
			Total %	T	B	"Null"	
				% of lymphocytes			
Control patients (n - 7)	25.7 ± 4.1	84 ± 1.6	12 ± 2.3	41.2	6.3	52.5	4 ± 1.2
P.KP-45 treated patients (n - 10)	38.3 ± 7.2	87 ± 1.9	11 ± 1.7	35.0	5.1	59.9	2 ± 1.3

TABLE 4

Function of alveolar macrophages obtained from lung cancer patients treated with Propionibacterium

	Cell adherence %	Phagocytosis %	% bacterial killing 1 h	2 h
Control (7)	100	100	32.1 ± 4.1	43.2 ± 5.9
P.KP-45 (10)	153.1 ± 17.4	178.3 ± 27.3	70.6 ± 6.2	83.6 ± 11.3

cytotoxicity of macrophages observed *in vitro* does not play any significant role in control of tumour growth *in vivo* or it does play a role but only until such time as the number of tumour cells does not exceed the boarder value limiting the capacity of this defence mechanism. The significance of other antitumour immunological mechanisms based mainly on T lymphocyte function was omitted from our consideration, because the number of these cells in the lung is very small.

If the second of the above mentioned conceptions is true the theoretical chance of the expression of propionibacteria efficiency may occur only in the case of non-advanced tumours or in adjuvant treatment of operable lung cancer. Surgical treatment of lung cancer is combined with particularly high risk of local progression or clinical demonstration of pre-existing distant micrometastases.

It encouraged us to undertake clinical investigations to evaluate the influence of immunotherapy with P.KP-45 applied prior to surgery. In this study patients are receiving intravenous infusions of the bacteria 7 days before surgical intervention. After surgery, an analysis of cellular content of resected tumours is performed.

The results obtained until now show that general immunotherapy does not significantly influence the cellular composition of the primary lung tumours and intensity of their inflammatory infiltration (Fig. 4). This agrees with the results obtained by us in the treatment of advanced tumours and with the results of our experiments performed on transplantable lung tumour in animals (Roszkowski *et al.*, 1981b).

How propionibacterial immunotherapy influences clinically undetectable secondary tumours remains unclear as such a short time of observation does not enable us to establish the frequency of local recurrences or appearance of distant

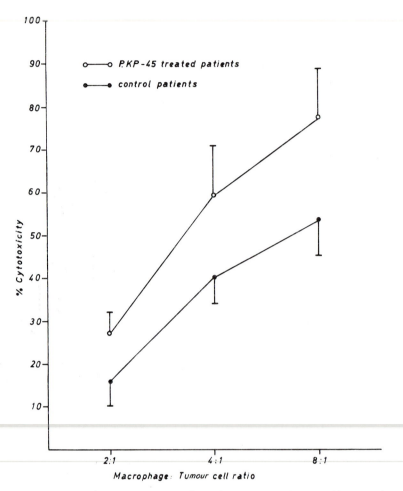

FIG. 3 *Cytotoxic effect of alveolar macrophages directed against established line of human lung cancer cells LC-1. Cytotoxicity measured by inhibition of ^{3}H-thymidine incorporation in 48 h tumour cell cultures.*

metastases. Likewise we cannot yet evaluate antitumour activity of P.KP-45 in conditions of small tumour load.

B. Chemoimmunotherapy of Breast Cancer

The study was performed on 82 patients with advanced disseminated breast cancer. The characteristics of this group of patients are shown in Table 5. All patients were receiving chemotherapy according to regimen FAC (fluorouracil,

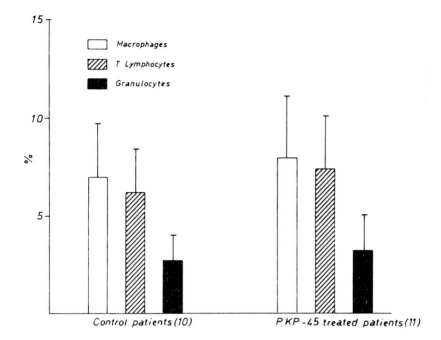

FIG. 4 *Cellular composition of resected lung tumours from patients treated with P.KP-45 and untreated controls. Results are expressed in mean percentage ± S.D.*

adriablastin, cyclophosphamide — Blumenschein *et al.*, 1974). Forty of them were additionally treated with general immunotherapy with P.KP-45 (once a week, intravenous infusions of 10 mg of bacteria). Tables 6 and 7 present the results of this study. A higher percentage of objective remissions and the prolonged mean time of their duration was observed in the group of patients receiving chemoimmunotherapy. Prolonged survival was also noted in this group of patients (Fig. 5). The results of this study were analysed taking into consideration predominant metastatic organs. Significant differences between both groups of patients (treated with chemo- and chemoimmunotherapy) were noticed. Patients with metastases in lungs and liver responded more frequently to chemoimmunotherapy than others. These differences were not observed among patients with other localization of metastases. In the case of metastases to the central nervous system regressions were not seen. It must be emphasized that the antitumour effect of chemoimmunotherapy was most pronounced in organs (lungs, liver) possessing a highly developed reticuloendothelial system. We can reject possible accidental difference

TABLE 5

*Clinical characteristics of patients with
metastatic breast cancer*

Group	FAC	FAC + P
Number of patients	42	40
Age	50 (31 - 66)	54 (39 - 62)
Time from detection of metastases	6 (2 - 8)	7 (3 - 12)
Karnofsky index	65 (50 - 90)	70 (50 - 90)
Previous treatment		
VCMFP	9	11
Radiotherapy	10	4
Hormonal treatment	12	14
No therapy	11	11
Predominant organs of metastases		
Lung	14	17
Osseus	17	13
Liver	4	6
Mediastinal lymph nodes	3	1
Peripheral nodes	1	2
Brain	3	1

FAC = fluorouracil, adriablastin, cyclophosphamide.
P   = *Propionibacterium granulosum* KP-45

in size of metastases among both analysed groups, as the size
of measurable tumour lesions was evaluated and both groups of
patients were similar in this respect at the beginning of
treatment.

Attention should also be drawn to the fact that transient
liver enlargement was observed in patients receiving P.KP-45
which may suggest stimulation of the reticuloendothelial system in this organ. This observation corresponds to the
results of experimental studies performed on animals showing
increased weight of liver as well as its infiltration with
monocytes or macrophages and lymphocytes (Roszkowski *et al.*,
1980).

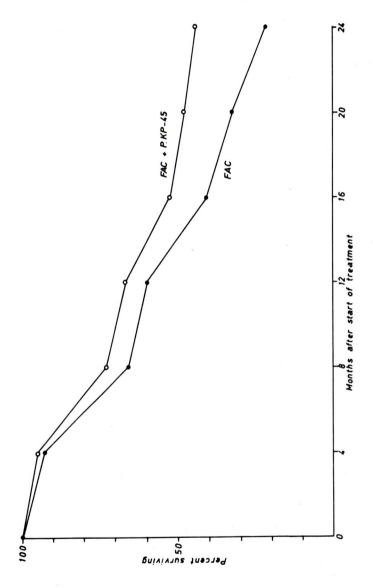

FIG. 5 Survival of patients with breast cancer treated with chemo- and chemoimmunotherapy.

TABLE 6

Results of FAC versus FAC + P.KP-45 treatment of patients with breast cancer

	Number of patients	CR	PR	SD	PD	Duration of remission (months)
FAC	42	4	23	7	8	11 (4 - 26)
FAC + P.KP-45	40	6	26	3	5	16 (7 - 22)

CR = complete response; PR = partial response; SD = stable disease; PD = progressive disease; FAC = fluorouracil, adriabastin, cyclophosphamide.

C. Chemoimmunotherapy of Malignant Melanoma

The study was performed on 34 patients of both sexes with disseminated malignant melanoma. Characteristics of this group are shown in Table 8. All patients received chemotheray with bleomycin, vincristine and cyclochloronitrosourea (De Wash, 1976) and 18 of these patients were additionally treated with P.KP-45 as intravenous infusions at weekly intervals. The results of this study are presented in Tables 9 and 10. The patients undergoing chemoimmunotherapy showed higher percentages of responses and longer duration of remission time as compared to those receiving only oncostatics. The observation of survival time has not been completed.

The analysis of results in respect to localization of metastases has shown pronouncedly higher efficacy of chemoimmunotherapy in the case of tumour lesions localized in lungs and liver than in other organs. However, due to considerable dissemination of melanoma, predominant metastatic organs have not been defined and analysis was performed on individual metastatic sites.

D. Chemoimmunotherapy of Intestinal Carcinoma

The study was performed on 33 patients with metastasizing cancer of sigmoid or rectum. The patients were treated with chemotherapy according to regimen MACF (melphalan, adriamycin, CCNU, 5-fluorouracil — Roszkowski *et al.*, 1982) and 15 of these patients received, additionally, general immunotherapy with P.KP-45.

Low percentage of responses not exceeding 22% was observed in both groups of patients receiving chemo- and chemoimmuno-

TABLE 7

*Results of FAC versus FAC + P.KP-45 treatment
of patients with breast cancer
according to predominant metastatic organs*

	Number of patients	CR	PR	SD	PD
		Lungs			
FAC	14	1	11	–	2
FAC + P.KP-45	18	3	14	1	–
		Osseus			
FAC	17	2	8	5	2
FAC + P.KP-45	13	–	8	1	4
		Liver			
FAC	4	–	1	2	1
FAC + P.KP-45	6	1	4	1	–
		Mediastinal lymph nodes			
FAC	3	1	2	–	–
FAC + P.KP-45	1	1	–	–	–
		Peripheral lymph nodes			
FAC	1	1	–	–	–
FAC + P.KP-45	1	1	–	–	–
		Brain			
FAC	3	–	–	–	3
FAC + P.KP-45	1	–	–	–	1

CR = complete response; PR = partial response; SD = stable disease; PD = progressive disease; FAC = fluorouracil, adriablastin, cyclophosphamide.

therapy. Neither did we observe any other objective beneficial therapeutic effects of P.KP-45 application. However, it has to be emphasized that the examined group consisted of patients with highly advanced malignant disease.

TABLE 8

Clinical characteristics of patients with malignant melanoma

Group	BVC	BVC + P.KP-45
Number of patients	16	18
Age	38 (23 − 52)	42 (34 − 56)
Karnofsky index	70 (50 − 90)	70 (50 − 90)
Distribution of metastases		
liver	6	8
skin	12	9
lung	3	6
lymph nodes	7	4
brain	−	1
bone	3	2

BVC = bleomycin, vincristine, cyclochloronitrosourea.

TABLE 9

Results of BVC versus BVC + P.KP-45 treatment in malignant melanoma patients

	Number of patients	CR	PR	SD	PD	Duration of remission (months)
BVC	16	1	5	1	9	3 (2 − 6)
BVC + P.KP-45	18	3	7	3	5	7 (4 − 11)

CR = complete response; PR = partial response; SD = stable disease; PD = progressive disease; BVC = bleomycin, vincristine, cyclochloronitrosourea.

TABLE 10

Results of BVC versus BVC + P.KP-45 treatment according to localization of metastatic sites of malignant melanoma

Metastatic organ	CR + PR	
	BVC	BVC + P.KP-45
Liver	2/6	6/8
Skin	7/12	6/9
Lung	1/3	4/6
Lymph nodes	4/7	3/4
Brain	–	0/1
Bone	2/3	1/2

CR = complete response; PR = partial response; BVC = bleomycin, vincristine, cyclochloronitrosourea.

II. LOCAL IMMUNOTHERAPY

A. Local Immunotherapy of Digestive Tract Cancer

The study was performed on 20 patients with advanced cancer of the upper and lower part of the digestive tract or with recurrences after surgical treatment (9 patients with stomach cancer, 1 patient with duodenal cancer, 3 patients with sigmoidal cancer and 7 patients with rectal cancer). Surgical treatment according to clinical indications and tumour advancement was performed in 8 cases of 10 patients with upper part of digestive tract cancer and in 8 cases of 10 patients with cancer of the lower part of the digestive tract. Eight patients were treated with chemotherapy (5-fluorouracil) and in these cases local immunotherapy was applied at least 6 weeks after completion of treatment. Characteristics of this group of patients are presented in Table 11.

The general clinical state of patients was evaluated according to 5-range subjective scale. The criteria of the scale are summarized in Table 12. Control of tumour size was performed on the basis of photographic documentation of endoscopic and X-ray examinations.

Ten mg of lyophilized P.KP-45 microorganisms were suspended in 1 ml of 1% lidocaine in saline and injected intratumourally with the help of drains introduced through the

TABLE 11

Patients with adenocarcinoma treated with intratumoural injections of Propionibacterium KP-45

	Stomach		Duodenum	Rectosigmoid	Total
	Infiltrating (N = 5)	Solid (N = 4)	Solid (N = 1)	(Rectum, N = 7) (Sigmoid, N = 3)	
Sex					
Male (age)	4 (56–65 yr)	2 (66,71 yr)	1 (47 yr)	6 (52–72 yr)	13 (47–72 yr)
Female (age)	1 (69 yr)	2 (54,72 yr)	–	4 (43–56 yr)	7 (43–72 yr)
Time from diagnosis of cancer	12–36 mth	7–24 mth	14 mth	12–18 mth	7–36 mth
Therapy before application of Propionibacterium					
Resection or hemicolectomy	2	–	–	5	8
Ventriculointestinal anastomosis	–	–	1	–	–
Resection or hemicolectomy + FU[1]	1	–	–	3	8
Laparotomy + FU	2	2	–	–	–
No therapy	–	2	–	2	4
Size of tumour or infiltration					
2– 6 cm	3	2	–	3	8
6–10 cm	1	1	1	6	9
10–20 cm	1	1	–	1	3
General clinical state					
2	2	1	1	2	6
3	2	2	–	6	10
4	1	1	–	2	4

[1]FU = 5-fluorouracil.

TABLE 12

Scale for evaluation of general clinical state of patients with digestive tract cancer

Grade	Criteria
0	Excellent clinical condition, no complaints, normal body weight, no pain, normal biochemical test results.
1	Good condition, temporary local pain without need for analgetics, body weight normal or slightly (10-15%) subnormal, no difficulty in alimentation and passage, normal or slightly changed results of laboratory tests
2	Weakness, frequent pain with temporary need for analgetics, lowered (10-15%) body weight, slight anaemia with increased (up to 40 mm) erythrosedimentation rate (ESR), abnormal results of laboratory tests without marked hypoproteinaemia.
3	Marked weakness with only temporary improvement, progressive loss of weight, markedly lowered (25-30%) body weight; troubles with alimentation and/or passage; frequent pain with continuous need for analgetics; anaemia with increased (40-90 mm) ESR, abnormal results of biochemical tests with hypoproteinaemia and increased serum alkaline phosphatase.
4	Cancerous cachexia, no appetite, serious troubles with alimentation and/or passage up to occlusion, strong and very frequent pain with continuous need for analgetics; anaemia with high ESR, highly elevated serum alkaline phosphatase (above 800 IU/ml), blood protein and ion abnormalities.

bioptic canal of the endoscope (all injections were done under visual control). Intratumoural injections of P.KP-45 were applied on day 1, 2, 3, 10 and 17 of observation. The time of observation lasted from 2 to 12 months from the first injection of P.KP-45. During this period, 3 patients, all of them with very advanced disease and large tumours, died from cancerous cachexia or intestinal occlusion after 3, 4 and 8 months of observation. In the remaining 17 cases, improvement of general clinical state and marked decrease of tumour size were observed.

In all patients treated with intratumoural injections of
P.KP-45 endoscopic observations revealed increasing swelling
and redness of tumorous tissue during the first 3 to 5 days
after injection. After 7 to 10 days tumour size returned to
its initial value and this was followed by regression. After
2 months of observation the tumour size diminished markedly
in 9 of 20 patients while after 3 months this was observed in
14 of 20 patients. In addition to the 3 patients who did not
react to the therapy and finally died, in another 3 patients
despite improvement of general clinical state only slight
involution of tumour was observed. In all these cases, the
tumour size diminished after 4 months of observation and this
was accompanied by further clinical improvement. The reduction of tumour size occurs during the first 3 to 4 months
after a course of intratumoural injections of P.KP-45. Only
slight changes were observed later. However, after 6, 8 and
12 months of observation no worsening of the clinical conditions of the patients or enlargement of tumour size were
found. The mean tumour size (Fig. 6) calculated for all
treated patients reflects to some extent this phenomenon,
except for the eighth and twelfth months of observation
because of only single cases at that time. The misleading
decrease in mean tumour size was due to the death or shorter
observation period of patients (cases 1 and 2) with large
tumours.

Improvement of the general clinical state was accompanied
by normalization of blood picture and basic immunological
parameters (Fig. 7).

B. Local Immunotherapy of Digestive Tract Cancer — "Twin-Pairs" Study

This study was performed on 14 patient-pairs with advanced
metastatic stomach and colo-rectal adenocarcinoma. The diagnosis was based on the same criteria as described above (Section A). Single or multiple metastases were found in all subjects. Surgical intervention according to indications and
advancement of the disease was performed in 3 of 5 patient-
pairs with stomach cancer and in 6 of 9 patient-pairs with
sigmoidal and rectal cancer (hemicolectomy). Intratumoural
injections of propionibacteria were applied on days 1,7,14,
30, and later once each month. All the 14 patients treated with
P.KP-45 were matched with their control "twins"-patients suffering from the same form of cancer, of as similar as possible
parameters like age, advancement of disease, application of
surgery, former history of the disease and its therapy. During
observation the "twin" controls did not receive anticancer
therapy, except symptomatic treatment and analgetics. This

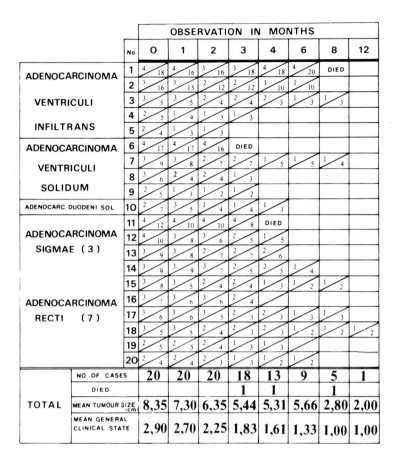

FIG. 6 *Local immunotherapy of digestive tract cancer. General clinical state (upper left number in each compartment) and tumour size (lower right numbers).*

allowed us to analyse statistically the differences between the twin-pairs and to use Wilcoxon's test of pairs for analysis of differences between the whole groups. Evaluation of the general clinical state of patients was based on criteria presented in Table 12. The observation of patients lasted up to 30 months from the first injection of P.KP-45 for the subjects who survived this period of time. In all the observed cases the tumour size diminished after 4-5 months of therapy and only slight changes were observed later. Survival of the 14 "twin"-pairs of cancer patients is presented in Table 13. In all cases the P.KP-45-treated patients survived longer than their control "twins" and 4 of the patients are still

DAYS	0	5	15	25	35	45	55	65
WHITE BLOOD CELLS [cells ×10³/mm³]	6,5±1,4	9,7±2,1	10,3±2,3	10,8±3,7 S	11,2±3,1 S	11,0±4,3 S	10,2±2,9 S	7,1±3,0
BLOOD LYMPHOCYTES [cells ×10³/mm³]	1,6±0,5	1,75±0,57	2,8±0,53 S	3,8±1,1 S	3,1±1,3 S	2,4±1,0	2,7±1,2	2,1±0,9
BLOOD LYMPHOCYTES T [cells ×10³/mm³]	0,74±0,18	0,95±0,2	1,7±0,4 S	2,4±0,6 S	1,8±0,8 S	1,6±0,9	1,7±0,7 S	1,3±0,5 S
BLOOD LYMPHOCYTES B [cells ×10³/mm³]	0,42±0,1	0,41±0,09	0,5±0,13	0,63±0,27	0,47±0,2	0,32±0,09	0,56±0,11	0,56±0,08
IMMUNOGLOBULIN G [mg%]	1300±270	1050±310	1150±250	1100±360	900±290	1010±310	1050±220	1200±410
IMMUNOGLOBULIN A [mg%]	241±70	196±74	208±82	290±69	196±93	282±101	210±83	238±67
IMMUNOGLOBULIN M [mg%]	168±100	156±53	179±62	148±71	164±51	166±78	138±84	141±66
α-2-MACROGLOBULIN [mg%]	326±118	351±97	431±124	350±71	382±83	291±97	403±92	316±107
COMPLEMENT C_3 [mg%]	189±71	230±69	260±52	210±41	190±62	240±76	230±68	220±92
DELAYED HYPERSENSITIVITY TO PHA [mm]	5,6±2,4	5,3±2,6	9,4±3,0 S	9,8±4,1 S	11,0±1,9 S	10,6±3,7 S	10,4±2,8 S	9,2±2,9 S
DELAYED HYPERSENSITIVITY TO PPD [mm]	4,1±1,6	4,3±2,0	8,6±2,7 S	7,4±3,1 S	10,9±2,9 S	9,2±4,2 S	9,7±3,1	8,9±2,6 S
HAEMOGOBLIN [g%]	11,6±3,9	11,2±4,2	10,8±3,6	11,8±2,7	12,3±2,1	13,1±3,2	13,6±3,0	14,2±2,4 S
RED BLOOD CELLS [cells ×10⁶/mm³]	3,9±1,2	3,7±1,3	3,8±1,1	4,1±0,9	4,5±1,3	4,4±2,0	4,7±1,6	4,8±1,8 S
ERYTHROCYTES SEDIMENTA-TION RATE [mm/h]	37±11	49±14	36±13	98±21 S	72±32 S	61±18	47±21	22±7

FIG. 7 Local immunotherapy of digestive tract cancer. Bloodpictures and basic immunological parameters.

TABLE 13

General clinical state and survival of patients treated with intratumoural injections of P.KP-45 and their untreated "twin"-matched controls

Diagnosis	Pair No.	Sex	Age	General clinical state	Survival (months) Control	Survival (months) P.KP-45	Difference in survival (months)
Adenocarcinoma ventriculi	1	M	37–41	2.5	4	25+	21+
	2	M	58–61	2.0	8	24+	16+
	3	M	62–64	4.0	2	5	3	14.0 ± 7.2
	4	M	57–61	4.0	3	14	11
	5	M	47–50	2.5	9	28	19	$t = 1.3$
Adenocarcinoma recti/sigmae	6	F	46–50	3.5	7	29+	22+	$p = N.S.$
	7	F	65–67	3.0	7	16+	9+
	8	F	41–44	2.0	4	32	28
	9	F	44–47	3.5	2	12	10
	10	F	47–50	2.0	11	34	23	18.88 ± 6.11
	11	M	61–64	2.5	3	21	18
	12	M	54–60	2.0	13	33	20
	13	M	54–57	3.0	8	26	18
	14	M	56–59	3.0	9	31	22
				Mean	6.4 ± 3.4	23.5 ± 8.8	17.1 ± 6.7

$t = 9.5$
$p < 0.01$

alive after 16-29 months of therapy.   The mean survival time
of patients treated with P.KP-45 was established to be 23.5
months while that of "twin"-controls 6.4 months, the difference in survival time being about 17 months.

C. Local Immunotherapy of Pleural Maligant Effusion

Local immunotherapy in the form of intrapleural instillation
of 10 mg of P.KP-45 was applied in 7 patients with advanced
breast cancer and with the presence of pleural malignat
effusion combined with neoplastic processes.   All patients
received chemotherapy according to regimen VCMFP (vincristine,
cyclophosphamide, methotrexate, 5-fluorouracil, prednison —
Ansfield *et al.*, 1971).   During the therapy recurrent pleural
effusions were observed in these patients.   The control group
consisted of 8 patients receiving intrapleurally 400 mg of
cyclophosphamide instead of P.KP-45.   Dynamics of increase of
pleural effusion before treatment was similar in both groups,
this was evaluated by X-ray examination and by aspiration of
pleural fluid.
     In 5 cases of 7 patients treated with P.KP-45 increases in
pleural effusion were stopped after only a single intrapleural instillation of bacterial suspension.   Therapeutic
success was achieved only in 2 patients out of 8 in the control group treated with cyclophosphamide.

III. SIDE-EFFECTS

No complications and serious side-effects were observed which
would cause us to give up the application of immunotherapy
with P.KP-45.   However, evaluation of P.KP-45 side-effects
was rather difficult as a majority of patients received simultaneously a chemotherapy with a combined high risk of toxic
symptoms such as nausea, vomiting, diarrhoea, myelosuppression, disturbances in blood coagulation and in the case of
FAC regimen the possibility of cardiomyopathy.   Therefore,
we assumed that only those symptoms which appeared during the
first 24 hours after application of the vaccine were side-
effects of P.KP-45 treatment.   To assess possible late side-
effects or complications, all patients underwent biochemical
monitoring to follow up liver and kidney function.
     Fever, in some cases ranging up to 40°C was the most com-
mon side-effect observed in almost all patients receiving
general immunotherapy (intravenous infusions).   The elevated
body temperature dropped to the normal value in a few hours
after completing P.KP-45 injection.   Chills also accompanied
high fever.   Moderate elevation of blood pressure (about 20-
30 mm Hg) was noted in 30% of all patients immediately after

TABLE 14

Toxicity of chemo- and chemoimmunotherapy of advanced breast, lung, colorectal cancer and malignant melanoma

Side-effects and complications	Chemotherapy	Chemo-immunotherapy
Alopetia	82 (70%)	73 (66%)
Nausea (vomiting)	56 (48%)	72 (65%)
Diarrhoea	12 (10%)	16 (14%)
Leukopaenia 2000	45 (38%)	21 (19%)
Trombocytopaenia 100,000	20 (17%)	6 (5%)
Bone pain	4 (3%)	7 (6%)
Headache	8 (7%)	15 (13%)
Infective complications	18 (15%)	3 (3%)
Total number of patients	117	111

injection of P.KP-45. Nausea and/or vomiting were observed
in 9% of all cases shortly after vaccine infusions.
We also noticed sporadic headache and transient pain in
bones and joints in a few patients. All these symptoms passed
after 24 hours. A comparison of the frequency of toxic symptoms during chemo- and chemoimmunotherapy is summarized in
Table 14. It is significant that myelosuppression and infective complications were more seldom observed in patients
receiving chemoimmunotherapy than in those undergoing chemotherapy alone. In the case of local immunotherapy with P.KP-
45, general side-effects were limited to moderate elevation
of body temperature in 8 of 20 patients receiving intratumoural injections of P.KP-45 and in all 7 patients after
intrapleural instillation of bacterial suspension. Intratumoural injections of P.KP-45 also resulted in local pain
being relieved by suspending the bacteria in lidocaine solution.

IV. CONCLUSIONS

Our results demonstrate that the efficiency of immunotherapy
with P.KP-45 differs considerably, being dependent on factors
such as type of tumour, the organ localization, the form of
immunotherapy as well as the accompanying chemotherapy.
In our clinical material, malignant melanoma and breast
cancer seem to be the most susceptible tumours for general

immunotherapy. In patients with these tumours, undoubtedly
a more pronounced antitumour effect was obtained after chemoimmunotherapy than in parallel groups receiving only cytostatics.

Direct antitumour effect expressed by tumour regression or
inhibition of pleural malignant effusion production was also
observed in patients treated with local injections of P.KP-45.
Breast cancer and malignant melanoma metastases in lungs were
most susceptible to immunotherapy whereas primary small cell
lung carcinoma did not react to this treatment. It is not
clear whether this phenomenon is related to different biology
of the tumours coming from their histological structure or
from differences between primary and secondary origin. In
addition to the above mentioned antitumour effects of P.KP-45
immunotherapy was undoubtedly beneficial also in cases where
these antitumour effects were not noticed. Most important
seems to be the diminishment of dangerous side-effects of
oncostatic treatment. This was expressed by decreased frequency of myelosuppression in patients receiving chemoimmunotherapy as compared to those treated only with oncostatics.
This could be the reason for the rarely observed infective
complications in the former group of patients. However, activation of nonspecific organ defence mechanisms may also be
responsible for this phenomenon.

The results of these studies point to the usefulness of
further clinical trials on the application of P.KP-45 in other
malignant diseases as well as in combination with other forms
of antitumour therapy.

At present we conclude that P.KP-45 can be efficiently
applied as a routine adjuvant form of antitumour therapy with
particularly low toxicity.

REFERENCES

Ansfield, F.J., Ramirez, G., Korbitz, B.C. and Davis, H.L.
(1971). Five drug therapy for advanced breast cancer. A
phase I study. *Cancer Chemother. Rep.* **55**, 183-190.

Blumenschein, R., Cardenas, J., Freireich, E. and Gottlieb,
J. (1974). FAC chemotherapy for breast cancer. *Proc. Am.
Soc. Clin. Oncol.* **15**, 193-197.

De Wasch, G. (1976). Combination chemotherapy with three
marginally effective agents. *Cancer Treat. Rep.* **60**, 1273-
1276.

Halpern, B.N., Prevot, A.R., Biozzi, G., Stiffel, G., Mouton,
D., Morard, J.C., Bouthillier, Y. and Decreusefond, C.
(1964). Stimulation de l'activité phagocytaire du système
reticuloendothelial provoquée par *Corynebacterium parvum*.
Res. J. Reticuloendothel. Soc. **1**, 77-96.

Halpern, B.N., Biozzi, G., Stiffel, G. and Mouton, D. (1966).
Inhibition of tumour growth by administration of killed
Corynebacterium parvum. *Nature (London)* **212**, 853-854.

Karnofsky, D.A. and Burchenal, J.H. (1949). "The Clinical
Evaluation of Chemotherapeutic Agents". (Ed. C.M. MacLeod)
pp. 191-223, Combia University Press, New York.

Roszkowski, K., Gil, J., Roszkowski, W., Szmigielski, S.,
Nowakowski, W., Ko, H.L., Pulverer, G. and Jeljaszewicz, J.
(1981a). Immunostimulation by systemic administration of
Propionibacterium granulosum in patients with primary or
secondary lung tumours. *Eur. J. Respir. Dis.* **62**, 425-433.

Roszkowski, K., Krzakowski, M., Nowakowski, W., Piatkowski,
Z. and Roszkowski, W. (1982). Combination chemotherapy of
advanced colorectal cancer. *In* Chemiotherapia Oncologia.
(In press).

Roszkowski, K., Roszkowski, W., Ko, H.L., Pulverer, G. and
Jeljaszewicz, J. (1981b). Macrophage and T lymphocyte
content of tumours in mice treated with *Propionibacterium*.
Oncology **38**, 334-339.

Roszkowski, K., Roszkowski, W., Ko, H.L., Szmigielski, S.,
Pulverer, G. and Jeljaszewicz, J. (1980). Changed murine
lymphocyte trapping after treatment with *Propionibacterium
granulosum*. *Cancer Immunol. Immunother.* **10**, 33-37.

Woodruff, M.F.A. and Boak, J.L. (1966). Inhibitory effect of
injection of *Corynebacterium parvum* on the growth of tumour
transplants in isogeneic hosts. *Br. J. Cancer* **20**, 345-355.

BACTERIAL PRODUCTS IN THE IMMUNOTHERAPY OF HUMAN TUMOURS

Joseph G. Sinkovics

*Department of Virology and Epidemiology,
Baylor College of Medicine, Texas Medical Centre,
Houston, Texas 77030, USA*

I. INTRODUCTION: SPECIFIC HUMAN TUMOUR ANTIGENS AND IMMUNOINCOMPETENCE OF THE HOST

Many basic questions concerning the immunological relationship between tumours and the human host remain unanswered. Some ideas have been forcefully presented but were not confirmed. The major questions relate (1) to specific antigenicity of malignant human neoplasms, and (2) to the dependence of tumour growth on the immunoincompetence of the host.

Highly antigenic (in particular virally induced or artificially "xenogenized") (Austin and Boone, 1979; Kobayashi, 1979) animal tumours elicit strong and specific immune responses and thus often are immunologically rejected. Even though the likelihood is strong that some human tumours are virally induced and exert a strong antigenic stimulus on the host (Burkitt's lymphoma, nasopharyngeal carcinoma; Kaposi's sarcoma; mycosis fungoides and T-cell lymphomas and leukaemias), in most cases of human tumours, tumour-specific antigens and tumour-specific immune responses could not so far be convincingly demonstrated. Analysis with monoclonal antibodies of human tumour-associated antigens so far failed to document strictly tumour-specific antigens (Sinkovics, 1983).

When lymphocyte-mediated cytotoxicity was first tested against human tumours in the colony inhibition assay, the preoccupation with tumour specificity was so overwhelming that the activities of natural killer cells were ignored (or eliminated from the tabulated results). Since tumour-specific lymphocyte-mediated cytotoxicity could not be documented in the case of human tumours by colony inhibition or other cytotoxicity assays (Kay *et al.*, 1976), attention has now been turned toward natural killer cells which attack target

cells with abnormal cell membrane composition, such as virally
infected cells or tumour cells with the hope that stimulation
of these cells by bacterial products or interferon may lead
to tumour rejection. The fact that human lymphocytes can
lyse tumour cells in the autologous setting, i.e. antigenicity
of tumour cells can be recognized and reacted to in the human
host (Sinkovics *et al.*, 1971), provides the rationale for
further current trials along these lines.

Another notion has been the association of immunoincompetence and tumour growth, envisioning that "correction" of this
immunoincompetence will enable the tumour-bearing human host
to survive longer or even to recover from its malignancy.
However, in most impartially conducted trials, defects of
delayed hypersensitivity as tested with recall skin test antigens or with *in vitro* assays of lymphocyte blastogenesis did
not reflect prognosis, except that patients extensively
treated for advanced tumours became anergic and had a poor
prognosis. As a continuation of this simplistic line of
thought, the early results of nonspecific immunostimulation
of patients with various malignant tumours, in particular
with BCG, were grossly exaggerated and this attitude still
prevails in the clinical evaluation of interferon. Conse-
quential to highly publicized but unconfirmed claims, a profound disbelief permeates the field of human tumour immunology
at the present time.

Against a background in turmoil, this reviewer will en-
deavour to present a critical but balanced view with the
utilization of selected recent data reflecting primarily, but
not exclusively, the effect of bacterial products on the
human host-tumour relationship. Some important modalities of
immunotherapy (levamisole, thymosin, interferon, transfer
factor, immune RNA) will not be evaluated in this review, but
were reviewed recently elsewhere (Sinkovics, 1978).

II. IMMUNOLOGICAL EFFECTS OF BACTERIAL PRODUCTS

The major effects of bacterial products on the immune system
in animals and humans include the well known activation of
B- and T-lymphocytes, natural killer cells and macrophages
and the more recently documented induction of tumour necrosis
factor production (Kiger *et al.*, 1978; Haranaki and Satomi,
1981), abrogation of suppressor cell induction (Pehambarger
and Knapp, 1981), intensification of antibody-dependent cell-
mediated cytotoxicity (Richman *et al.*, 1981), and induction
of immune interferon in non-T lymphoid cells (Catalona *et
al.*, 1981). All these immune activities are known to be
directly or indirectly cytotoxic to tumour cells. However,

a tumour-bearing host, while displaying many of these immune functions as tested *in vitro*, may still fail to reject its tumour.

III. TUMOUR GROWTH IN THE IMMUNOCOMPETENT HOST

Generation of various subsets of suppressor cells (monocytes and T-lymphocytes) by bacterial products (corynebacteria, mycobacteria) (Druker *et al.*, 1981; Wepsic *et al.*, 1981; Koyama *et al.*, 1981; Lichtenstein *et al.*, 1981) can lead to subtle to profound immune defects either generalized, or confined to immune reactions to a specific antigen, thus allowing for the growth of antigenic tumours.

Another mechanism of tumour growth occurs against the background of decreased suppressor cell and increased helper cell activity. In this situation, excessive antibody production ensues. A tumour shedding soluble substances (antigens) may escape immune rejection by preempting the receptors of mononuclear cytotoxic effector cells due to immune complex formation in blood and tissue fluids.

Not only soluble tumour antigens shed into the microenvironment but molecular mediators of tumour cell origin may exert immunosuppressive effect (Hudig *et al.*, 1981; Renk *et al.*, 1981; Sinkovics *et al.*, 1972a,b). These molecular mediators are probably similar to those released by the foetus in order to escape maternal rejection directed to paternal antigens. Among soluble substances of foetal origin, chorionic gonadotropin and α-fetoprotein are known to be immunosuppressive (Gershwin, 1980).

Still another mechanism of tumour growth in an immunocompetent host would be that of antigenic modulation, i.e. cessation of target antigen expression by tumour cells under immunologic attack (Sinkovics *et al.*, 1976).

IV. INFECTIOUS COMPLICATIONS OF HUMAN CANCER

Naturally occurring infections may be expected to influence the host-tumour relationship in favour of the host. However, this is usually not the case. Patients with hairy cell leukaemia (leukaemic reticuloendotheliosis) frequently succumb to infection with mycobacteria (*M. kansasii*), while in experimental systems mycobacteria have been successfully used to intensify immune reactions against tumours. Gram negative sepsis and endotoxin shock is a leading cause of death in the cancer hospital (Sinkovics, 1979). Tumour necrosis leading to remission seldom if ever happens during endotoxin shock. It is possible that immunostimulation with bacterial products may protect against the infectious complications of cancer

and prolongation of survival may be due to the reduced rate
of infectious complications and not due to antitumour effects
(Hewitt, 1980).

V. CONTRADICTORY EXAMPLES OF IMMUNOTHERAPY OF HUMAN TUMOURS

A. Acute Leukaemia

In 4 recent trials for acute lymphoblastic leukaemia of childhood (Table 1), one suggests benefits from BCG vaccination
(Eremeev, 1981); 3, in conformity with most clinical trials
conducted in the past 10 years, show no benefit (Ekert *et
al.*, 1980; Haghbin *et al.*, 1980; Yu *et al.*, 1981).

TABLE 1

*Recent chemoimmunotherapy regimens
for acute lymphoblastic leukaemia*

Reference	Clinical trial and results			
Australia Ekert *et al.*, 1980	196 pts with 117 CR 83 pts chemotherapy 82 pts chemotherapy BCG	No difference in remission duration and survival		
MSKCC Haghbin *et al.*, 1980	3 y chemotherapy No maintenance BCG maintenance	No difference in relapse rate		
China Yu *et al.*, 1981	64 pts Chemotherapy Chemoimmunotherapy	CR/Pts 13/16 41/48 54/64	% 81 85	
Soviet Union Eremeev *et al.*, 1981		Remission duration	Survival	CR >6 y
	50 pts chemo-therapy	13 mth	23 mth	1
	25 pts chemo-therapy BCG	36 mth	45 mth	6

pts = patients; MSKCC = Memorial Sloan-Kettering Cancer
Centre; BCG = Bacille Calmette-Guerin; CR = complete remission.

TABLE 2

Recent chemoimmunotherapy regimens
for acute nonlymphoblastic leukaemia

Reference		Clinical trial and results				
England Summerfield et al., 1979			1st CR	2nd CR	CNS relapse	Survival
	18 pts	chemotherapy	7	1	5	171 d
	32 pts	chemotherapy BCG	8	2	0	388 d
England Zuhrie et al., 1980	51 pts	No maintenance		18 wk remission		
		BCG allogen L cells		35 wk duration		
England Whittaker et al., 1980	182 pts 81 CR	chemotherapy chemotherapy BCG allog L cells		survival 2nd remission prolonged 21/34 62%		
England Lister et al., 1980	86 pts 45% CR 5 pts CR	At 3 y chemotherapy chemotherapy BCG allog L cells CR No CR	Remission duration 8 mth		8 mth 19 mth 2 mth	

Netherlands Storms et al., 1980	116 pts 45 CR		Remission duration	Survival	2nd remission
		23 pts chemotherapy	5 mth	19 mth	2
		18 pts chemotherapy BCG allog L cells	6 mth	34 mth	10
		without 2nd CR with 2nd CR		4 mth 13 mth	

New York Bekesi and Holland, 1979		Remission duration
	chemotherapy	243 d
	chemotherapy NL cells	686 d
	chemotherapy NL cells BCG MER	336 d

Los Angeles Baehner et al., 1979	163 pts		Remission duration	Relapse/pts
		chemotherapy	11 mth	25/47
		chemotherapy L cells BCG	10 mth	26/39

China Yu et al., 1981	157 pts		CR/pts	%
		ANLL chemotherapy	13/42	31
		chemoimmunotherapy	62/115	54
		AML chemotherapy BCG MV		79
		chemotherapy BCG		45
		chemotherapy MV		46

Japan Maekawa et al., 1979	31 pts 14 CR		Survival	Duration of survival from treatment	from relapse	CR
		6 pts chemotherapy	1	240 d	90 d	115 d
		8 pts chemotherapy PS-K	4	590 d	210 d	150 d

N = neuraminidase-treated; L = leukaemic; MV = measles vaccine; CNS = central nervous system; d = day; mth = month; MER = methanol extract residue.

The balance of failure and success in the case of acute nonlymphocytic (acute myeloblastic and myelomonocytic) leukaemia favours immunotherapy (Table 2). Of the 4 British trials (Lister *et al.*, 1980; Summerfield *et al.*, 1979; Whittaker *et al.*, 1980; Zuhrie *et al.*, 1980), 3 claim prolonged remission duration and/or survival. In one of these trials,

prolonged survival was attributed to the high incidence of
second remission induction in relapsed patients who previously
received chemoimmunotherapy (Whittaker *et al.*, 1980). The
same principle emerged in the Dutch trial (Storms *et al.*,
1980). In the New York trial, active tumour-specific immunization with neuraminidase-treated leukaemic cells prolonged
remission duration but this effect was abrogated by BCG-MER
application (Bekesi and Holland, 1979). The Los Angeles
trial failed to demonstrate benefit from the use of leukaemic
cells and BCG (Baehner *et al.*, 1979). In the Chinese trial,
the combined use of BCG and measles vaccine was more effective than BCG or measles vaccine used separately (Yu *et al.*,
1981). In the Japanese trials, the use of polysaccharide PS-K
from *Coriolus versicolor* increased survival and remission
duration in small numbers of patients treated (Maekawa *et
al.*, 1979). BCG cell wall skeleton and irradiated autologous
leukaemic cells added to chemotherapy resulted in temporary
prolongation of survival (Ohno and Kato, 1980).

B. Lymphoma

Earlier reports claimed delayed progression of disease in
patients receiving immunostimulation with bacterial products.
More recently, BCG added to chemotherapy lengthened the survival of patients with nodular lymphocytic lymphomas (Jones,
1979). On the other hand, BCG-MER added to standard treatment regimens of Hodgkin's disease was of no benefit (Vinciguerra *et al.*, 1981), or it was considered to be toxic (fever)
and unacceptable (pain, ulcerations) (Cooper *et al.*, 1980).

C. Sarcoma

Earlier trials with *Corynebacterium parvum* or irradiated
tumour cells and BCG were reviewed (Leventhal, 1981; Sinkovics, 1978). More recent trials used sarcoma cell viral
oncolysates and BCG in patients with metastatic tumours
treated with combination chemotherapy. A delay in progression
of metastatic sarcomas was observed in comparison with patients receiving combination chemotherapy only (Sinkovics *et
al.*, 1978) and patients with circulating lymphocytes cytotoxic to sarcoma cells and/or unblocking serum factors experienced the longest tumour-free survival (Sinkovics, 1980).

D. Melanoma

In Canada, 12 of 64 patients (19%) receiving BCG after surgical removal of primary melanoma relapsed and 5 of the relapsed
patients died; 2 of the 12 patients relapsed locally and 10

developed distant metastases. Of 74 patients treated with
surgery only, 20 relapsed (26%) and 9 of these died; 12 relapsed locally and 8 developed distant metastases (Paterson
et al., 1980). The Sloan-Kettering Cancer Centre recorded no
benefit from the use of BCG, *Corynebacterium parvum* or pseudomonas vaccine in postoperative adjuvant trials for patients
with excised regional lymph node metastases (Pinsky *et al.*,
1980; Hilal *et al.*, 1981). At the Roswell Park Memorial
Institute, recurrence rates of stages I and II postoperative
patients were with surgery only 9 of 28 patients with 6 deaths;
with surgery, dacarbazine and Estracyt 8 of 27 patients with
5 deaths; with surgery and BCG 8 of 29 patients with 4 deaths
(Holtermann, 1980) (Table 3). An Italian trial with postoperative BCG claims significant prolongation of disease-free
interval (Canaletti *et al.*, 1981).

The SWOG treated postoperatively stage III and IV (regional
lymph node or solitary distant metastases excised) patients
with either carmustine, hydroxyurea and dacarbazine or with
these drugs and BCG without significant difference in relapse
rate or survival (Quagliana *et al.*, 1980). In another trial
treatment of stages II and III patients with dacarbazine or
with dacarbazine and BCG-MER gave identical results (Sterchi
et al., 1980). Stage IV patients did not benefit from immunostimulation with bacterial products. In England, vincristine,
dacarbazine and BCG gave a response rate of 38% with 8.5% CR
rate and median survival of 6.7 months for responders
(Thatcher *et al.*, 1981). In Australia, response rates and
survival were the same for patients treated with dacarbazine
or with dacarbazine and *Corynebacterium parvum* (Clunie *et al.*,
1980). At Hershey, Pennsylvania, 1 of 11 patients treated
with dacarbazine and BCG achieved CR and in 2 patients the
disease stabilized with median survival 10 months. Of 9
patients treated with dacarbazine, BCG and *Corynebacterium
parvum*, none entered CR, none had stabilization of disease
and the median survival was 7 months (Wynert *et al.*, 1980).
(Table 3). At M.D. Anderson Hospital dacarbazine, BCG and
transfer factor were given to 35 patients with a 23% response
rate and 7 months median survival; when melphalan was added,
the response rate of 29 patients dropped to 17% and the survival to 4 months (Schwarz *et al.*, 1980). Another clinica-
trial at M.D. Anderson Hospital combined viral oncolysates
and BCG. Patients with excised regional lymph node metastases
received chemotherapy (vincristine, dacarbazine and actinomycin D or dacarbazine and semustine) and BCG; or chemotherapy, BCG and melanoma cell lysates ("viral oncolysates")
for 2 years. Relapse rates were identical in the first 3
years in both groups but in the 4th and 5th years no further
relapses with melanoma occurred in the group treated with

TABLE 3

Recent chemoimmunotherapy trials for melanoma with BCG

Reference	Clinical trial and results
MSKCC Pinsky *et al.*, 1980-1	BCG, *C. parvum*, Pseudomonas vaccine and dacarbazine. Postop: no benefits
Roswell Park Holtermann *et al.*, 1980	Surgery / Surgery DTIC Estracyt / Surgery BCG — Recurrence at 22 mth: no difference
Canada Paterson *et al.*, 1980	64 pts Surgery BCG / 74 pts Surgery — Recurred 12, 20; Died 5, 9
SWOG Quagliana *et al.*, 1980	Surg BCNU Hy DTIC / Surg BCNU Hy DTIC BCG — Relapse/pts 35/86 41%, 37/75 44%; Med. NED time 99 wk, 60 wk
Sterchi *et al.*, 1980	DTIC vs DTIC BCG-MER: no difference
Hershey Wynert *et al.*, 1980	DTIC BCG / DTIC BCG Cp — Pts 11, 9; CR 1, 0; Med. survival 10 mth, 7 mth

BCNU = carmustine; DTIC = dacarbazine; Postop = postoperatively; NED = no evidence of disease; Hy = Hydrea; MER = methanol extraction residue; Cp = *Corynebacterium parvum*.

chemotherapy, BCG and melanoma cell lysates, whereas relapses with melanoma continued to occur in the group treated with chemotherapy and BCG. Those patients with stage IV disease who received chemotherapy, BCG and melanoma lysates experienced the longest survival (12 months) in comparison with patients receiving chemotherapy or chemotherapy and BCG (5-6 months) (Sinkovics, Plager, Papadopoulos, in preparation).

E. Head and Neck Carcinomas

Patients with some of these tumours may express herpes virus-coded or associated antigens in their tumours and may respond immunologically to these antigens. On the other hand, many of these patients undergo profound immunosuppression.

Chemoimmunotherapy or radioimmunotherapy of these tumours remain controversial, since early reports favouring BCG vaccination could not be confirmed. While BCG exerted a weak effect in reducing relapse rate (54% in treated and 76% in control group), neuraminidase-treated autologous tumour cells and BCG together had no beneficial effect (Taylor *et al.*, 1980). *Corynebacterium parvum* added to radiotherapy also failed to improve disease-free state or survival (Cheng *et al.*, 1982). Of 36 patients treated after surgery and radiotherapy with *Nocardia rubra* cell wall skeleton preparation, 75% at 6 months, 61% at 12 months and 50% at 18 months remained tumour-free. Of 35 control patients without immunostimulation, 71% at 6 months, 41% at 12 months and 25% at 18 months remained tumour-free (Itoh *et al.*, 1980). More recently, human leukocyte interferon injected peritumourally was claimed to have reduced tumour size and regional or distant spread (Ikic *et al.*, 1981).

F. Breast Carcinoma

The effects of immunostimulation with BCG remain controversial in advanced disease. As with acute myelogenous leukaemia, melanoma, and colon carcinoma, a group at the M.D. Anderson Hospital claimed prolonged survival for those stage IV patients who received FAC chemotherapy (5-fluorouracil, adriamycin and cyclophosphamide) and BCG versus historical control patients receiving FAC chemotherapy only (Hortobagyi *et al.*, 1979). However, the responders in this historical control group had an unusually short survival of 16 months versus 24 months for the group receiving also BCG. In clinical trials using adriamycin-containing combinations without immunotherapy, median survival in excess of 20 months is expected. In France, the comparison of vincristine and FAC with vincristine, FAC and BCG gave 73 and 74% response rates and 20

months median survivals without any difference in favour of
BCG (Pouillart *et al.*, 1979). At the Mayo Clinic BCG-MER
added to cyclophosphamide, 5-fluorouracil and prednisone
failed to improve remission rate or prolong time to progression (Britell *et al.*, 1979). In North Carolina BCG-MER was
added to cyclophosphamide, adriamycin, vincristine, 5-fluoro-
uracil and prednisone or to cyclophosphamide, adriamycin,
methotrexate and 5-fluorouracil without additional benefit
(Muss *et al.*, 1981).

In the adjuvant setting, the addition of tamoxifen to the
CMF regimen (cyclophosphamide, methotrexate and 5-fluoroura-
cil) improved the response rate of women with estrogen recep-
tor-positive tumours but further addition of BCG provided no
more benefit (Hubay *et al.*, 1980). However in Montpellier,
France, postoperative patients with T_{1-3} N_{0-1} disease received
either no further treatment or BCG. Of 160 patients, relapse
rates were 58% versus 26% suggesting benefits derived from
BCG vaccination (Serrou *et al.*, 1981) (Table 4).

G. Gastrointestinal Carcinomas

Immunostimulation with bacterial products did not make a
major impact on the natural course of gastric carcinoma. In
Toronto, Canada (Makowka *et al.*, 1980) and Vienna, Austria
(Dittrich *et al.*, 1981), postoperative chemotherapy versus
chemoimmunotherapy resulted in no improvement of survival.
In Japan treatment with streptococcal polysaccharide OK-432
or with mycelial extract PS-K lengthened survival from 7
months in the untreated group and 11 months in the group
receiving chemotherapy (mitomycin C, 5-fluorouracil and cytarabine) to 14 months for those receiving mitomycin C, fluorouracil and OK-432 (Fujimoto *et al.*, 1979; Nio *et al.*, 1980).
Immunostimulation with BCG or *Corynebacterium parvum* or
immunorestoration with levamisole was ineffective in Dukes B
and C colon carcinoma (Bancewicz *et al.*, 1980; McPherson *et
al.*, 1980; Souter *et al.*, 1979). However, one trial claims
median disease-free time of 35 and 34 months for those patients with resected Dukes C disease who received either BCG or
5-fluorouracil and BCG versus 21 months for resected historical controls. In the BCG and 5FU-BCG group relapse rates
were 18 of 52 patients with 10 deaths and 30 of 69 patients
with 16 deaths, respectively (Mavligit, 1980). BCG-MER was
also effective in prolonging disease-free time and survival
(Robinson *et al.*, 1980).

TABLE 4

Recent chemoimmunotherapy trials for breast carcinoma

Reference	Clinical trial and results

MDAH Hortobagyi et al., 1979

Treatment	Pts	% PR CR	Duration of response and survival	
FAC	44	73	9 mth	16 mth
FAC BCG	105	76	14 mth	24 mth

France Pouillart et al., 1979

Treatment	Response	% PR CR	Median survival
VCR FAC	54/74	73	20 mth
VCR FAC BCG	51/69	74	20 mth

Mayo Britell et al., 1979

Treatment	% CR	% PR	% stable	Time to progression
CFP	13	39	39	248-261 d
CFP BCG MER	6	32	50	159 d

N. Carolina Muss et al., 1981

Treatment	% Response	Duration of response and survival	
CDVFP	59	16 mth	25 mth
CDVFP BCG MER	54	14 mth	23 mth
CDMF	43	12 mth	26 mth
CDMF BCG MER	43	15.5 mth	25.6 mth

Ohio, Hubay et al., 1980, 1981

CMF TXF vs CMF TXF BCG: Additional benefit from TXF to ER+ women; no additional benefit from BCG.

France Serrou et al., 1981

160 pts		% relapse at 1 y
T_{1-3} N_{0-1}	Surgery BCG	26
	Surgery ----	58

PR = partial remission; FAC = 5-fluorouracil, adriamycin, cyclophosphamide; VCR = vincristine; CFP = cyclophosphamide, 5-fluorouracil, prednisone; CDV = cyclophosphamide, doxorubicin, vincristine; FP = 5-fluorouracil, prednisone; ER = estrogen receptor; CMF = cyclophosphamide, methotrexate, 5-fluorouracil; TXF = tamoxifen.

H. Bronchogenic Carcinomas

Small cell undifferentiated (oat cell) carcinoma of the lung
is highly responsive to chemo- and radiotherapy but immunostimulation with BCG or BCG-MER provided no additional benefit (Aisner and Wiernik, 1980; Holoye *et al.*, 1978; McCracken
et al., 1980; Paschal *et al.*, 1980). However, patients not
responding well to chemotherapy survived slightly longer when
BCG was added to the treatment regimen (Holoye *et al.*, 1978)
(Table 5). At M.D. Anderson Hospital radio-chemotherapy
(etoposide, ifosfamide, vincristine and adriamycin) plus
Corynebacterium parvum resulted in 11 CR of 14 patients (79%)
with limited, and in 7 CR of 19 patients (37%) with extensive
disease. CNS relapse of those patients who did or did not
receive prophylactic radiotherapy to the brain was 1 of 15
and 5 of 17, respectively. Survival in weeks was 63 for all
patients, 71 for limited and 56 for extensive disease, 71 for
those in complete and 50 for those in partial remission, 101
for those with, and 48 for those without brain irradiation.
Corynebacterium parvum did not improve these results (Valdivieso *et al.*, 1981). In Japan, the addition of OK-432 to
radiotherapy and ifosfamide increased survival (Miyamoto *et
al.*, 1980) (Table 5).

Non oat cell bronchogenic carcinomas have been treated
after resection of stages I and II disease with intrapleural
injection of BCG or *Corynebacterium parvum*. The trial in
Albany, New York, continues to favour the treated patients:
in 1980 12 of 30 BCG-treated (40%) and 24 of 36 control
patients (66%) relapsed (Bennett and McKneally, 1980). In
1981, of 169 patients entered, 33% in the BCG-treated and 62%
in the nontreated group relapsed (McKneally *et al.*, 1981).
Similar but not identical clinical trials tend to confirm
(Edwards, 1979; Jansen *et al.*, 1980; Perlin *et al.*, 1980) and
refute (Lowe *et al.*, 1980) these results (Table 6). When BCG
was given intrapleurally and levamisole by mouth, the tendency of BCG to reduce recurrence rate was lost and an increased recurrence rate occurred (Wright *et al.*, 1980). Side-
effects of intrathoracic BCG injection are hyperpyrexia and
BCG pneumonia; these patients require isoniazid therapy for
12 weeks (Gail *et al.*, 1981). Patients with advanced squamous cell carcinoma receiving both BCG and levamisole in
addition to radiotherapy experienced less metastases and
longer survival than patients receiving no immunotherapy
(Pines, 1980). In patients with stage III squamous cell carcinoma treated by radiotherapy and chemotherapy, intrapleural
BCG shortened survival (Ruckdeschel *et al.*, 1980).

Killed *Corynebacterium parvum* vaccine was injected intrapleurally in 7 patients after resection of stage I tumour:

TABLE 5

Recent trials of chemoimmunotherapy for oat cell carcinoma

Reference	Clinical trial and results

MDAH Holoye et al., 1978

Med. survival (wks)

Response	VAC	VAC BCG
CR	50	48
PR	37	52
No response	9	19

SWOG McCracken et al., 1980

Treatment	% response	Duration of response and survival	
XRt chemotherapy	46	23 wk	85 wk
XRt chemotherapy BCG	50	20 wk	60 wk

Baltimore Aisner and Wiernik, 1980

Chemotherapy vs chemotherapy BCG MER: no additional benefit.

Paschal et al., 1980

Treatment	% PR CR	Duration of response and survival	
XRt chemotherapy	63	5 mth	9 mth
XRt chemotherapy BCG MER	70	7.6 mth	10.4 mth
XRt chemotherapy Extensive	30 pts		8.6 mth
XRt chemotherapy Localized	19 pts		10.5 mth
XRt chemotherapy Extensive	28 pts		8.5 mth
BCG MER Localized	15 pts		12.9 mth

Japan Miyamoto et al., 1980

Treatment	Pts	% survival	1 y	2 y	3 y
XRt Ifosf	39	6.7 mth	24	13	7
XRt Ifosf OK 432	10	19.2 mth	40	40	20

VAC = vincristine, adriamycin, cyclophosphamide; XRt = radiotherapy.

TABLE 6

Recent immunotherapy trials for bronchogenic carcinoma

Reference	Clinical trial and results			
Albany, N.Y. Bennett and McKneally, 1980; McKneally *et al.*, 1981	169 pts Stage I	Recurrence 1980		At 3 y (1981)
	Control	24/36		62%
	BCG	12/30		33%
Ruckdeschel *et al.*, 1980	Stage III		Med. survey (wks)	
	XRt chemotherapy		46.5	
	XRT chemotherapy BCG		23.3	
F. Hutchinson C C Seattle Wash. Wright *et al.*, 1980	Stages I-II BCG intrapleurally, levamisole p.o.: increased recurrence rate			
England Pines, 1980	Stages III-IV BCG and levamisole reduced rate of metastases and improved survival			
Fox *et al.*, 1980	Treatment	Relapse		
	Postop. Cp levamisole	1/17		
	Postop. ——— ——	5/21		
Mitcheson and Castro, 1979	Treatment	Survival	Deaths	
	XRt ———	10/22	12 3.3 mth	
	XRt Cp	10/22	12 4.4 mth	

Hollinshead, 1981	Treatment	Pts	Died	4 y survival %
	Surgery	16	8	} 50
* high dose	Surgery chemotherapy*	8	4	
methotrexate	Surgery immunotherapy**	15	4	} 80
leucovorin	Surgery chemo*immunotherapy**	13	4	
rescue		Pts		3 y survival %
	Surgery	27		35
**tumour-associ- ated antigen in Freund's adjuvant	Surgery immunotherapy**	24		85
	Surgery immunotherapy**	29		90

Cp = *Corynebacterium parvum.*

6 patients remained alive tumour-free at 2.5 years (Sloan *et al.*, 1980). Post resection administration of *Corynebacterium parvum* resulted in 1 relapse of 17 patients, whereas in the control group receiving no immunotherapy, 5 of 21 patients relapsed (Fox *et al.*, 1980). Of 22 and 22 nonresectable patients receiving radiotherapy and radiotherapy and *Coryne-*

bacterium parvum, respectively, 12 and 12 died at 12 to 14 months, respectively (Mitcheson and Castro, 1979) (Table 6).

Intralesional BCG placed through fiberoptic bronchoscopy in nonresectable tumours, in addition to chemotherapy, appeared to reduce tumour size (Miller *et al.*, 1980). Percutaneous injection of BCG into pulmonary tumours may result in increase in size of tumours as granulomas are formed. Nine of 21 patients developed pneumothorax during this procedure (Fon *et al.*, 1981). The lymphocyte population of tumours injected with BCG consisted of activated T- and natural killer cells (Niitsuma *et al.*, 1981).

Japanese oncologists claim that OK-432 when added to chemotherapy prolonged the survival of resected patients (Watanabe *et al.*, 1980).

Newer developments of a clinical trial started in Canada by T.H.M. Stewart, A.C. Hollinshead and J.E. Harris continue to show prolonged survival for those patients who after resection of stages I-II tumours received immunotherapy with tumour-associated antigen in Freund's adjuvant with or without methotrexate (Table 6) (Hollinshead, 1981).

I. Genitourinary Carcinomas

Small superficial transitional cell carcinomas of the urinary bladder may regress after repeated intravesical instillations of BCG (Camacho *et al.*, 1980; Lamm *et al.*, 1980; Gunther *et al.*, 1979; Morales, 1980; Morales and Ersil, 1979; Rodriguez-Netto *et al.*, 1979), but not all of the clinical trials are confirmatory (Stober and Peter, 1980) (Table 7).

Patients with carcinoma of the prostate in all stages appear to have benefited from BCG vaccination added to conventional treatment modalities (Guinan *et al.*, 1979) (Table 8).

The SWOG chemoimmunotherapy (adriamycin, cyclophosphamide and BCG) trial for advanced ovarian carcinoma resulted in 57% and 43% response rates for chemoimmunotherapy and chemotherapy, respectively, but CR occurred at higher rate in the chemoimmunotherapy group and the 22 months median survival duration of this group of patients was significantly better than 14 months for the chemotherapy group (Alberts, 1979).

Intraperitoneal *Corynebacterium parvum* for the treatment of ascitic ovarian carcinoma resulted in disappearance of ascites in 3 of 8 patients without reduction in size of solid tumours. Patients receiving intraperitoneal *Corynebacterium parvum* did not show activation of macrophages or natural killer cells (Mantovani *et al.*, 1981). At Roswell Park Memorial Institute, cyclophosphamide with high dose methotrexate and citrovorum factor rescue (MECY regimen) resulted in

TABLE 7

*Immunotherapy with BCG after standard treatment
for carcinoma of urinary bladder*

Reference	Clinical trial and results		
	Treatment	Recurrence	
San Antonio, Tx.	Controls	8/19	42%
Lamm *et al.*, 1980	BCG	3/18	17%
MSKCC N.Y.	Controls	10/22	45%
Camacho *et al.*, 1980	BCG	4/22	18%
Brazil	TUR	145 pts	100%
Rodriguez-Netto	TUR Thio TEPA	20 pts	35%
et al., 1979	TUR BCG	21 pts	5%
Germany	Controls	36 pts	61%
Stober and Peter,	BCG	30 pts	57%
1980			

TUR = transurethral resection; TEPA = triethylene phosphora-
mide; MSKCC = Sloan Kettering Cancer Centre.

TABLE 8

Immunotherapy with standard treatment for prostatic carcinoma

Reference	Clinical trial and results		
Guinan *et al.*,	92 pts		
1979	Stage	Survival (mth)	
		BCG	Control
	A	38.4	24.4
	B	59	33
	C	22.7	13.5
	D	35	22.9
	Average	37.5	21

TABLE 9

Newer microbial preparations for the immunostimulation of tumour-bearing patients

Preparation	Reference
Mycobacterium smegmatis oil-attached cell wall skeleton with trehalose dimycolate	Vosika *et al.*, 1979
BCG cell wall skeleton with trehalose dimycolate attached to oil microdroplets	Vosika *et al.*, 1979; Richman *et al.*, 1981
Nocardia rubra cell wall skeleton	Azuma *et al.*, 1979; Ogura *et al.*, 1979; Itoh *et al.*, 1980
Pseudomonas vaccine	Snyder *et al.*, 1981
C1740 Klebsiella pneumoniae glycoprotein	Lang *et al.*, 1980
OK-432 streptococcus polysaccharide (Picibanil)	Watanabe *et al.*, 1980; Nio *et al.*, 1980; Miyamoto *et al.*, 1980
PS-K (polysaccharide Kreskin) from Coriolus versicolor	Maekawa *et al.*, 1979; Mugitani *et al.*, 1979; Kasamatsu *et al.*, 1980
Bestatin from Streptomyces olivoreticuli	Serrou *et al.*, 1981
Cyclomunine (Fusarium equiseti)	Serrou *et al.*, 1981
Schizophyllan polysaccharide from Schizophyllum commune	Takahashi *et al.*, 1979 Kuramoto *et al.*, 1979

48-67% CR of stages III-IV ovarian carcinoma but addition of *Corynebacterium parvum* did not improve the regimen (Barlow *et al.*, 1980).

Carcinoma of the kidney is resistant to radio- and chemo-therapy. Of 11 patients immunized with their own tumour cells, 2 "improved" (Prager *et al.*, 1980). Of 14 patients immunized with autologous irradiated tumour cells mixed with

Corynebacterium parvum, 4 showed "objective response", but different metastatic tumours in the same patient displayed divergent responses (McCune *et al.*, 1981).

J. New Microbial Products

Table 9 summarizes those biological response modifiers that are of microbial origin and have already been given to patients. Because of the abuse some earlier preparations (BCG, interferon) have been subjected to (gross exaggeration of efficacy; reckless publicity), these newer, more refined and promising products will require decency and great care in their clinical evaluation.

VI. ABBREVIATIONS

CNS	=	central nervous system
BCG	=	Bacille Calmette-Guerin
CR	=	complete remission
CMF	=	cyclophosphamide, methotrexate, 5-fluorouracil
FAC	=	5-fluorouracil, adriamycin, cyclophosphamide
MDAH	=	M.D. Anderson Hospital
MSKCC	=	Memorial Sloan-Kettering Cancer Centre
MER	=	methanol extraction residue
PR	=	partial remission
PS-K	=	polysaccharide Kreskin
SWOG	=	Southwest Oncology Group
N	=	neuraminidase-treated
L	=	leukaemic
MV	=	measles vaccine
DTIC	=	dacarbazine
BCNU	=	carmustine
VCR	=	vincristine
CDVFP	=	cyclophosphamide, doxorubicin, vincristine, 5-fluorouracil, prednisone
CDMF	=	cyclophosphamide, doxorubicin, methotrexate, 5-fluorouracil
TXF	=	tamoxifen
CFP	=	cyclophosphamide, 5-fluorouracil, prednisone
XRt	=	radiotherapy
TUR	=	transurethral resection

VII. REFERENCES

Aisner, J. and Wiernik, P.H. (1980). *Cancer* **46**, 2543-2549.

Alberts, D.S. (1979). Abstr. 5th Ann. Meeting Med. Oncol. Soc. Nice, France, Dec. 1979, p.52.

Austin, F.C. and Boone, C.W. (1979). *Adv. Cancer Res.* **30**, 301-345.

Azuma, I. and Yamamura, Y. (1979). *Gann Monogr. Cancer Res.*
   24, 121-141.

Baehner, R.L., Bernstein, I.D., Sather, H., Higgins, G., Mc-
   Credie, S., Chard, R.L. and Hammond, D. (1979). *Med. Ped.
   Oncol.* **7**, 127-139.

Bancewicz, J., Macpherson, S.G., McVie, J.G., Calman, K.C.,
   McArdle, C.S. and Soukop, M. (1980). *J. Roy. Soc. Med.*
   73, 137-139.

Barlow, J.J., Piver, M.S. and Lele, S.B. (1980). *Cancer* **46**,
   1333-1338.

Bekesi, J.G. and Holland, J.F. (1979). *Haematol. Blood
   Transfus.* **23**, 79-87.

Bennett, J.A. and McKneally, M.F. (1980). *Proc. Am. Assoc.
   Cancer Res.* **21**, 249 (abstr. 1000).

Britell, J.C., Ahmann, D.L., Bisel, H.F., Frytak, S., Ingle,
   J.N., Rubin, J. and O'Fallon, J.R. (1979). *Cancer Clin.
   Trials* **2**, 345-350.

Camacho, P., Pinsky, C., Kerr, D., Whitmore, W. and Oettgen,
   H. (1980). *Proc. Am. Soc. Clin. Oncol.* **21**, 359 (abstr.
   C-160).

Canaletti, R., Nouvenne, R., Marola, G., Rossi, G., Sturba,
   F., Bachetti, G., Piva, G., Pezzarossa, G., Moroni, P. and
   Bonsignori, M. (1981). *Tumori* **67**, 89 (suppl. A).

Catalona, W.J., Ratliff, T.L. and McCool, R.E. (1981).
   Nature **29**, 77-79.

Cheng, V.S.T., Suit, H.D., Wang, C.C., Raker, J., Kaufman, S.,
   Rothman, K., Walker, A. and McNulty, P. (1982). *Cancer*
   49, 239-244.

Clunie, G.J., Gough, I.R., Dury, M., Furnival, C.M. and
   Bolton, P.M. (1980). *Cancer* **46**, 475-479.

Cooper, M.R., Pajak, T.F., Gottlieb, A. and Holland, J.F.
   (1980). *Proc. Am. Soc. Clin. Oncol.* **21**, 467 (abstr. C-
   583).

Dittrich, C., Fritsch, A., Hofbauer, F., Jakesz, R., Miksche,
   M., Moser, K. and Rainer, H. (1981). *Excerpta Med. Int.
   Congr. Ser.* **542**, 389-395.

Druker, B.J., Wepsic, H.T., Alaimo, J. and Murray, W. (1981).
   Cancer Immunol. Immunother. **10**, 227-237.

Edwards, F.R. (1979). *Thorax* **34**, 801-806.

Ekert, H., Waters, K.D., Matthews, R.N., Tauro, G.P., Rice,
   M.S., Seshadi, R., Mauger, D.C., Tiernan, Y.R., McWhirter,
   W.R., O'Reagan, P., Olsen, T.E. and Mathews, J.D. (1980).
   Med. Ped. Oncol. **8**, 353-360.

Eremeev, V.S., Rumyantsev, A.G., Akimova, G.V., Osipov, S.G.,
   Kislyak, N.S. and Bergolz, V.M. (1981). *Neoplasma* **28**,
   219-221.

Fon, G.T., Bein, M.E., Holmes, E.C. and Huberman, R.P. (1981).
   A.J.R. **137**, 269-275 (I.C.R.D.B. Dec. 1981, abstr. 2).

Fox, R.M., Woods, R.L., Tattersall, M.H. and Basten, A.
    (1980). *Int. J. Rad. Oncol. Biol. Phys.* **6**, 1043-1045.
Fujimoto, S., Takahashi, M., Minami, T., Ishigami, H.,
    Miyazaki, M. and Itoh, K. (1979). *Japan. J. Surg.* **9**, 190-
    196.
Gail, M.H., Oldham, R.K., Holmes, E.C., Wright, P.W., McGuire,
    W.P., Moutain, C.F., Lukeman, J.M., Feld, R., Hill, L.D.,
    Eagan, R.T. and Pearson, F.G. (1981). *Cancer Immun.
    Immunother.* **10**, 129-137.
Gershwin, M.E. (1980). *Cancer Immun. Immunol.* **8**, 1-2.
Guinan, P., Baumgartner, G., Totonchi, E., John, T., Crispen,
    R., Rao, R. and Ablin, R. (1979). *Urology* **14**, 561, 565.
Gunther, M., Seyfarth, M., Hudemann, B., Riedel, H., Schutt,
    C., Jense, H.L., Werner, H., Nizze, H., Friedrich, A.,
    Nimmich, W., Budde, E. and Erdmann, T. (1979). *Zentralbl.
    Chirurg.* **104**, 1610.
Haghbin, M., Cunningham-Rundles, S., Thaler, H.T., Gupta, S.,
    Hecht, S., Murphy, M.L. and Oettgen, H.F. (1980). *Cancer*
    46, 2577-2586.
Haranaka, K. and Satomi, N. (1981). *Jap. J. Exp. Med.* **51**,
    191-194.
Hewitt, B.H. (1980). *Mod. Med. Canada* **35**, 1352-1361.
Hilal, E.Y., Pinsky, C.M., Hirshaut, Y., Wanebo, H.J., Hansen,
    J.A., Braun, D.W., Fortner, J.G. and Oettgen, H.F. (1981).
    Cancer **48**, 245-251.
Hollinshead, A.C. (1981). *Cancer Detect. Prev.* **3**, 419-448.
Holoye, P.Y., Samuels, M.L., Smith, T. and Sinkovics, J.G.
    (1978). *Cancer* **42**, 34-40.
Holtermann, O.A., Karakouris, C.P., Berger, J. and Constan-
    tine, R.L. (1980). *Proc. Am. Soc. Clin. Oncol.* **21**, 400
    (abstr. C-319).
Hortobagyi, G.N, Gutterman, J.U., Blumenschein, G.R.,
    Tashima, C.K., Burgess, M.A., Einhorn, L., Buzdar, A.U.,
    Richman, S.P. and Hersh, E.M. (1979). *Cancer* **44**, 1955-
    1962.
Hubay, C.A., Pearson, O.H., Marshall, J.S., Rhodes, R.S.,
    DeBanne, S.M., Rosenblatt, J., Mansour, E.G., Hermann, R.
    E., Jones, J.C., Flynn, W.J., Eckert, C. and McGuire, W.L.
    (1980). *Cancer* **46**, 2805-2808.
Hubay, C.A., Pearson, O.H., Marshall, J.S., Stellato, T.A.,
    Rhodes, R.S., DeBanne, S.M., Rosenblatt, J., Mansour, E.G.,
    Hermann, R.E., Jones, J.C., Flynn, W.J., Eckert, C. and
    McGuire, W.L. (1981). *Breast Cancer Res. Treatm.* **1**, 77-82.
Hudig, D., Djobodze, M., Redelman, D. and Mendelsohn, J.
    (1981). *Cancer Res.* **41**, 2803-2808.
Ikic, D., Brodarec, I., Padovan, I., Knazevic, M. and Soos,
    E. (1981). *Lancet* **1**, 1025-1027.

Itoh, M., Sakai, S., Murata, M. and Sasaki, R. (1980). *Jibi. Inkoka. Rinsho* **73** (suppl. 1), 415-421 (I.C.R.D.B. April, 1981, abstr. 3).

Jansen, H.M., The, T.H. and Orie, N.G. (1980). *Thorax* **35**, 781-787.

Jones, S.E. (1979). Abstracts 5th Ann. Meeting Med. Oncol. Soc. Nice, France, Dec 1-3, 1979, p. 52.

Kasamatsu, T. (1980). *Exc. Med. Int. Congr. Ser.* **512**, 1140-1142.

Kay, H.D., Thota, H. and Sinkovics, J.G. (1976). *Clin. Immun. Immunopathol.* **5**, 218-234.

Kiger, N., Schulz, J., Khalil, A., Caillou, B. and Mathe, G. (1978). *Cancer Immun. Immunother.* **5**, 207-210.

Kobayashi, H. (1979). *Adv. Cancer Res.* **30**, 279-299.

Koyama, S., Fujimoto, S., Tada, T. and Sakita, T. (1981). *Int. J. Cancer* **27**, 829-835.

Kuramoto, M., Okamura, K., Yajima, A., Higashiiwai, H. and Suzuki, M. (1979). *Gann To Kagaku Ryoho* **6**, 843-848 (abstr. 19 and 20) I.C.R.D.B., Oct. 1980.

Lamm, D.L., Thor, D.E., Harris, S.C., Reyna, J.A., Stogdill, V.D. and Radwin, H.M. (1980). *J. Urol.* **124**, 38-42.

Lang, J.M., Giron, C., Oberling, F., Marchiani, C. and Zalisz, R. (1980). *Cancer Immunol. Immunother.* **8**, 273-274.

Leventhal, B.G. (1981). *In* "Sarcomas of Soft Tissue and Bone in Childhood". *NCI Monogr.* **56**, 183-187.

Lichtenstein, A., Murahata, R., Sugasawara, R. and Zighelboim, J. (1981). *Cell. Immun.* **5**, 257-268.

Lister, T.A., Whitehouse, J.M., Oliver, R.T., Bell, R., Johnson, S.A., Wrigley, P.F., Ford, J.M., Cullen, M.H., Gregory, W., Paxton, A.M. and Malpas, J.S. (1980). *Cancer* **46**, 2146-2148.

Lowe, J., Iles, P.B., Shore, D.F., Langman, M.J. and Baldwin, R.W. (1980). *Lancet* **1**, 11-14.

Maekawa, I., Kawamura, T., Okabe, M., Imamura, M. and Shira-ishi, T. (1979). *Gan To Kagaku Ryoho* **6**, 611-617 (I.C.R.D. B. Nov. 1980, abstr. 7).

Makowka, L., Falk, R.E., Ambus, U., Bugala, R. and Landi, S. (1980). *Canad. J. Surg.* **23**, 429-431.

Mantovani, A., Sessa, C., Peri, G., Allavena, P., Introna, M., Polentarutti, N. and Mangioni, C. (1981). *Int. J. Cancer* **27**, 437-446.

Mavligit, G.M. (1980). *Prog. Exp. Tumour Res.* **25**, 275-292.

McCracken, J.D., Heilbrun, L., White, J., Reed, R., Samson, M., Saiers, J.H., Stephens, R., Stuckey, W.J., Bickers, J. and Livingston, R. (1980). *Cancer* **46**, 2335-2340.

McCune, C.S., Schapera, D.V. and Henshaw, E.C. (1981). *Cancer* **47**, 1984-1987.

McKneally, M.F., Maver, C., Lininger, L., Kausel, H.W.,
    McIlduff, J.B., Older, T.M., Foster, E.D. and Alley, R.D.
    (1981). *J. Thorac. Cardiovasc. Surg.* **81**, 484-492.
McPherson, T.A., Young, D. and Hratiuk, L. (1980). *Proc. Am.
    Soc. Clin. Oncol.* **21**, 402 (abstr. C-328).
Millar, J.W., Hunter, A.M., Wrightman, A.J. and Horne, N.W.
    (1980). *Eur. J. Resp. Dis.* **61**, 162-166.
Mitcheson, H.D. and Castro, J.E. (1979). *Br. J. Cancer* **40**,
    823.
Miyamoto, H., Abe, S., Inowe, S., Tsuneta, I., Osaki, Y.,
    Murao, M. and Tsuji, H. (1980). *Gan Rinsho* **26**, 146-153
    (I.C.R.D.B. Sept. 1980, abstr. 7).
Morales, A. (1980). *Cancer Immunol. Immunother.* **9**, 69-72.
Morales, A. and Ersil, A. (1979). *In* "Cancer of the Genito-
    urinary Tract". (Eds D.E. Johnson and M.L. Samuels),
    p. 330, Raven Press, New York.
Mugitani, H., Naruki, Y., Matsuo, K., Yoshida, M., Ohtsuka,
    K., Satake, K., Motoki, N., Kakihara, U., Sugawara, M.,
    Kaneda, K., Shinbo, T. and Yata, J. (1979). *Nippon Gan
    Chiryo Gakkai Shi* **14**, 812-818 (O.C.R.D.B. April 1980,
    abstr. 8).
Muss, H.B., Richards, F., Cooper, M.R., White, D.R., Jackson,
    D.V., Stuart, J.J., Howard, V., Shore, A., Rhyne, A.L.
    and Spurr, C.L. (1981). *Cancer* **47**, 22-5-2301.
Niitsuma, M., Golub, S.H., Edelstein, R. and Holmes, E.C.
    (1981). *J. Natl. Cancer Inst.* **67**, 997-1003.
Nio, Y., Nakamoto, K., Tanaka, A., Kiruchi, S. and Henmi, K.
    (1980). *Nippon Gan Chiryo Gakkai Shi* **15**, 161-165)
    (I.C.R.D.B. Feb. 1981, abstr. 4).
Ogura, T., Hirao, F. and Azuma, I. (1979). Proc. 38th Ann.
    Meeting Japanese Cancer Assoc., Tokyo, Japan, 1979) (cited
    by I.C.R.D.B. Oct. 1980 Series CT06 80/10, p.3.
Ohno, R. and Kato, Y. (1980). *Rinsho Ketsueki* **21**, 1324-1327.
    (I.C.R.D.B. June 1981, abstr. 1).
Oshimi, K., Wakasugi, H., Seki, H. and Kano, S. (1980).
    Cancer Immunol. Immunother. **9**, 187-191.
Paschal, B., Richards, F., Muss, H.B., Cooper, M.R., White,
    D.R., Spurr, C.L. and Bearden, J.D. (1980). *Proc. Am.
    Soc. Clin. Oncol.* **21**, 461 (abstr. C-564).
Paterson, A.H.G., Williams, D., Jerry, L.M. and McPherson,
    T.A. (1980). *Proc. Am. Soc. Clin. Oncol.* **21**, 403 (abstr.
    C-333).
Pehambarger, H. and Knapp, W. (1981). *J. Invest. Dermatol.*
    76, 502-505.
Perlin, E., Oldham, R.K., Weese, J.L., Heim, W., Reid, J.,
    Mills, M., Miller, C., Blom, J., Green, D., Bellinger, S.,
    Cannon, G.B., Law, I., Connor, R. and Herberman, R.B.
    (1980). *Int. J. Rad. Oncol. Biol. Phys.* **6**, 1033-1039.

Pines, A. (1980). *Int. J. Radiat. Oncol. Biol. Phys.* **6**, 1041-1042.

Pinsky, C.M., Wittes, R.E., Linvingston, P.O., Krown, S.E., Hirshaut, Y. and Oettgen, H.F. (1980). *Proc. Am. Assoc. Cancer Res.* **21**, 186 (abstr. 746).

Pouillart, P., Jouve, M., Palangie, T., Giralt, E.C. and Asselin, B. (1979). Abstr. 5th Ann. Meet. Med. Oncol. Soc. Nice, France, Dec. 1979, p.52.

Pouillart, P., Palangie, T., Jouve, M., Garcia-Giralt, E., Magdelenat, H. and Asselain, B. (1981). *Bull. Cancer Paris* **68**, 678.

Prager, M.D., Peters, P.C., Baechtel, F.S. and Brown, G. (1980). *Proc. Am. Soc. Med. Oncol.* **21**, 213.

Quagliana, J., Tranum, J., Neidhardt, J. and Gagliano, R. (1980). *Proc. Am. Soc. Clin. Oncol.* **21**, 399 (abstr. C-316).

Renk, C.M., Gupta, R.K. and Morton, D.L. (1981). *Cancer Immunol. Immunother.* **11**, 7-16.

Richman, S.P., Chism, V.T. and Murphy, S.G. (1981). *Cancer Immunol. Immunother.* **11**, 233-238.

Robinson, E., Bartal, A. and Mekori, T. (1980). *Rec. Res. Cancer Res.* **75**, 80-87.

Rodriguez-Netto, N., Napoli, F. and Caserta Lemos, G. (1979). *Rev. Paulista Med.* **94**, 51-53.

Ruckdeschel, J.C., McKneally, M.F., Devore, C., Baxter, D., Sedransk, N., Caradonna, R. and Horton, J. (1980). *Proc. Am. Soc. Clin. Oncol.* **21**, 374 (abstr. C-218).

Schwarz, M.A., Gutterman, J.U., Burgess, M.A., Heilbrun, L.K., Murphy, W.K., Bodey, G.P., Stone, E., Turner-Chism, V. and Hersh, E.M. (1980). *Cancer* **45**, 2506-2515.

Serrou, B., Pujol, H., Gary-Bobo, J. and Cupissol, D. (1981). Abstr. 4th Ann. San Antonio Breast Cancer Symp. Nov. 1981.

Sinkovics, J.G. (1978). *Pathobiol. Ann.* **8**, 241-284.

Sinkovics, J.G. (1979). *J. InterAm. Med.* **4**, 10-14.

Sinkovics, J.G. (1980). *In* "Tumour Progression". (Ed. R. Crispen), pp. 316-331, Elsevier/North-Holland, New York, Amsterdam.

Sinkovics, J.G. (1982). *Rev. Infect. Dis.* (In press).

Sinkovics, J.G., Shirato, E., Martin, R.G., Cabiness, J.R. and White, E.C. (1971). *Cancer* **27**, 782-793.

Sinkovics, J.G., Cabiness, J.R. and Shullenberger, C.C. (1972a). *Front. Rad. Ther. Oncol.* **7**, 141-154.

Sinkovics, J.G., Reeves, W.J. and Cabiness, J.R. (1972b). *J. Natl. Cancer Inst.* **48**, 1145-1149.

Sinkovics, J.G., Campos, L.T., Loh, K.K., Cormia, F., Velasquez, W. and Shullenberger, C.C. (1976). *In* "Neoplasm Immunity: Mechanisms". (Ed. R.G. Crispen), pp. 193-212, Univ. Chicago.

Sinkovics, J.G., Plager, C. and Romero, J. (1978). *In* "Neo-plasm Immunity: Solid Tumour Therapy". (Ed. R.G. Crispen), pp. 211-219, Franklin Institute Press.

Sinkovics, J.G., Plager, C. and Papadopoulos, N.E. (1981). *In* "Advances of Comparative Leukaemia Research". (Eds D.S. Yohn and J.R. Blakeslee), pp. 613-615, Elsevier/North-Holland, New York, Amsterdam.

Snyder, R.D., Hortobagyi, G.N., Bodey, G.P., Gutterman, J.U. and Hersh, E.M. (1981). *J. Surg. Oncol.* **16**, 87-92.

Solan, A.J., Vogl, S.E., Zaravinos, T., Salazar, C. (1980). *Cancer Immunol. Immunother.* **8**, 263-264.

Souter, R.G., Gill, P.G. and Morris, P.J. (1979). *Br. J. Surg.* **66**, 901.

Sterchi, J.M., Cole, W., Richards, F., Muss, H.B., White, D.R., Cooper, M.R. and Spurr, C.L. (1980). *Proc. Am. Soc. Clin. Oncol.* **21**, 363 (abstr. C-177).

Stober, U. and Peter, H.H. (1980). *Therapiewoche* **30**, 6067-6073.

Storms, G.E., Burghouts, J. and Haanen, C. (1980). *Netherland J. Med.* **23**, 54-58.

Summerfield, G.P., Gibbs, T.J. and Bellingham, A.J. (1979). *Br. J. Cancer* **40**, 737-743.

Taguchi, T., Nakano, Y., Urushizaki, I., Kure, S., Ishitani, K., Saito, T., Wakui, A., Takahashi, K., Koyama, Y., Kiumura, T., Furue, H., Kusama, S., Toyama, J., Ito, I., Tomigana, T., Majima, H., Kondo, T., Kamei, H., Terabe, K., Kosaki, G., Mori, T., Yayoi, K., Masaoka, T., Kimura, I., Ohnoshi, T., Inokuchi, K. and Kumashiro, R. (1979). *Gan To Kagaku Ryoho* **6**, 619-626 (I.C.R.D.B. Nov. 1980, abstr. 19).

Takahashi, H., Okamoto, M., Suzuki, T., Saito, A., Yao, K., Ikegami, A., Furukawa, K. and Yagi, T. (1979). *Jibi To Rinsho* **25**, 436-442.

Taylor, S.G., Sikora, P., Sisson, G.A. and Bytell, D.E. (1980). *Proc. Am. Soc. Clin. Oncol.* **21**, 370 (abstr. C-200).

Thatcher, N., Blackledge, G., Palmer, M.K. and Crowther, D. (1981). *Eur. J. Cancer* **17**, 465-469.

Valdivieso, M., Tenczynski, T.F., Rodriguez, V., Burgess, M.A., Mountain, C.F., Barkley, H.T., Hersh, E.M. and Bodey, G.P. (1981). *Cancer* **48**, 238-244.

Vinciguerra, V., Coleman, M., Pajak, T.F., Rafla, S., Stutzman, L., Gomez, G., Weil, M., Brunner, K., Cuttner, J., Nissen, N., Leventhal, B. and Gottlieb, A. (1981). *Cancer Clin. Trials* **4**, 99-105.

Vosika, G., Schmidtke, J., Goldman, A., Parker, R., Ribi, E. and Gray, G.R. (1979). *Cancer Immunol. Immunother.* **6**, 135-142.

Watanabe, Y., Yamada, T., Muranaka, Y., Hayashi, T., Urayama, H., Fujino, S. and Iwa, T. (1980). *Gan Kagau Ryoho* **7**, 1256-1263 (I.C.R.D.B. 1981 March, abstr. 5).

Wepsic, H.T., Alaimo, J., Druker, B.J., Murray, W. and Morris, H.P. (1981). *Cancer Immunol. Immunother.* **10**, 217-225.

Whittaker, P.W., Hill, L.D., Peterson, A.V., Bagley, C.M., Johnson, L.P. and Morgan, E.H. (1980). *Proc. Am. Soc. Clin. Oncol.* **21**, 452 (abstr. C-524).

Wynert, W., Harvey, H., Gottlieb, R., Dixon, R., White, D. and Lipton, A. (1980). *Proc. Am. Soc. Clin. Oncol.* **21**, 372 (C-209).

Yu, Z., Lin, M., Wang, B. and Lin, X. (1981). *Chin. Med. J.* **94**, 31-34.

Zuhrie, S.R., Harris, R., Freeman, C.B., MacIver, J.E., Geary, C.G., Delamore, I.W. and Tooth, J.A. (1980). *Br. J. Cancer* **41**, 372-377.

Dr PRAGER: Dr Goldin, drug synergism is obviously an important issue but there can be difficulty trying to find such synergism. For example, initial studies showed that asparaginase blocked the therapeutic effect of methotrexate (MTX), but by careful scheduling Dr Capizzi showed that there was a time point following asparaginase administration when there was maximal sensitivity to MTX. Do you have a recommendation regarding a systematic approach to looking for drug synergism?

Dr GOLDIN: Attention can be focussed on the mechanism of action of the drugs in regard to kinetics of tumour cells and other characteristics. There have been several instances when attempts have been made to synchronize the tumour cell population and to treat with a second drug when the cells are traversing a portion of the cell cycle and when they will be most susceptible. Overall, it is important to investigate the influence of scheduling a drug combination activity.

Dr SEDLACEK: Dr Goldin, do you see any difference between the SRC assay and the s.c. xenograft of human tumours when you screen the various cytostatics? Do you know any substance which is effective on these tumours, but not effective in your mice tumour screening system, especially in L1210 and P-388?

Dr GOLDIN: In general, the drugs that are active in the human tumour xenograft systems are active in the mice tumour systems. But there are a few drugs active in the xenograft systems but not in our animal screenings and we are interested in them. We are also interested in drugs active in both animal tumour screens and xenograft screens as compared to drugs active only in the animal models.

Dr PULVERER: Dr Goldin, many substances mentioned by you are known to us bacteriologists as active inducers of lysogeny and many of them are also mutagenic.

Dr GOLDIN: The question of mutagenicity and carcinogenicity of antitumour agents is receiving increasing attention, as therapy becomes more successful and new antitumour agents are being tested in this regard. If an analogue of a human anti-tumour agent is less mutagenic, this would be considered as a positive feature.

Dr MITCHELL: The majority of studies on immunotherapy are uncontrolled and/or nonrandomized. This is why controversies remain, whereas definitive results could be obtained with attention to appropriate numbers of patients in each group, use of randomized control (untreated where appropriate) and evaluation of data statistically. While this may seem only too obvious a statement, remarkably few well-designed studies such as Dr Roszkowski's have been performed in immunotherapy. They are absolutely essential, otherwise we remain in the present state of uncertainty indefinitely.

Dr SEDLACEK: The analysis of clinical studies often gives survival or regression curves after immunotherapy which have a smaller slope compared to the respective control group. Any minor change of the various treatments may change the outcome so that the results are no longer significant. Dr Sinkovics, analysing the data you have reviewed, do you know of any studies where the survival curves after immunotherapy have not just a linear slope but show a plateau?

Dr SINKOVICS: Unfortunately, immunotherapy in patients is marginally effective at best. Therefore very careful analysis of data is needed. A 10–20% difference is very difficult to prove. Therefore, right now we must proceed step by step with careful analysis of data. Even with the best statistics and analysis a 10–30% difference may not be significant in the human situation.

Dr MITCHELL: While major differences could be detected without statistics or even controls, many of the studies made in chemotherapy were relatively small but significant. In acute lymphocytic leukaemia of children survival was improved step-wise with each advance in treatment and now many children are cured. This may well be the situation in immunotherapy: minor but significant changes in survival that we can build upon — but receiving careful analysis to be certain there really is a difference.

Dr SCHWAB: Several problems prevent a rational approach to immunotherapy: it is difficult to predict whether an immuno-modulator will have the net effect of enhancement or suppres-sion of the immune response, even with purified, well characterized agents. Many variables including dose, timing, physical state of agent, and host processing of the agent, determine whether the immunomodulator will produce suppression or enhancement. It is uncertain, at any stage of tumour growth, whether suppression or enhancement of the immune system is desirable. For example, we know that *C. parvum* can suppress many T-cell functions. BCG and measles vaccines can induce anergy, but these agents can be effective in some trials. Therefore, it is questionable to proceed on the premise that stimulation of the immune system is desirable.

Dr PULVERER: Because of great differences in immunomodulation activities from bacterial strain to strain, in future one should take special care on the propionibacterial strain used. The same seems to be necessary as concerns BCG.

Dr SINKOVICS: Dr Pulverer is absolutely right. We should standardize to the best of our ability the bacterial products we use for immunostimulation. This we can do. What we cannot easily do is to categorize our patients. Here there are tremendous individual differences both patient-related and tumour-related. Dr Roszkowski and associates made a most remarkable effort to subgroup and categorize their patients to reduce variables and render their results comparable and controlled. This is a good example to follow. We prefer to register our patients against established protocols, such as protocols of the South West Oncology Group. We still use BCG around the excised primary melanoma but seldom, if ever, use it against visceral metastases. It appears worthwhile to use BCG in nodular lymphocytic lymphoma or in stage III—IV ovarian carcinoma in addition to conventional chemotherapy. But one must retain flexibility and riskage. One should not be convinced that these modalities of immunotherapy truly constitute a significantly better modality than the best conventional treatment methods.

Dr SCHNEIDER: To be able to compare different bacterial vaccines and results from different clinical studies it seems necessary to have commonly acceptable standardization methods. Which methods can really be proposed? Is the mouse spleen weight test really such an acceptable one?

Dr BREDE: From my point of view the spleen weight test is not an ideal one. It is too wide and too vague. We should prefer biochemical standards, if possible, expressed in weight units. There should be standard-reference preparation, prepared from a defined strain lyophilized and handed out from a reference laboratory in small amounts, to enable other researchers to create their own "internal standard".

Dr ADLAM: As stressed in my talk, we do use a standard batch of *C. parvum* in all of our quality control procedures and we also work by means of the "seed lot" system, i.e. we have laid down and recorded the history of all our seeds for growing *C. parvum* batches. When one compares the *C. parvum* used in clinical studies throughout the world, it is fortunate that they appear to be similar in their activities. This cannot be said for BCG preparations where there are great differences in viability and method of production. Finally, although the spleen weight increase assay may at first sight look to be a strange choice for testing, in fact as stressed in both my talk and that of Dr Roszkowski, this simple and highly reproducible assay does parallel activity in several tumour models and activity in many other assays of immuno-stimulation, e.g. adjuvant activity and histamine sensitizing activity.

Dr BALDWIN: We have attempted to standardize BCG preparations from multiple sources using a range of rodent tumours. The general conclusion from these studies was that classification of BCG by antitumour properties would not be appropriate because of the wide number of variables. These include the composition of BCG preparations particularly with respect to numbers of viable/nonviable organisms. At the screening level the problem is even more complicated since agents such as BCG influence a broad spectrum of host responses, including specific (T- and B-cells, antibody) and nonspecific (NK-cells, macrophages).

Helen NAUTS: We must not only stress the need for stable preparations but also the optimum technique of administration of immunotherapeutic agents as regards site, dose, frequency and duration of these agents and to develop new ways of getting them into the tumour prior to surgery.

Dr BALDWIN: There is good evidence from many studies in the animal tumour system that localization of agents including the whole bacterial vaccine (BCG, *C. parvum*) and isolated products (cell wall, cord factor, etc.) at tumour deposit (regional therapy) produces an improved therapeutic response. This has been carried out by intralesional injection of agents and studies are now being reported in patients. There is, for example, one study showing that injection of BCG into lung tumours 24 h prior to resection increased the NK-cell activity within the tumour. These approaches have much to commend them and one hopes that more clinical work will be attempted. In the long term, however, intralesional injection will have limited value and alternative methods of trafficking must be sought. One approach being developed by Dr

Fiedler at the Frederick Cancer Centre, USA, is to use liposomes as carriers for tumour localizing muramyl dipeptide.
An alternative approach that we are pursuing is to use monoclonal antibodies as carriers for immunomodulating agents.
Antibody-coupled to interferon has already been used to generate NK-cells and similar approaches may be developed for other
immunomodulators.

Dr **KIRCHNER**: A general question to the people working with
different immunostimulants: what is the antibody status as
for example against *C. parvum* or BCG?

Dr **ROSZKOWSKI**: You mentioned quite an important problem. As
we know, prolonged immunotherapy with, for example, *C. parvum*,
often leads to humoral responses directed against the immunostimulant and with consequent dangerous side-effects (e.g.
impairment or kidney and/or liver function). In our studies
we carefully looked for such propionibacterial strains with
weak immunogenicity. Finally, we decided on a compromise
and selected strains which are moderately active (using spleen
test as a criterion) but with very low risk of the above-
mentioned side-effects.

   The only way to reduce side-effects to a minimum is to
make an effort to isolate and characterize "active" compounds
from whole bacteria and use them instead of vaccines. Dr
Chedid's work is a good example to follow.

BACILLUS CALMETTE-GUÉRIN AS A MODULATOR OF CELL-MEDIATED IMMUNITY

Malcolm S. Mitchell

*University of Southern California,
Comprehensive Cancer Centre,
Los Angeles, California, USA*

Bacillus Calmette-Guérin (BCG) has been known for many years to be an immunological "adjuvant" and has received considerable study during that time. We have reviewed that large body of work recently (Mitchell and Murahata, 1979) and will concentrate here on some of the work done by my colleagues and myself over the past decade. We will specifically examine the influence of BCG on the efferent (effector) arm of cell-mediated immunity and on afferent functions, both helper and suppressor. Finally, some recent unpublished work bearing upon the effects of BCG on the cell membrane of macrophages will be described. All of these effects exemplify properties that might be sought in future studies of potential biomodulators (biological response modifiers). Thus, regardless of its own role, BCG can serve as a useful prototypical modulatory agent.

I. ACTIVATION OF NONADHERENT EFFECTOR CELLS

In some of our earliest experiments we showed that intravenous BCG not only potentiated the effects of suboptimal doses of alloantigen on spleen cell-mediated immunity (CMI) but that BCG by itself evoked a high level of CMI (Mitchell *et al.*, 1973 (Fig. 1). Originally we assumed that T-cells were the effectors stimulated by BCG, since nonadherent cells were the only killer cells. Later we returned to this issue, after the demonstration of other types of nonadherent effector cells and more particuarly the discovery by Dr Richard Murahata that the effector cells generated by BCG and alloantigen (L1210 cells into C57Bl/6 mice) were not active in a 4 h ^{51}Cr release assay (Murahata and Mitchell, 1976). The

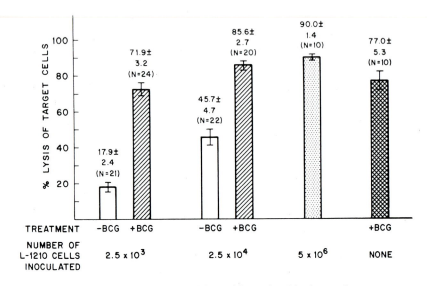

FIG. 1 *Effect of BCG on cell-mediated alloimmunity, expressed as percentage killing of P815Y(H-2^d) target mastocytomas cell in a 48 h growth-inhibition assay (referred to as "lysis" in this figure from an early paper). Means ± standard errors are depicted, with numbers of mice in each group shown in parentheses. Statistical significance in the 2-tailed t-test, was demonstrable for augmented immunity in both groups treated with BCG and L1210 compared with those controls given L1210 alone (p<0.001). The reactivity resulting from treatment with mycobacteria alone, the double hatched bar, was only slightly less than from optimal immunization with 5 × 10^6 L1210 cells. (From Mitchell et al., 1973.)*

effector cells are neither classical T-cells nor classical macrophages. As summarized in Table 1, their properties suggest a myeloid cell, perhaps an immature macrophage or a granulocyte, with an atypical lymphocytic cell least likely. The cells have a specificity and lack of reactivity in 4 h cytolytic assays that make them certainly not natural killer cells. Specificity may be conferred by sensitized T-cells in the spleen, since after removal of CMI activity by treatment with carbonyl iron and a magnet, activity could be replaced by adding normal peritoneal macrophages or more accurately peritoneal cells treated with anti-Thy 1 and complement (Table 2). If the spleen cells were instead treated with anti-Thy 1 and complement, the cell—cell cooperation was abolished (Table 3). We have not been successful in identifying soluble arming factors or cytophilic antibodies

TABLE 1

Effector cells induced by mycobacteria and alloantigen

1. Lack Thy-1 antigen.

2. Nonadherent to glass or plastic. Only lightly adherent to nylon wool.

3. Appear to be *specific* for the alloantigen used for immunization.

4. May be immunoglobulin-bearing cells.

5. Relatively radioresistant: resistant to 1000 rads but sensitive to 3000.

6. Reduced by carbonyl iron and magnet but not affected by silica and so may not be truly phagocytic.

7. Require sensitized splenic T-cells for activity. Perhaps "armed" by sensitized T-cells.

TABLE 2

Restoration of activity to nonphagocytic cells by normal PEC

	Percent inhibition of target cell growth Mean ± S.E.M.			
BCG	Unfrac- tionated	Non- phagocytic	Nonphagocytic + PEC	Nonphagocytic + killed PEC
−	27.2 ± 3.0	2.4 ± 5.4	59.8 ± 1.2	6.3 ± 2.8
+	40.9 ± 6.5	6.5 ± 4.5	66.9 ± 3.0	7.2 ± 2.9

All mice received the standard immunizing dose of L1210 on day 0. Mice receiving BCG were injected i.v. with 0.5 mg on day (-10). All mice were sacrificed on day 14, and the spleens assayed for activity in the MCA before of after depletion of phagocytic (and adherent) cells by treatment with carbonyl iron and magnet. Five x 10^5 normal PEC were added to the indicated cultures. In some cases, the PEC were killed by heating to 56°C for 30 min before addition to the assay.

PEC: peritoneal exudate cells.

TABLE 3

Effect of ATS treatment of immune spleen cells on their ability to cooperate with normal PEC

BCG	ATS	Percent inhibition of target cell growth Mean ± S.E.	
		Spleen	Spleen + PEC
−	−	43.9 ± 2.8	68.8 ± 2.7
−	+	52.2 ± 7.3	59.5 ± 6.3
+	−	51.4 ± 4.8	65.1 ± 2.0
+	+	73.7 ± 3.2	61.7 ± 9.7

All mice received 2×10^7 irradiated L1210 cells on day 0. They were sacrificed on day 14, and their spleens were assayed for activity in the microcytotoxicity assay before or after treatment with ATS and complement. Five x 10^5 normal PEC were added to the indicated cultures. The effector cells were ATS-resistant, but ATS treatment abolished cooperation with normal PEC. ATS: Antithymocyte serum; PEC: Peritoneal exudate cells.

TABLE 4

Effector cells induced by mycobacteria alone

1. Lack Thy-1 antigen. Produced in thymectomized, irradiated bone marrow-reconstituted mice. Therefore are not T-cells and do not require T-cells for induction.

2. Nonadherent to glass or plastic. Only lightly adherent to nylon wool.

3. *Nonspecific.* Kill several allogeneic targets (but not fibroblasts).

4. Exist concomitantly with BCG-induced suppressor cells.

that might explain how the cells interacted, leaving a more direct cell-cell interaction as most likely.

Dr Barbara Kinder (unpublished data) has also investigated the nonadherent cells elicited by BCG alone, and has found great similarities, with the notable exception that those cells lacked specificity for the alloantigens (Table 4). These cells were not only non-T-cells, but also could be

TABLE 5

*BCG antagonizes immunosuppression caused by Cytarabine
(ara-C) measured by 48 h growth inhibition assay*

Treatment		% specific growth inhibition
BCG	Ara-C	
−	+	19.75 ± 3.7
+	−	70.2 ± 3.6
+	+	60.0 ± 2.5
−	−	50.4 ± 10.1

All groups of mice received 20 x 10^6 killed L1210 cells i.p.
on day 0. Those receiving BCG were given 2 x 10^7 viable
units on day (-10). Ara-C was injected i.p. at 20 mg/kg/day,
days 3—7. All mice were sacrificed and their spleen cells
assayed on day 14 for cell-mediated immunity to $H-2^d$ alloantigens. Values shown are the means of at least 5 determinations and the standard errors of the means. Effector:target
ratio, 100:1.

generated in thymectomized, irradiated mice reconstituted
with anti-Thy 1 treated bone marrow. In other words, they
were neither T-cells nor dependent upon T-cells for their
generation.
    It is worth noting, for its possible therapeutic implica-
tions, that BCG can antagonize the immunosuppressive effects
of the antimetabolite cytarabine (ara-C) if one measures
killer cells active in a 48 h growth inhibition (microcytotoxicity) assay (Table 5) (Murahata and Mitchell, submitted
for publication, 1982). BCG pretreatment 10 days before
alloantigen completely abrogated the suppressive effect of
subsequent cytotoxic therapy with ara-C given at the moderate
dose of 20 mg/kg/day for 5 days ($p<0.001$). There was no
significant difference in the CMI of mice given the combined
therapy from that of normal controls receiving saline with
the immunization. BCG was partially protective against
immunosuppression by ara-C even at a higher dose of 40 mg/kg
/day, with CMI greater than that of the ara-C treated group
($p<0.05$) but less than that of the immunized controls (data
not shown). Reversal of immunosuppression by BCG given on
days 10 and 13 after alloantigen, with ara-C again on days
3—7 at 20 mg/kg/day, was achievable transiently but was much
less successful than antagonism by pretreatment. Doses in

the range of 1 to 4 x 10^7 viable units of BCG (Tice) were
protective and the relative immunity among the 4 experimental
groups remained consistent over effector:target cell ratios
of 25:1 to 250:1. BCG thus appears to augment the number or
potentiate the activity of these predominantly non-T effector
cells. Yet, in direct contrast, when cytolytic T-cells are
measured as an indicator of immunity, BCG acts with cytarabine
to further diminish the activity of such cells. The likely
involvement of suppressor macrophages elicited by BCG in
depressing the generation of cytolytic T cells will be considered in detail shortly.

The precise nature of the cells elicited by BCG and alloantigen or by BCG alone has not been determined but is under
investigation, through cell depletions and cell sorting with
monoclonal antibodies to macrophage antigens in a collaboration between our group and Dr Timothy Springer at Harvard. The
intriguing possibility that the cells may be granulocytes has
also been entertained, particularly since large numbers of
those cells enter the spleen after intravenous injection of
BCG.

II. STIMULATION OF INTERLEUKIN I

When Gery and Waksman described "lymphocyte activating factor
(LAF) as a stimulant of *in vitro* T-cell blastogenesis (Gery
and Waksman, 1972), it was uncertain whether this mediator
had a role *in vivo*. Since BCG given i.v. to mice had augmented the cytotoxic activity of their nonadherent spleen
cells, which were thought to be T-cells at that time, it
seemed appropriate to investigate whether LAF production was
increased in those mice. As shown in Table 6, this mediator,
now known as interleukin I (IL-I) was produced in amounts
2—3-fold higher in BCG-treated spleens than in normal spleens
(Mitchell *et al.*, 1973). Splenic adherent cells were the
source of the interleukin, and the production of IL-I was
approximately proportional to the increased percent of macrophages in the BCG-treated spleen. BCG-treated spleens contained approximately 25% macrophages, as opposed to 5 to 10%
in normal spleens, as estimated from Wright's stained cytological preparations. Yet when BCG was added *in vitro* to
cultures of unfractionated or adherent splenic cells, the
amount of IL-I per cell was increased (Mokyr and Mitchell,
1975). Approximately 20-fold more IL-I was produced per
culture compared with the baseline output of that mediator
(Fig. 2). The presence of T-cells increased the output of
IL-I, further indicating that while T-cells were not necessary for the production of IL-I, their presence and presumably interaction with macrophages through lymphokines

TABLE 6

Effect of BCG on Interleukin I (lymphocyte-activating factor) production. (From Mitchell et al., 1973.)

Volume of supernatant Fluid added	Source of spleen cells yielding supernatant fluid*	
	Normal mice	BCG-treated mice
A. Unfractionated spleen cell population as source of supernatant fluid		
0.5 ml	1585 ± 123[+]	3227 ± 66
0.1 ml	524 ± 11	623 ± 96
B. Purified adherent spleen cells ("macrophages") as source of supernatant fluid		
0.5 ml	2073 ± 84	7181 ± 299
0.1 ml	403 ± 36	1283 ± 65

*Cultures used for production of supernatant fluids were composed of 12 x 10^6 unfractionated spleen cells (A) or the monolayers of adherent cells obtained from 2 x 10^7 spleen cells (B) in 4 ml of minimal essential medium.

[+]Mean values (± range) of CPM of ^{3}H-thymidine incorporated into 5 x 10^6 thymocytes *in vitro* after incubation with supernatant fluid. 0.1 ml of a 1/200 dilution of PHA was added to all thymocyte cultures. Control cultures without any supernatant fluid incorporated 291 ± 1 CPM.

improved the production. This is only one of several examples of the interdependence of lymphocytes and macrophages in which each cell is equally important. For BCG to act as a mitogen for T-cells (thymocytes) *in vitro*, a small admixture of macrophages is an absolute requirement, probably through IL-I or a similar macrophage-made mediator. As few as 0.25% to 0.5% macrophages, which is the percentage ordinarily present in the thymus, are sufficient whereas their "complete" removal (to less than 0.25%) by adherence to plastic completely prevented the mitogenic effect of BCG (Table 7). There was no effect on the viability of the T-cells as measured by the response to Concanavalin A, which was essentially unchanged. Replacement of purified peritoneal macrophages (to 0.5%) completely restored the response to BCG. BCG was more potent in stimulating IL-I than any subcellular

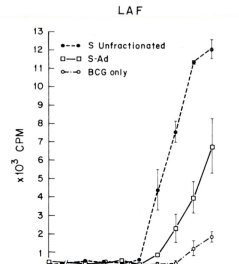

FIG. 2 *Activity of Interleukin I (Lymphocyte-activing Factor, LAF) in supernatant fluids from unfractionated spleen cells and splenic adherent cells treated with BCG* in vitro. *(From Mokyr and Mitchell, 1975.)*

fraction derived from it, including methanol extraction residue (MER) (Fig. 3) (Mitchell *et al.*, 1975).

IL-I is elicited by a wide variety of substances, and in fact is now known to be identical with the "endogenous pyrogen" described by Beeson many years before the studies of Gery and Waksman. There is evidence that IL-I is also important in regulating B-cells and probably other types of cell including perhaps haematopoietic cells. Certainly macrophages make a family of mediators, including colony-stimulating factor augmenting white blood cell precursors. It is not inconceivable that IL-I may prove to be identical with other mediators now described only by their functions in very different, often nonimmunological, systems.

III. ACTIVITY ON NATURAL SUPPRESSOR CELLS

What seems to us a particularly good example of how "applied science" with a biomodulator can lead to a fairly fundamental observation in immunology is the finding that BCG can activate "natural" suppressor cells. These experiments were

TABLE 7

*Effect of removal of macrophages from thymus cells
on the mitogenic response to BCG in vitro
(From Mokyr and Mitchell, 1975.)*

BCG (viable units x 10^{-5})	Index of stimulation* (Mean ± S.E.)	
	Thymus cells	Nonadherent thymus cells
250.0	0.67 ± 0.16	1.03 ± 0.23
83.3	12.87 ± 2.45	1.80 ± 0.33
27.8	24.58 ± 3.15	2.21 ± 0.56
9.3	24.19 ± 4.25	1.37 ± 0.19
3.1	8.66 ± 1.48	0.70 ± 0.07
1.0	2.11 ± 0.38	0.87 ± 0.13
0.3	1.11 ± 0.24	0.87 ± 0.17
0.1	0.96 ± 0.09	0.92 ± 0.11

*Index of stimulation = ratio of the uptake of ^{3}H-TdR in presence of BCG *in vitro* to the uptake in the absence of BCG. Means of 3 experiments are shown.

performed by Dr James A. Bennett coincidentally with very similar work by Klimpel and Henney (Bennett *et al.*, 1978; Klimpel and Henney, 1978).

While attempting to improve the immune response to H-2^d alloantigens generated by the *in vitro* immunization of C57 Bl/6 (H-2^b) spleen cells, we found paradoxically that cells from BCG treated mice could not be immunized at all. Only relatively large doses of intravenously administered BCG elicited suppressor macrophages in numbers sufficient to inhibit the immunization of spleen cells. As shown in Table 8, 1×10^7 viable units of Tice strain BCG or more caused inhibition of the ability of spleen cells to be allo-immunized *in vitro*, as measured by the generation of cytolytic T-cells, and the generation of both T and non-T cells reflected by a 48 h growth inhibition assay. Below 1×10^7 viable units there was no discernible effect of BCG pretreatment on immunity. A dose of 5×10^7 viable units was lethal to the mice. The existence of suppressor cells within the spleen was indicated by mixing experiments, which have shown a clear dose-response relationship between the number of BCG treated cells added and the suppression achieved (Fig. 4).

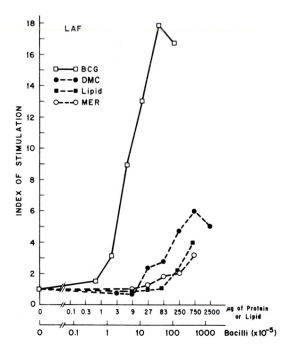

FIG. 3 *Effect of whole BCG and its fractions* in vitro *on the production of interleukin I (LAF). DMC = "delipidated" (lipid-free) mycobacterial cells; MER = methanol extraction residue of BCG. See Mitchell* et al., *1975 for details.*

Exploration of this phenomenon further revealed that it was the nylon wool-adherent cells from the BCG treated spleen that prevented the development of normal spleen cells into specifically allosensitized T-cells as identified in a 4 h ^{51}Cr release assay (Bennett *et al.*, 1978) (Table 9). The splenic adherent cells appear to be fairly classical macrophages, from their characterization by us and others (Table 10). In particular, their suppressor activity is very sensitive to indomethacin, indicating a prostaglandin-mediated effect, which we will shortly discuss. Suppressor adherent cells had in fact been suggested in our earliest work (Mitchell *et al.*, 1973). Whereas the mitogenic responses to phytohemagglutinin (PHA), Con A and lipopolysaccharide (LPS) were markedly reduced with spleen cells from BCG-treated mice, the response to the 2 T-cell mitogens was considerably improved when glass-adherent cells were removed.

Of perhaps most interest biologically is that the suppressor cells, or their relatives, antedate BCG administration, and are in fact "natural" suppressor cells found in normal

TABLE 8

Effect of intravenous BCG on the ability of spleen cells to be alloimmunized in vitro. (From Bennett et al., 1981a.)

No. of BCG organisms	Spleen weight (gm)	% specific ^{51}Cr release	% specific growth inhibition
None	0.08 ± 0.01	65	80
1×10^6	0.07 ± 0.01	66	75
5×10^6	0.08 ± 0.01	63	76
1×10^7	0.19 ± 0.05	15	26
2×10^7	0.31 ± 0.04	5	0
5×10^7	lethal dose	–	–

20×10^6 C57B1/6 (H-2^b) spleen cells were cultured with 2×10^5 irradiated (4000 r) P815 mastocytoma cells (H-2^d) for 4 days. Both 4 h ^{51}Cr release from labelled P815 targets and 48 h growth inhibition of unlabelled targets were used to determine immunization. Effector target ratios were 100:1 for the former and 10:1 for the latter.

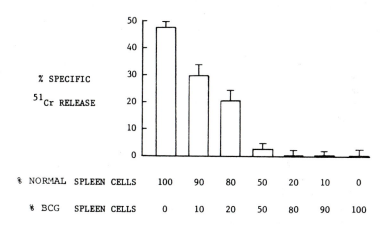

FIG. 4 *Cytotoxicity of C57B1/6 spleen cells (cytolytic T cells) to H-2^d alloantigens generated in vitro: suppressor influence of spleen cells from BCG-treated mice. 2×10^7 normal spleen cells and various numbers of mycobacteria-treated spleen cells were mixed, and cultivated with allo-antigenic P815Y tumour cells for 4 days.*

TABLE **9**

Splenic suppressor cells from BCG-treated mice are adherent
(From Bennett and Mitchell, 1980.)

Cell population	% specific ^{51}Cr release	% specific 48 h growth inhibition
Normal unfractionated spleen	69 ± 2	83 ± 8
BCG unfractionated spleen	4 ± 4	0
Normal Nonadh + Normal Adh	62 ± 5	80 ± 11
Normal Nonadh + BCG Adh	9 ± 3	12 ± 7
BCG Nonadh + BCG Adh	5 ± 4	0
BCG Nonadh + Normal Adh	54 ± 4	70 ± 8

Cells derived from spleens of C57B1/6 mice treated with 2 x 10^7 viable units of Tice BCG i.v. 8 days earlier, fractionated over nylon wool. *In vitro* alloimmunization of 2 x 10^7 total spleen cells against H-2^d was attempted. Mixtures of 30% nonadherent and 70% adherent cells (optimal for normal spleens) were made with cells of each type indicated. Mean ± S.E. shown.

TABLE 10

Splenic suppressor cells stimulated by mycobacteria

1. Nonspecifically suppress the *in vitro* generation by spleen cells of allogenic and syngeneic antitumour cytotoxicity.

2. Adherent to glass, plastic and nylon wool.

3. Lack Thy-1 antigen. Produced in thymectomized, lethally irradiated mice reconstituted with anti-Thy 1-treated bone marrow cells. Therefore, are not T-cells and do not require T-cells for induction.

4. Resistant to 1000 rads X-irradiation.

5. Removed by carbonyl iron and magnet treatment and are sensitive to silica. Therefore, are truly phagocytic cells.

6. Sensitive to indomethacin and other prostaglandin inhibitors.

7. Exist in the spleen concomitantly with mycobacteria-induced effector cells.

TABLE 11

Effect of various types of cell on immunization:
influence of BCG. (From Bennett and Mitchell, 1980.)

Days after BCG (2 x 10^7 viable units)	% specific ^{51}Cr release with spleen cells immunized in the presence of:		
	Thymocytes	Bone marrow cells	Spleen cells
No BCG	62	38	66
2	66	9	64
7	59	5	45
11	67	7	33
16	64	3	31

C57B1/6 mice were given BCG and at the times indicated various cells were tested for suppressor activity by admixture of 6 x 10^6 cells with 20 x 10^6 normal spleen cells. Allo-immunization to H-2^d tumour cells was attempted and on day 4 cytolytic T-cells were assayed by a 4 h ^{51}Cr release test at 100:1, effect target cells. Mean values indicated.

bone marrow (Table 11). Mixture of 6 x 10^6 normal bone marrow cells with 20 x 10^6 normal spleen cells inhibited the development of cytotoxic T cells by 30% (relative to normal controls). Normal spleen cells or thymocytes at this dose never suppressed alloimmunization. Bone marrow cells from BCG-treated mice were 2 to 4-fold more active as the same number of normal bone marrow cells. That is, at all concentrations of bone marrow cells added, BCG therapy of the mice at least doubled the degree of suppression achieved (Bennett *et al.*, 1981a). Spleen cells derived from thymectomized, irradiated mice reconstitued with normal bone marrow cells had significant suppressor activity. Moreover, when BCG was administered to these reconstituted mice, the suppressor activity of the spleen cells was increased still further (Table 12 (Bennett and Mitchell, 1980). These experiments also proved that T-cells were not required for the induction of suppressor macrophages by BCG, nor did the presence of T-cells lead to enhanced effects of BCG here. Comparisons of the activity of adherent cells isolated from either the reconstituted spleen or the bone marrow, from untreated or BCG treated mice, have failed to establish conclusively

TABLE 12

Effect of nylon-adherent cells from T-cell deficient
("B") mice on immunization: influence of BCG
(From Bennett and Mitchell, 1980.)

Adherent cells added	% specific ^{51}Cr release (E:T = 100:1)	% specific growth inhibition (E:T = 10:1)
None	64 ± 2	84 ± 6
Normal spleen (1 x 10^7)	63 ± 2	80 ± 5
"B" spleen (6 x 10^6)	67 ± 3	78 ± 4
"B" spleen (1 x 10^7)	51 ± 3	60 ± 8
BCG treated "B" spleen (6 x 10^6)	15 ± 3	30 ± 5
BCG treated "B" spleen (1 x 10^7)	7 ± 2	0

Thymectomized, lethally irradiated mice were reconstituted
with normal bone marrow cells treated with anti-Thy 1 + complement. These were called "B" mice. 2 x 10^7 viable units
of BCG was injected 10 days previously into the groups indicated. Means ± standard deviations from 2 experiments are
indicated.

whether the BCG augments the activity per cell, or creates
additional suppressors. However, we favour the former explanation of "activation", rather than the recruitment of additional suppressor cells from precursors.

Circumstantial evidence suggests that the natural suppressor cells in the bone marrow are induced by intravenous BCG
to migrate to the spleen, where they proliferate (Fig. 5).
The number of cells in the bone marrow decreases within 2 to
3 days after intravenous BCG, remaining low for several weeks.
During the first week after BCG there is a steady increase in
the number of spleen cells, with a large increase in the 2nd
and 3rd weeks. The latter can be inhibited by antiproliferative agents such as cyclophosphamide or cytarabine (Bennett
and Mitchell, 1979). An interesting but unexplained decrease
in thymocytes also occurs soon after BCG administration.
There is a fall of 90% in the number of thymocytes, with a
return to normal in 3 weeks. Where these thymocytes migrate,
why they go there, and most importantly, whether there is
any relationship to the cellular changes occurring at the
other sites are all tantalizingly uncertain.

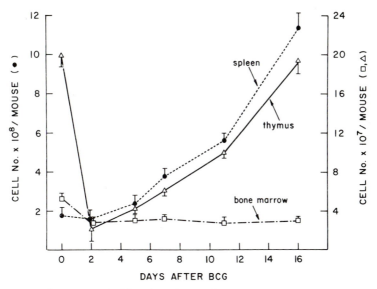

FIG. 5 *Numbers of cells in the spleen, thymus and bone marrow after the intravenous injection of BCG. (From Bennett et al., 1981a.)*

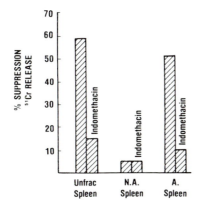

FIG. 6 *Effect of indomethacin on splenic suppressor cells. (From Bennett et al., 1981b.)*

Suppressor cells stimulated by BCG in the spleen are very sensitive to indomethacin, at concentrations of 10^{-7} M or greater *in vitro* (Bennett *et al.*, 1981b) (Fig. 6). In contrast the bone marrow suppressor cells are a heterogeneous population consisting not only of indomethacin-sensitive and -insensitive cells, but of both adherent and nonadherent cells too (Dorshkind *et al.*, 1980; Bennett *et al.*, 1981a,b).

TABLE 13

Effect of indomethacin on suppressor cells from the bone marrow. (From Bennett et al., *1981b.)*

Treatment of bone marrow donors	Population of cells	% suppression of immunization	
		Indomethacin (10^{-6}M)	
		−	+
Untreated	Unfractionated	36	23
	Nonadherent	30	18
	Adherent	38	20
BCG	Unfractionated	46.5	35.5
	Nonadherent	31	34
	Adherent	51	28.5

6×10^6 putative suppressor cells and 20×10^6 normal spleen cells were mixed and immunization was attempted against H-2^d alloantigenic tumour cells. % suppression was calculated as the difference between spleen cells immunized in the absence or presence of putative suppressor cells. Data shown are means from 2 experiments.

Indomethacin (10^{-6} M) significantly but incompletely inhibited the activity of nylon wool adherent bone marrow natural suppressor cells from normal or BCG-treated mice (Table 13). Nonadherent suppressor cells from normal bone marrow, which may be lymphocytes, were also partially inhibited by indomethacin. However, the nonadherent cells from BCG treated bone marrow were unaffected by 10^{-6} M indomethacin or higher. BCG may thus have induced an indomethacin resistant (i.e. non-prostaglandin-mediated) suppressor mechanism in the same nonadherent cells, or evoked a different population of indomethacin-resistant nonadherent cells. A final, nonexclusive, possibility, is that BCG caused the emigration from the bone marrow of all indomethacin-sensitive nonadherent suppressor cells.

  BCG stimulates myeloid (macrophage-granulocyte) colony forming units in the bone marrow and spleen of the same mice in which suppressor cells are generated (Bennett and Marsh, 1980; Bennett *et al.*, 1981a). As suppressor cell activity in the marrow and spleen increased, colony forming units increased in parallel (Fig. 7), and was likewise concentrated

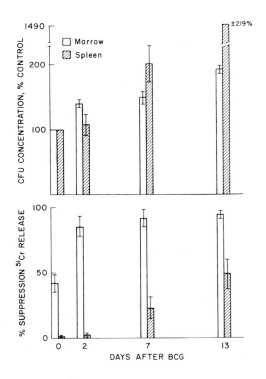

FIG. 7 *Generation of colony-forming units and suppressor cells in the spleen and bone marrow of mice given BCG. (From Bennett* et al., *1981a.)*

TABLE 14

Density gradient centrifugation of cells in the BCG-treated spleen

	% suppression of immunization	Colony-forming units /2.5 x 10^4 cells
Unfractionated spleen cells	51	38.3 ± 4.8
"light" fraction	8	5.5 ± 0.5
"medium" fraction	52	31.8 ± 4.8
"heavy" fraction	63	34.5 ± 6.8

Ficoll-400 discontinuous density gradient (12%, 16%, 21%, 25%) centrifuged at 20,000 x g for 60 min. See Bennett *et al.*, 1981a for further details. Spleen cells were obtained 2 weeks after 2 x 10^7 viable units of BCG.

in the nylon wool-adherent spleen cell fraction. Colony-forming units were also found in the same medium and heavy fractions of a Ficoll discontinuous density gradient as were suppressor cells (Table 14). It thus appears that the colony-
forming unit and the macrophage-like suppressor cell are in a
very similar subpopulation, if not identical with each other.
Widely discordant results obtained by a group of investigators
at the University of Illinois working with *in vitro* immunization to the syngeneic plasmacytoma MOPC-315 and our group,
studying alloimmunization, led to an exchange of materials in
a collaborative trial. The results indicated that the discrepancies were due to the BCG and not to the test system.
In that study we found that different strains of BCG and even
different lots of the same strain behaved differently in their
ability to induce suppressor cells (Mokyr *et al.*, 1980).
Tice BCG consistently abrogated the ability of suppressor
cells to be immunized whereas several lots of the Phipps
strain, especially NC-724 and NC-734, consistently led to
augmented in vitro reactivity. A third lot of the Phipps
strain, batch A-8, consistently caused strong induction of
suppression (Table 15). Intravenous or intraperitoneal administration both led to suppression of ability of suppressor
cells to be immune in our experiments, but the demonstration
of suppressor macrophages by administration was more consistent by the i.v. route. Suppression of immunity was uniformly noted at intervals 10—14 days or longer after BCG,
whereas augmentation was frequently noted 2—7 days after the
same suppressive BCG strain. Which component of the BCG
organism is responsible for induction of suppressor cell is
uncertain but could be approached directly by fractionation
into lipids, cell walls, and cell wall skeletons, or perhaps
by the chemical construction of various muramyl dipeptides.

It should be emphasized for perspective that the splenic
macrophage-like suppressor cells exist concomitantly with the
nonadherent splenic killer cells described above, approximately 14 to 30 days after BCG. It is unclear whether the
suppressor function we have measured *in vitro* is in fact only
one manifestation of the activation state of splenic macrophages. Other properties of these activated macrophages such
as nonspecific cytotoxicity might well be helpful to the host
against foreign cells. However, equally plausibly, there may
well be separate subpopulations of macrophages, with different regulatory effects on immunity, akin to the various sub-
classes of T-cells. We have not yet noted deleterious consequences *in vivo* ascribable to the presence of suppressor
macrophages in the spleen or marrow of BCG-treated mice. As
a working hypothesis it seems reasonable to consider that the
overall effect of BCG on the immune response at a particular

TABLE 15

Effect of different strains and batches of BCG on in vitro
immunization of BALB/c spleen cells to MOPC-315

Expt.	BCG strain	Batch No.	% specific ^{51}Cr release* at E/T ratio of: 100/1	10/1
1	None	–	56	21
	Phipps (5 x 10^7)	NC-734	84	37
	Tice (5 x 10^7)	–	11	3
2	None	–	23	11
	Phipps (1 x 10^7)	NC-734	88	34
	Phipps (1 x 10^7)	NC-724	83	47
	Phipps (1 x 10^7)	SC-NC-724$^+$	60	32
	Phipps (1 x 10^7)	A - 8	9	3

*Mean values are shown; $^+$Subculture of NC-724.

route and schedule is always the vector sum of the positive
and negative ("helper" and "suppressor") influences it exerts.
That in fact is why "immunomodulation" rather than "adjuvanticity" best describes BCG's actions.

IV. RING ("DOUGHNUT") POLYMORPHONUCLEAR LEUKOCYTES

A recent set of morphological observations may bear upon the
killer and suppressor activities of BCG spleen cells. Dr
James Hengst has found that a significant increase of spleen
cells with a ring nucleus, which we have termed "ring" or
"doughnut" cells, occurs after intravenous BCG (Fig. 8).
Pels and his colleagues (1980) found similar cells in the
peritoneal exudate of C57Bl/6 mice alloimmunized 10 days previously with SL-2 (H-2^d) tumour cells. Cytochemical staining
has revealed that some of the cells are macrophages while
others are polymorphonuclear leukocytes. The function of
these cells is uncertain but through sorting based on surface
markers, light scattering and physiochemical properties we

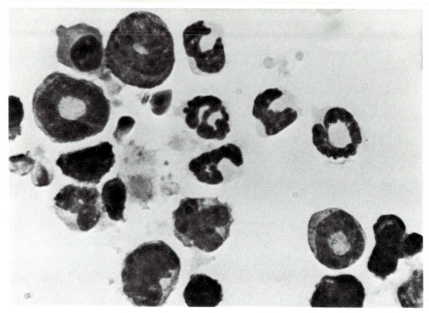

FIG. 8 *Macrophage- and granulocyte-like cells with a ring-shaped nucleus ("doughnut" cells) in the spleen of mice that received BCG 2 weeks previously.*

intend to compare the activities of doughnut cells with the functions detected in BCG treated spleens, to see if any might be subserved by morphologically homogeneous cells.

V. EFFECTS ON MACROPHAGE MEMBRANE PROTEINS

Thus far we have discussed the modulation of immune responses where macrophage-made mediators or macrophage mediated reactions were measured. A more fundamental question of *how* BCG affects the macrophage might be addressed in part through a biochemical approach such as one being explored by Dr June Kan-Mitchell. This consists of radiolabelling of externally disposed membrane proteins with a nonenzymatic, solid phase iodinating agent, Iodogen, solubilizing the membrane selectively with a nonionic detergent, NP-40, and subsequently performing polyacrylamide gel electrophoresis and autoradiography (Kan-Mitchell *et al.*, 1980). As illustrated in Fig. 9, the "profiles" of various types of macrophage differ from one another and appear to fall into 3 broad classes. Normal resident peritoneal macrophages and those elicited by proteose-peptone have similar membrane protein patterns. These

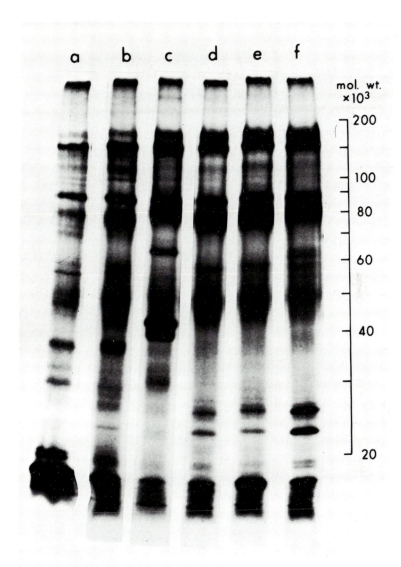

FIG. 9 *Autoradiographic profiles of solubilized macrophage membranes whose surface proteins were labelled with* [125]I *and subjected to SDS-polyacrylamide gel electrophoresis. Note the differences among the normal and proteose-peptone groups (colums a and b), the thioglycollate-stimulated macrophages (column c) and the "activated" macrophages from mice receiving the immunomodulators BCG (d),* Corynebacterium parvum *(e) or pyran (f).*

differ from the profile of thioglycollate-stimulated macrophages, which in turn are different from that of the activated
macrophages induced by BCG, *Corynebacterium parvum* or pyran.
Three characteristic additional membrane proteins are found
on thioglycollate stimulated macrophages, and both deletions
and additions are found with activated macrophages. Resolu-
tion of which specific membrane proteins are added or deleted,
particularly if they are receptors or ectoenzymes, may afford
new insights into the immunological functions of these macrophages. This biological investigation goes well beyond simply
helping to explain the mechanisms by which such microbes as
BCG activate those cells.

VI. CONCLUSION

If there is any principle that our work with BCG has indicated
to us, it is that intrinsic regulation of immune responses
can be elucidated through their perturbation by extrinsic
modulators. Such studies must necessarily begin as "phenomen-
ology" but can lead to interesting examinations of immunologically active cells and their interrelationships, through
direct interactions or mediators. We may yet obtain some
understanding of how to modify immunity in the host *predictably*, to achieve significant therapeutic advances. When
tumour-associated antigens isolated by monoclonal antibodies
are combined with predictably stimulatory doses of immunomodulators such as BCG, or its components, rational active
immunotherapy of cancer will then have replaced the well-
intentioned but somewhat misguided earlier efforts. This
aspect of biomodulation will then truly have reached its
maturity.

VII. REFERENCES

Bennett, J.A. and Mitchell, M.S. (1979). Induction of sup-
    pressor cells by intravenous administration of Bacillus
    Calmette-Guerin (BCG) and its modulation by cyclophospha-
    mide. *Biochem. Pharmacol.* (Arnold Welch Festschrift) **28**,
    1947-1952.
Bennett, J.A. and Marsh, J.C. (1980). Relationship of Bacil-
    lus Calmette-Guerin induced suppressor cells to hemato-
    poietic precursor cells. *Cancer Res.* **40**, 80-85.
Bennett, J.A. and Mitchell, M.S. (1980). Systemic administra-
    tion of BCG activates natural suppressor cells in the bone
    marrow and stimulates their migration into the spleen. *In*
    "Neoplasm Immunity: Experimental and Clinical". (Ed. R.G.
    Crispen), pp. 61-83, Elsevier/North-Holland, New York,
    Amsterdam, Oxford.

Bennett, J.A., Rao, V.S. and Mitchell, M.S. (1978). Systemic bacillus clamette-Guerin (BCG) activates natural suppressor cells. *Proc. Nat. Acad. Sci. USA* **75**, 5142-5144.

Bennett, J.A., Marsh, J.C. and Mitchell, M.S. (1981a). Suppressor macrophages: their induction, characterization and regulation. *In* "Mediation of Cellular Immunity in Cancer by Immune Modifiers". (Eds M.A. Chirigos, M.S. Mitchell, M.J. Mastrangelo and M. Krim), pp. 9-26, Raven Press, New York.

Bennett, J.A., Rao, V.S., Kemp, J.D. and Mitchell, M.S. (1981b). Selectivity in the effects of indomethacin on BCG-activated suppressor cell populations. *J. Immunopharm.* **3**, 221-239.

Dorshkind, K., Klimpel, G.R. and Rosse, C. (1980). Natural regulatory cells in murine bone marrow: inhibition of *in vitro* proliferative and cytotoxic responses to alloantigens. *J. Immunol.* **124**, 2594-2588.

Gery, I. and Waksman, B.H. (1972). Potentiation of the T-lymphocyte response to mitogens. II. The cellular source of potentiating mediator(s). *J. Exp. Med.* **136**, 143-155.

Kan-Mitchell, J., Rao, V.S. and Mitchell, M.S. (1980). Membrane proteins of peritoneal macrophages in different states of activation. *Proc. Reticuloendothel. Soc.* **17**, 113.

Klimpel, G.R. and Henney, C.S. (1978). BCG-induced suppressor cells. I. Demonstration of macrophage-like suppressor cell that inhibits cytotoxic T cell generation *in vitro*. *J. Immunol.* **120**, 563-569.

Mitchell, M.S., Kirkpatrick, D., Mokyr, M.B. and Gery, I. (1973). On the mode of action of BCG. *Nature New Biol.* **243**, 216-218.

Mitchell, M.S., Mokyr, M.B. and Kahane, I. (1975). Stimulation of lymphoid cells by components of BCG. *J. Natl. Cancer Inst.* **55**, 1337-1343.

Mitchell, M.S. and Murahata, R.I. (1979). Modulation of immunity by Bacillus Calmette-Guerin (BCG). *Pharmacol. Ther.* **4**, 329-353.

Mokyr, M.B., Bennett, J.A., Braun, D.P., Hengst, J.C.D., Mitchell, M.S. and Dray, S. (1980). Opposite effects of different strains or batches of the same strain of BCG on the *in vitro* generation of syngeneic and allogeneic antitumour cytotoxicity. *J. Natl. Cancer Inst.* **64**, 339-344.

Mokyr, M.B. and Mitchell, M.S. (1975). Activation of lymphoid cells by BCG *in vitro*. *Cell. Immunol.* **15**, 264-273.

Murahata, R.I. and Mitchell, M.S. (1976). Antagonism of immunosuppression by BCG. *In* "The Macrophage in Neoplasia". Ed. M.A. Fink, pp. 263-265, Academic Press, New York.

Pels, E., de Groot, J.W., Mullink, R., Van Unnik, J.A.M. and den Otter, W. (1980). Identification of 2 different types of mouse peritoneal exudate cells with ring shaped nuclei. *J. Reticuloendothel. Soc.* **27**, 367-376.

CLINICAL EVALUATION OF ANTICANCER ACTIVITY OF A STREPTOCOCCAL PREPARATION OK-432 (PICIBANIL)

Motoharu Kondo[1], Haruki Kato[1], Toshikazu Yoshikawa[1],
Mitsuo Katano[2] and Motomichi Torisu[2]

[1]*1st Department of Medicine,
Kyoto Prefectural University of Medicine, Kyoto 602, Japan;*

[2]*Department of First Surgery,
Kyushu University School of Medicine, Fukuoka 812, Japan*

I. HISTORY

In 1968 Busch reported a case of a 19-year old female with sarcoma in the face and neck, whose tumour disappeared when she accidentally suffered from erysipelas during her hospitalization. Since then, accumulating evidence has shown that erysipelas seemed to have an antitumour effect in man.

Fehleisen (1882) demonstrated that erysipelas is caused by the inoculation with pure culture of the *Streptococcus erysipelatis*. He inoculated 7 patients with inoperable malignant diseases, and of these 6 reacted to the new trial. In one case in particular, the tumour disappeared within one week after the inoculation of streptococci.

Bruns (1888) described 14 cases of malignant diseases in which erysipelas occurred either accidentally or by inoculation. Of the 5 cases of sarcoma, 3 cases were fully and permanently cured, and the rest showed reduction in size of their tumours.

Coley (1891) added 9 cases of malignant diseases, and inoculation of erysipelas had a significant effect upon the disappearance or reduction of tumours. He speculated that the effect of erysipelas on malignant tumour might be due to:

1) direct destruction of tumour by streptococci;

2) the high temperature due to erysipelas caused degeneration of the tumours; and

3) streptococci had a direct antagonistic effect upon the cancer bacillus.

Coley's toxin was then widely used and favourable anticancer effects have been reported (Nauts *et al.*, 1953).

II. STREPTOCOCCAL PREPARATION OK-432 (PICIBANIL[R])

In 1940, Okamoto reported that streptolysin S is produced by haemolytic streptococci. Koshimura *et al.* (1955, 1958) found that degradation and loss of Ehrlich carcinoma cells occurred together with streptolysin S production when a suspended mixture of tumour cells and living *Streptococcus hemolyticus* were incubated at 37°C for 90 min. Okamoto *et al.* (1966, 1967) further studied the correlation between the anticancer effect and streptolysin S production of the streptococcal cell suspension, and reported that cell suspension of a low virulent strain, *Su*, of *Streptococcus hemolyticus* in Bernheimer's basal medium (BBM), attenuated by the addition of penicillin G and by the heating to 45°C, exhibited a strong inhibitory effect on the growth of Ehrlich carcinoma transplanted intraperitoneally into mice. The antitumour effect of this preparation was enhanced about 20-fold from that of the original streptococcal cell suspension. It was also found that the attenuated preparation was completely devoid of streptolysin S production and no sign of systemic infection occurred when the preparation was given to experimental animals. However, these streptococci still provide glycogenesis and certain other enzymatic activities, therefore, they are not necessarily a dead bacterial preparation. Streptococcal cell suspension in phyiological saline instead of BBM showed a slight antitumour effect, and an active preparation lost its antitumour effect when heated to 100°C.

The lyophilized preparation is designated "OK-432" (Picibanil[R], Chugai Pharm. Co., Tokyo), and its clinical unit is expressed by KE (Klinische Einheit). 1 KE corresponds to 0.1 mg dry weight of bacilli. Since the preparation contains a trace amount of penicillin, it cannot be used for patients allergic to penicillin.

III. ANTITUMOUR EFFECT OF OK-432

OK-432 exerts an antitumour effect against transplantable tumours in experimental animals by itself (Sakurai *et al.*, 1972) or in combination with chemotherapeutic agents (Mashiba *et al.*, 1979), and it suppresses the development of some mouse leukaemias (Suzuki *et al.*, 1975; Aoki *et al.*, 1976). The cytotoxic effect of this preparation is still obscure, but it may be explained by:

1) the direct action of OK-432 to tumour cells; and

2) the potentiation of host-immune responses.

A. Direct Action of OK-432

OK-432 has been shown to have antitumour effect against
Yoshida sarcoma, ascites hepatoma AH13 and Ehrlich ascites
tumour *in vitro* (Sakurai *et al.*, 1972). Since OK-432 inhibited DNA and RNA synthesis of certain tumour cells in proportion to the dose of preparation applied *in vitro* (Ono *et al.*,
1973), OK-432 seemed to have a direct action against malignant tumour *in vivo* when it is systematically administered.

B. Host-Mediated Indirect Action of OK-432

Peritoneal exudate cells activated with intraperitoneal injection of OK-432 had either cytostatic or cytocidal activity
against Ehrlich ascites tumour cells, and this effect was
more prominent when OK-432 was used in combination with a
chemotherapeutic agent (Mashiba *et al.*, 1979). Ishii *et al.*
(1976) showed that peritoneal macrophages from rats treated
with intraperitoneal OK-432 inhibit the growth of syngeneic
tumours. It is interesting that macrophages thus activated
by OK-432 developed a cytotoxic effect against other tumour
cells, indicating that OK-432 activates macrophages nonspecifically.

Activation of T-lymphocytes by OK-432 was demonstrated by
Kai *et al.* (1979). He demonstrated that pretreatment of mice
with OK-432 resulted in the increase of cytotoxicity of
spleen cells against primed target cells. He also showed
that OK-432 enhanced the hapten-specific antibody response,
with which hapten-specific B-cell and carrier-specific helper
T-cell are known to cooperate.

Adjuvant effect of OK-432 was reported by Ryoyama *et al.*
(1979). L-1210 leukaemic cells are less affected by OK-432
when compared with Yoshida sarcoma or Ehrlich ascites tumour
cells. When mice were immunized with mitomycin C-treated
L-1210 cells and then injected with OK-432, the life span of
the mice after transplantation of L-1210 was markedly prolonged. Activation of T-cell in peritoneal exudate cells by
OK-432 confirmed that the antitumour effects decreased when
the cells were pretreated by anti-T-cell antibody, and not
by anti-B-cell antibody (Natsu-ume Sakai *et al.*, 1976).
Aoki *et al.* (1975) showed that OK-432 enhanced the synthesis
of natural antibodies to specific cell-surface antigens.

OK-432 also activates natural killer (NK) cells in peritoneal exudate cells in mice (Oshimi *et al.*, 1980a). Although
the mechanism by which OK-432 enhances NK cell activity is
not known, it seems likely that OK-432-activated macrophages

produce interferon which subsequently activates NK cells.
The production of interferon in OK-432-inoculated mice was
suppressed by pretreatment with the macrophage-inhibitor
carageenan (Matsubara *et al.*, 1979).

Protection of infections by OK-432 was demonstrated by
Shiraishi (1979) who reported that infection of *Candida albicans* was inhibited probably by activating macrophages, and
Yokota (1978) showed that bacterial infections were protected
by activated phagocytosis of macrophages and polymorphonuclear
leukocytes.

Besides the activation of cellular elements by OK-432, it
was demonstrated that OK-432 provided activation of the complement system mainly through its alternative pathway (Kondo
et al., 1975; Natsu-ume Sakai *et al.*, 1976). It should also
be noted that serum complement level CH50, measured by the
lysis of sensitized sheep RBC, as well as ACH50, measured by
the lysis of rabbit RBC which represents the alternative
pathway activity, were increased by subcutaneous injections
of OK-432 in guinea pigs (Kato *et al.*, 1979). Although it
is difficult to explain whether the anticancer effect of OK-432 *in vivo* is partly due to the complement system, the evidence that serum complement level is elevated by the use of
OK-432 might be of importance in clarifying the role of the
complement in tumour destruction.

IV. APPLICATION OF OK-432 TO AGED IMMUNE-DEFICIENT SUBJECTS

Based on various experimental data in animals, it was suggested that OK-432 might be useful in the treatment of human
patients with malignant tumours. However, since the host-immune response in patients with malignancy is modified by
the disease or by various forms of anticancer therapy, the
potentiating effect of the immune response by OK-432 was first
examined in aged individuals without malignancy who have
shown decreased cellular immunity as determined by PPD (tuberculin) and PHA (phytohemagglutinin) skin test as well as by
blastoid transformation of lymphocytes against PHA (Kondo *et
al.*, 1977). Ten aged individuals with decreased cellular
immunity were selected from 290 subjects without malignancy,
and they were treated by OK-432 for 6 months. Daily intramuscular injection of 1 KE of OK-432 for an initial 3 month
period followed by 2 KE weekly for 3 months enhanced the skin
responses and blastoid transformation of lymphocytes during
the therapy except for one case. In the other 9 cases, their
restored immune responses gradually decreased to the original
levels when OK-432 was discontinued (Fig. 1).

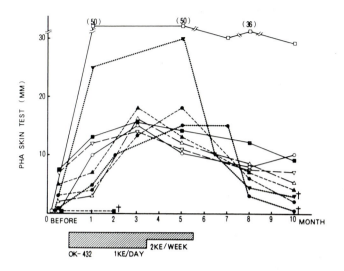

FIG. 1 *Effect of OK-432 on PHA skin test in aged immune-deficient subjects without malignancy.*

Thus, OK-432 was found to enhance the previously impaired immune response in aged subjects when it was administered in doses less than 1 KE daily. Although it is not known whether the amount used in aged individuals is sufficient to restore immune responses in patients with malignancy, it is suggested that OK-432 might be useful as an immunopotentiator to treat cancer patients.

V. ROUTE FOR ADMINISTRATION OF OK-432 IN PATIENTS WITH MALIGNANCY

A. Intramuscular Injection

This is the most common route for administering OK-432 to treat patients with malignancy. In general 1 to 5 KE is given daily or several times a week as maintenance doses, but no established schedule has been determined. At present, it is difficult to analyse the therapeutic effect of OK-432 alone on malignant diseases since it had only a slight effect in clinical trials of single administration in advanced cancer. Most investigators have tried to treat patients in combination with chemotherapy or irradiation.

Kimura *et al.* (1976) demonstrated the prolongation of survival as well as initiation of T-cell activity in stages III and IV lung cancer patients, when OK-432 was administered at maintenance levels with conventional inductive chemotherapeutic agents. Oshimi *et al.* (1980b) reported that OK-432 augmented the cytotoxic activity of natural killer (NK) cells in patients with malignancy, and the maximum level of cytotoxicity was evoked in all the patients on the third day after the beginning of intramuscular OK-432.

Interestingly, administration of repeated intramuscular OK-432 resulted in the elevation of serum complement levels in malignant diseases, and in addition, complement components C4, C3 and Factor B appeared in the gastric carcinoma tissue and on the leukaemic cell surface as determined by the fluorescent antibody technique (Kondo *et al.*, 1978) (Fig. 2).

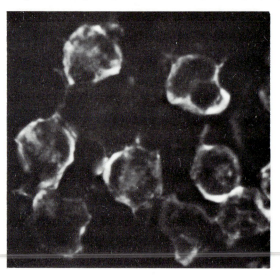

FIG. 2 *Localization of complement component factor B on CML cells appeared after OK-432 therapy.*

Results of the combination of OK-432 and chemotherapy upon advanced lung cancer patients receiving curative, relative-curative or non-curative resection are shown in Fig. 3.

B. Intradermal Injection

Uchida and Hoshino (1980) reported daily intradermal injections of OK-432 ranging from 1 to 10 KE to the forearm of 40 patients with advanced gastric cancer for over a period of 4 weeks in combination with 5-fluorouracil. Survival rates were significantly longer in the OK-432 group than those of the matched control patients treated by chemotherapy alone. These investigators showed that the improvement of *in vivo* and *in vitro* T-cell function tests achieved by the intradermal

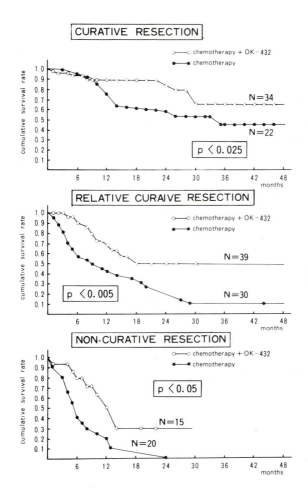

FIG. 3 *Survival of lung cancer patients treated by the combination of OK-432 and chemotherapy (Watanabe, 1980).*

injection of OK-432 was more significant than the improvements in patients given OK-432 by other routes such as intramuscular or intravenous injection. They also showed that intradermal injection of OK-432 was effective in reducing suppressor T-cell function.

Intradermal injection of OK-432 permitted much higher doses of the preparation with minimal side-effects than other routes, although it is still unclear whether the larger amount of OK-432 results in a more significant potentiation of host-immune responses.

C. Intravenous Injection

Because of severe side-effects such as fever and chill, this
route for administration of OK-432 is limited to only when
intramuscular or intradermal injection produces adverse reaction such as local bleeding or ulceration.

D. Intratumoural Injection

Intratumoural injection of OK-432 has been reported in patients with bladder carcinoma (Kagawa *et al.*, 1979) and malignant brain tumour (Tanaka *et al.*, 1980). In bladder tumour,
5 to 10 KE of OK-432 was first injected into the tumour and
then intramuscular injection of maintenance doses followed.
In brain tumour, patients received 3 to 9 KE of intramuscular
OK-432 as preimmunization, and then 0.25 to 1 KE of intratumoural injection and systemic maintenance doses of intramuscular injection followed. Clinical effects of intratumoural injection were evaluated by endoscopic examination
in the former, and by computerized tomography in the latter,
and a regression of tumour was observed in some cases. In
most of the cases receiving intratumoural injection of OK-432
histological examination revealed significant inflammatory
responses of mononuclear cell infiltrations and necrosis of
the tumour tissues.
    This intratumoural injection of OK-432 seems to have ad-
vantages over other routes of administration. One possibility
is the direct antitumour effect of OK-432. A more important
mechanism to be considered is that OK-432 might play a role
in inducing specific immune responses against the tumour-
specific antigen by its adjuvant effect, in addition to a nonspecific potentiation of host-immunity achieved by systemic
administration.

E. Intraperitoneal and Intrapleural Injection

Administration of OK-432 to the intraperitoneal cavity or
pleural cavity is a relatively new therapeutic procedure for
patients with cancer ascites or cancer pleurisy, that has
never been utilized for other bacterial preparations. Besides significant clinical results such as disappearance of
peritoneal or pleural fluid and tumour cells, this method
was helpful in clarifying the mode of action of OK-432 to
potentiate antitumour effect of the host.
    Ascitic fluids from 99 cases of advanced gastric cancer
were obtained by paracentesis for cytological examination
prior to OK-432 therapy. Six cell types were usually classi-
fied by Giemsa and Papanicolaou stains into: tumour cells,

TABLE 1

Survival of malignant ascites patients after
intraperitoneal injection of OK-432

Treatment	No. of patients	Survival time (month)[1]	50% Survival time (month)
OK-432			
Response group	63	$3.1-29.5(13.4 \pm 1.7)^2$	11.4
Non-response group	36	$0.3-7.8(\ 2.7 \pm 0.5)$	2.2
Total	99	$0.3-29.5(\ 8.7 \pm 1.2)^3$	5.6
Palliative group	40	$0.3-7.2(\ 2.6 \pm 0.4)$	2.1

[1]Mean ± SEM; [2]$P<0.001$ (unpaired Student's t-test): compared with that of palliative therapy. [3]$P<0.005$ (unpaired Student's t-test): compared with that of palliative therapy.

neutrophiles, lymphocytes, macrophages, eosinophiles and mesothelial cells. Patients then received intraperitoneal injections of 5 to 20 KE of OK-432 once or twice a week. By this method ascitic fluid disappeared in 59.9% and reduced in 4.1%, while 36.0% of the patients did not respond to the therapy. To obtain complete disappearance of ascitic fluid in the response group, 6 to 37 days (mean 22.5 ± 7.1 days) were required. Mean survival time of the patients after intraperitoneal OK-432 therapy was significantly prolonged, from 2.1 months in the palliative group to 11.4 months in the response group, while that in the non-response group was 2.2 months (Table 1). Survival rate of the patients is shown in Fig. 4.

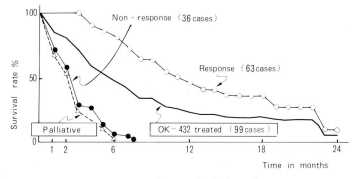

FIG. 4 *Effect of intraperitoneal injection of OK-432 on survival of patients with malignant ascites.*

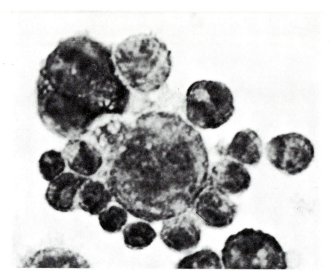

FIG. 5 *Attachment of lymphocytes and macrophages to the*
tumour cells in ascitic fluid after intraperitoneal injection
of OK-432.

By the intraperitoneal injection of OK-432, neutrophiles
appeared initially in the ascitic fluid within a few hours,
and the increase of macrophages and lymphocytes followed.
Tumour cells showed a gradual decrease with an attachment of
these cells to the tumour cell surface (Fig. 5).
    Usually, antitumour effect evoked by the use of OK-432 was
thought to be due to lymphocytes and macrophages. Subpopulation of lymphocytes in ascitic fluid and in circulating blood
was compared before and after the intraperitoneal OK-432
(Table 2). Although there was no change in peripheral blood,
E-rosette and stable E-rosette forming cells were significantly increased in ascitic fluid 5 to 10 days after OK-432
injection. Subpopulation of lymphocytes attached to tumour
cells in ascitic fluid was further investigated, and most of
the cells appearing after OK-432 injection were mainly E-
rosettes and stable E-rosette forming cells, namely T-cells
(Fig. 6, Table 3).
    In addition, macrophages and neutrophiles were also found
to attach to the tumour cells in ascitic fluid after intraperitoneal OK-432 (Table 4). Among them, it should be noted
that neutrophiles seemed to be associated with the disappearance of the tumour cells.
    In a case of a 77-year old man with advanced gastric can-
cer, intraperitoneal injection of 10 KE of OK-432 resulted in

TABLE 2

Effect of intraperitoneal injection of OK-432 on lymphocytes in ascitic fluid and in peripheral blood

OK-432 injection	Number of patients	Ascites-derived lymphocytes[1]			Peripheral blood-derived lymphocytes[1]		
		E-rosette	EAC-rosette	Stable E	E-rosette	EAC-rosette	Stable E
Before	12	42.2 ± 6.8	44.4 ± 6.2	7.6 ± 3.5	41.0 ± 2.3	46.5 ± 1.6	0.4 ± 0.2
After	12	73.4 ± 4.3[2]	21.8 ± 5.5	32.6 ± 4.1[3]	42.5 ± 3.4	44.4 ± 3.2	0.4 ± 0.2

[1] Assays were performed 5 to 7 days after the first administration of OK-432. Mean ± SEM.

[2] $P < 0.05$.

[3] $P < 0.02$.

TABLE 3

Attachment of lymphocytes to tumour cells in ascitic fluid after intraperitoneal injection of OK-432

OK-432 injection	Number of patients	Lymphocyte-tumour cell attachment (No. of lymphocytes/100 tumour cells)[1]			
		Lymphocyte	E-rosette	EAC-rosette	Stable E-rosette
Before	12	6.8 ± 1.8	5.6 ± 3.4	0.8 ± 0.7	2.8 ± 1.1
After	12	58.0 ± 12.9^2	51.0 ± 30.2^2	7.2 ± 4.7	28.8 ± 13.1^2

[1] Assays were performed 5 to 7 days after the first administration of OK-432. Mean ± SEM.

[2] $p < 0.005$.

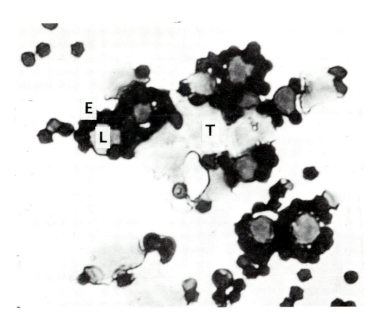

FIG. 6 *Attachment of E-rosette lymphocytes to the tumour cell in ascitic fluid after i.p. OK-432, demonstrated by double-rosette formation. T: Tumour cell, L: lymphocyte, E: Sheep erythrocyte.*

TABLE 4

Attachment of macrophages and neutrophiles to tumour cells in ascitic fluid after intraperitoneal injection of OK-432

OK-432 injection	No. of patients	No. of macrophages or neutrophils/ 100 tumour cells[1]	
		Macrophages	Neutrophils
Before	12	4.4 ± 3.0	1.8 ± 1.2
After	12	33.6 ± 20.0[2]	90.0 ± 20.0[3]

[1]Results were expressed as the maximum value of each patient. Mean ± SEM.

[2]$p < 0.01$.

[3]$p < 0.005$.

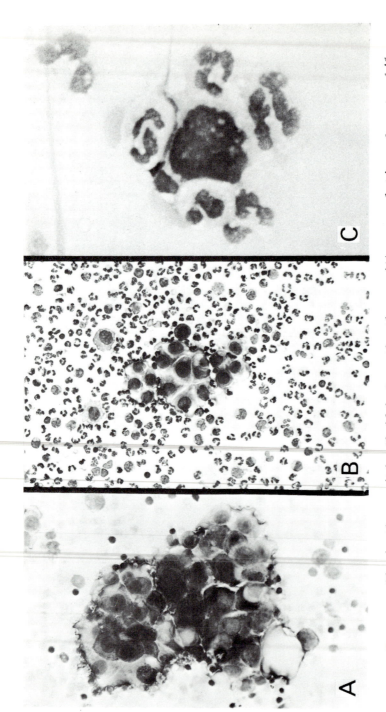

FIG. 7 Tumour cells in the ascitic fluid before the therapy (A), accumulation of neutrophiles in the ascitic fluid 12 h after the intraperitoneal injection of OK-432 (B), and attachment of neutrophiles to the tumour cells (C).

an increase of neutrophiles in the ascitic fluid. In 12 h
they surrounded the tumour cells, and after 48 h most of the
tumour cells disappeared. A few tumour cells remained and
revealed bulging and/or degeneration (Fig. 7). At that moment, neutrophiles, lymphocytes and macrophages in this ascitic fluid accounted for about 95%, 2% and 2% respectively.
When the patient was readmitted 3 months later because of
accumulation of ascites, he received the same schedule of
OK-432 therapy. It was confirmed that the increase of cellular elements in his ascitic fluid after the intraperitoneal
OK-432 was mainly neutrophiles, and tumour cells disappeared
shortly after OK-432 as ascites levels subsided. *In vitro*
cytotoxicity test of the neutrophiles to his own tumour cells
and tumour cells from other gastric cancer patients showed
that his neutrophiles had a cytotoxic effect only on his own
tumour cells, indicating that his neutrophiles obtained a
specific anticancer activity when OK-432 was given (Katano
and Torisu, 1982). It is not known whether the destruction
of the tumour cells is antibody-dependent cellular cytotoxicity (ADCC).

It is also unclear why these cellular elements appear in
ascitic fluid after intraperitoneal OK-432. Rapid increase
of neutrophiles in ascitic fluid may be explained by the fact
that OK-432 provides chemotactic activity to neutrophiles
(Table 5). This chemotactic factor may be derived from a
complement system since OK-432 activates the alternative
complement pathway (Kondo *et al.*, 1975; Natsu-ume Sakai *et
al.*, 1976).

Possible mechanism of anticancer effect of OK-432 in the
peritoneal cavity is shown in Fig. 8.

TABLE 5

Chemotactic activity of OK-432 to neutrophiles

OK-432 (KE/ml)	Chemotactic activity[1]
0	26 ± 2
0.001	48 ± 3
0.01	262 ± 50
0.1	220 ± 21

[1]Chemotactic activity is expressed as number of migrated
cells in 5 high power field. Mean ± SEM of 2 experiments.

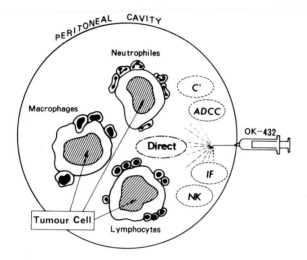

FIG. 8 *Proposed antitumour mechanism of OK-432 in the peritoneal cavity achieved after intraperitoneal injection.*

VI. SIDE-EFFECTS

Most of the patients complained of fever on the day when OK-432 was administered, but the grade of fever gradually decreased as the injections were repeated. Local redness and pain with induration were often noticed when the preparation was injected intramuscularly, and lidocaine was sometimes used with OK-432 to relieve the pain. Leukocytosis was observed in most of the cases. Loss of appetite, nausea and vomiting, liver dysfunction, anaemia and joint pain have been observed. However, their degrees were mild, and no fatal side-effects have been reported. The side-effects observed by the intraperitoneal injection of OK-432 are summarized in Table 6.

VII. ACKNOWLEDGEMENT

The authors express thanks to Mr O. Setoyama for his help.

VIII. REFERENCES

Aoki, T., Kveder, J.P., Kudo, T., Plate, E., Sendo, F. and Hollis, V.W. Jr. (1975). Enhancement of host immune response to cell surface antigens by a preparation of *Streptococcus hemolyticus*. *In* "Comparative Leukaemic Research". (Eds Y. Ito and R.M. Dutcher), pp. 277-280, University Tokyo Press, Tokyo/Karger, Basel.

TABLE 6

Side-effects of intraperitoneal injection of OK-432

Side-effect	Number of patients (%)
Fever : (<39.0)	64/99 (64.6)
(>39.0)	28/99 (28.2)
Chill	42/99 (42.4)
Nausea	22/99 (15.5)
Abdominal distension	29/99 (29.2)
Abdominal pain	24/99 (24.2)
Vomiting	7/99 (7.1)
Headache	6/99 (6.1)
Leukocytosis	94/99 (94.9)
Leukopaenia	5/99 (5.1)
Anaemia	6/99 (6.1)

Aoki, T., Kveder, J.P., Hollis, V.W. Jr. and Bushar, G.S. (1976). *Streptococcus pyogenes* preparation OK-432: immuno-prophylactic and immunotherapeutic effects on the incidence of spontaneous leukaemia in AKR mice. *J. Natl. Cancer Inst.* **56**, 687-690.

Bruns (1888). "Die Heilwirkung des Erysipel auf Geschwulste". Beitrage für klin. Chirurgie, pp. 443.

Busch, W. (1968). Niederrheinische Gesellschaft für Natur- und Heilkunde in Bonn. Aus der Sitzung der medicinischen Section vom 13, November 1867. *Berlin klin. Wschr.* **5**, 137-139.

Coley, W.B. (1891). Contribution to the knowledge of sarcoma. *Ann. Surg.* **14**, 199-220.

Von Fehleisen (1882). Über der Züchtung der Erysipelkokken auf künstlichem Nährboden und ihre Ubertragbarkeit auf dem Menschen. *Deutsch Med. Wschr.* **8**, 553-554.

Ishii, Y., Yomaoka, H., Toh, K. and Kikuchi, K. (1976). Inhibition of tumour growth *in vivo* and *in vitro* by macro-phages from rats treated with a streptococcal preparation OK-432. *GANN* **67**, 115-119.

Kagawa, S., Ogura, K., Kurokawa, K. and Uyama, K. (1979). Immunological evaluation of a streptococcal preparation OK-432 in treatment of bladder carcinoma. *J. Urol.* **122**, 467-470.

Kai, S., Tanaka, J., Nomoto, K. and Torisu, M. (1979).
    Studies on the immunopotentiating effects of a streptococ-
    cal preparation OK-432. I. Enhancement of T-cell-mediated
    immune responses of mice. *Clin. Exp. Immunol.* **37**, 98-105.
Katano, M. and Torisu, M. (1982). Neutrophile-mediated tumour
    cell destruction in cancer ascites. *Cancer* **50**, 62-68.
Kato, H., Yokoe, N., Matsumura, N., Nishida, K., Ikezaki, M.,
    Hosokawa, K., Hotta, T., Takemura, S., Yoshikawa, T. and
    Kondo, M. (1979). Effect of a streptococcal preparation
    on the complement system. *Jap. J. Exp. Med.* **49**, 343-350.
Kimura, I., Ohnishi, T., Yasuhara, S., Sugiyama, M., Urabe,
    Y., Fujii, M. and Machida, K. (1976). Immunotherapy in
    human lung cancer using the streptococcal agent OK-432.
    Cancer **37**, 2201-2203.
Kondo, M., Ikezaki, M., Imanishi, H., Nishigaki, I., Hosokawa,
    K. and Masuda, M. (1975). Streptococcal preparation as an
    activator of host-mediated immune response: cellular
    immunity and alternative pathway of complement. *GANN* **66**,
    675-678.
Kondo, M., Kato, H., Takemura, S., Yoshikawa, T., Yokoe, N.,
    Ikezaki, M. and Imanishi, H. (1977). Activation of cel-
    lular immunity and complement system in immune-deficient
    aged subjects by streptococcal preparation OK-432 (Pici-
    banil). *Jap. J. Clin. Oncol.* **7**, 21-26.
Kondo, M., Kato, H., Yokoe, N., Yoshikawa, T., Takemura, S.,
    Hotta, T., Matsumura, N. and Ikezaki, M. (1978). Activa-
    tion of complement system by streptococcal preparation
    OK-432 (Picibanil) in patients with malignancy. *Jap. J.
    Clin. Oncol.* **8**, 27-30.
Koshimura, S., Murasawa, K., Nakagawa, E., Ueda, M., Bando,
    Y. and Hirata, R. (1955). Experimental anticancer
    studies. Part III. On the influence of living haemolytic
    streptococci upon the invasion power of Ehrlich ascites
    carcinoma in mice. *Jap. J. Exp. Med.* **25**, 93-102.
Koshimura, S., Shimizu, R., Masusaki, T., Ohta, T. and Kishi,
    G. (1958). On the formation of streptolysin S by haemo-
    lytic streptococci acting on tumour cells. *Jap. J. Micro-
    biol.* **2**, 23-28.
Mashiba, H., Matsubara, K. and Gojobori, M. (1979). Effect
    of immunochemotherapy with OK-432 and yeast cell wall on
    the activation of peritoneal macrophages of mice. *GANN*
    70, 687-692.
Matsubara, S., Suzuki, F. and Ishiba, N. (1979). Induction
    of immune interferon in mice treated with a bacterial
    immunopotentiator, OK-432. *Cancer Immunol. Immunother.*
    6, 41-45.

Natsu-ume Sakai, S., Ryoyama, K., Koshimura, S. and Migita, S. (1976). Studies on the properties of a streptococcal preparation OK-432 (NSC-B116209) as an immunopotentiator. I. Activation of serum complement components and peritoneal exudate cells by Group A streptococcus. *Jap. J. Exp. Med.* **46**, 123-133.

Nauts, H.C., Fowler, G.A. and Bogatko, F.H. (1953). A review of the influence of bacterial infection and of bacterial products (Coley's toxins) on malignant tumours in man. *Acta Med. Scand.* **145** (Suppl.), 1-103.

Okamoto, H. (1940). Über die hochgradige Steigerung des Hämolysinbildungsvermögens des *Streptococcus hemolyticus* durch Nukleinsäure. *Jap. J. Med. Sci. (IV. Pharmacology)* **12**, 167-208.

Okamoto, H., Minami, M. and Shoin, S. (1966). Experimental anticancer studies. Part XXXI. On the streptococcal preparation having potent anticancer activity. *Jap. J. Med. Sci.* **36**, 175-186.

Okamoto, H., Shoin, S., Koshimura, S. and Shimizu, R. (1967). Studies on the anticancer and streptolysin S-forming abilities of haemolytic streptococci (a review). *Jap. J. Microbiol.* **11**, 323-336.

Ono, T., Kurata, S., Wakabayashi, K., Sugawara, Y., Saito, M. and Ogawa, H. (1973). Inhibitory effect of a streptococcal preparation (OK-432) on the nucleic acid synthesis in tumour cells *in vitro*. *GANN* **64**, 59-69.

Oshimi, K., Kano, S., Takaku, F. and Okumura, K. (1980a). Augmentation of mouse natural killer cell activity by a streptococcal preparation OK-432. *J. Natl. Cancer Inst.* **65**, 1265-1269.

Oshimi, K., Wakasugi, H., Seki, H. and Kano, S. (1980b). Streptococcal preparation OK-432 augments cytotoxic activity against an erythroleukaemic cell line in humans. *Cancer Immunol. Immunother.* **9**, 187-192.

Ryoyama, K., Murayama, T. and Koshimura, S. (1979). Effect of OK-432 on immunization with mitomycin-C treated L1210 cells. *GANN* **70**, 75-82.

Sakurai, Y., Tsukagoshi, S., Satoh, H., Akiba, T., Suzuki, S. and Takagaki, Y. (1972). Tumour-inhibitory effect of a streptococcal preparation (NSC-B116209). *Cancer Chemother. Rep.* **56**, 9-17.

Shiraishi, A., Mikami, Y. and Arai, T. (1979). Protective effect of OK-432 (a streptococcal preparation) on experimental candidasis. *Microbiol. Immunol.* **23**, 549-554.

Suzuki, S., Yamamoto, A. and Ogawa, H. (1975). Inhibitory effect of a streptococcal preparation (OK-432) on induction of splenomegaly by Friend luekaemia virus. *GANN* **66**, 455-456.

Tanaka, R., Sekiguchi, K., Suzuki, Y., Sobue, H. and Ueki, K. (1980). Preliminary evaluation of intratumoural injection of a *Streptococcus pyogenes* preparation in patients with malignant brain tumours. *Cancer* **46**, 1688-1694.

Uchida, A. and Hoshino, T. (1980). Clinical studies on cell-mediated immunity in patients with malignant disease. *Cancer* **45**, 476-483.

Watanabe, Y. (1980). Clinical value of immunotherapy by a streptococcal preparation OK-432 as an adjuvant for resected lung cancer (Japanese). *Cancer Chemother.* **7**, 1256.

Yokota, T., Nogaki, K., Ogawa, H. and Akiba, T. (1978). Stimulatory effect of the streptococcal preparation Picibanil on phagocytosis. *Current Chemother.* 324-326.

BACTERIAL INFECTIONS IN CANCER PATIENTS
AN OVERVIEW

Victor Fainstein and Gerald P. Bodey

*From the Department of Developmental Therapeutics,
Section of Infectious Diseases,
The University of Texas Cancer System Centre,
M.D. Anderson Hospital and Tumour Institute, Houston, Texas*

Bacterial infections in cancer patients are a common cause of morbidity and mortality. Several of the new bacterial organisms can cause infection in these patient populations. This chapter will only try to summarize some of the general concepts of infection in cancer patients and will primarily focus on these new organisms, their role in infection and the changing patterns of bacterial infections. It will finalize with a brief presentation of antibiotic therapy and several of the current procedures utilized for prophylaxis of bacterial infection.

I. GENERAL REMARKS

Multiple factors may increase the cancer patient's susceptibility to infection. Some of these factors are discussed in other chapters of this book. Cellular defects, such as neutropenia or monocytopenia, may be a consequence of the disease process itself. Impaired antibody production and gastrointestinal ulceration may develop as side-effects of treatment. A complex interrelationship of lymphocytes, neutrophils, macrophages, immunoglobulins, complement, along with physical barriers provide protection against infecting organisms. Any qualitative or quantitative defect in one of these factors may predispose to infection, although many patients frequently suffer from multiple defects simultaneously.

Neutropenia is probably one of the most important factors responsible for the increased frequency of infection. Both quantitative and qualitative defects have been recognized.

An inverse correlation exists between the number of circulating neutrophils and lymphocytes and the frequency of infection. For example, in a study of 52 patients with acute leukaemia, there were 43 episodes of major infection per 1000 days spent with a neutrophil count $<100/mm^3$ but only 4 episodes per 1000 days spent with a neutrophil count $>1000/mm^3$ (Bodey *et al.*, 1966). The longer the duration of neutropenia, the greater the risk of infection. Also the fatality rate from infection is dependent upon the trend in neutrophil count. Among patients with a neutrophil count $<1000/mm^3$ the fatality rate was 59% if the count decreased during the infection whereas it was 32% if the count increased.

The malignant process itself may also be responsible for increased susceptibility to infection. Often this is a consequence of tumour growth which causes obstruction or ulceration. Many infectious complications are a consequence of therapeutic measures. Cancer chemotherapeutic agents have multiple side-effects, such as gastrointestinal ulceration and myelosuppression. Obstruction to natural drainage may result from tumour growth and may predispose to local infection. For example, nearly 45% of patients with bronchogenic carcinoma die of infectious complications, and 80% of these infections are pneumonias. Most of these infections are caused by bronchial obstruction and atelectasis. Antibiotic therapy is often ineffective against these pneumonias even though the infecting organism is not resistant to the antibiotic used. Obstruction to urinary drainage commonly results in infection. The insertion of a catheter may relieve the obstruction, but chronic urinary tract infection is an inevitable consequence. Initially, the infecting organism is usually susceptible to multiple antibiotics, but eventually, following several courses of antibiotic therapy, superinfection with a resistant organism occurs.

Gastrointestinal ulceration may result from a variety of factors. Perhaps the most frequent cause of these ulcerations is the use of antitumour agents which destroy the proliferating epithelial cells of the gastrointestinal mucosa. Tumours of the gastrointestinal tract may undergo necrosis and ulcerations, serving as a portal for infection by enteric organisms. Infection is the proximate cause of death in about 50% of patients with gastrointestinal tumours. Septicaemia occurs during 45% of these fatal infections. In 70% of infections, the tumour itself is the major predisposing factor. Patients with gastrointestinal tumours are among those cancer patients who have a greater susceptibility to clostridial infections.

The skin and mucous membranes are the first line of defence against bacterial and fungal infections and are often breached in cancer by repeated venipunctures and finger sticks, infusion of drugs which may extravasate and cause phlebitis, the presence of decubitus ulcers and the use of intravenous catheters, needles, and hyperalimentation.

Microbial colonization is a special concern because many of the organisms acquired from the hospital environment are resistant to multiple antibiotics. The administration of oral absorbable antibiotics is associated with fungal colonization of the gastrointestinal tract. Both the proportion of patients colonized and the concentration of organisms in previously colonized patients increase during therapy. Colonization of the oropharynx by gram-negative bacteria during hospitalization occurs in a substantial proportion of patients, even without antibiotic administration. Major infections have been found to occur following marked increases in the concentration of gram-negative bacilli in the oropharynx. The respiratory tracts of patients with tracheostomies are usually colonized by gram-negative bacilli which often subsequently cause tracheitis and pneumonia.

A thorough understanding of all these underlying factors is necessary to identify likely sources of infection and permit more effective therapy. Bacterial infections are still the most common type of infection in cancer patients. While gram-positive infections, particularly with staphylococci, occur with increased frequency in patients with skin cancers and obstructing tumours, the gram-negative aerobic bacilli (particularly *E. coli*, *Klebsiella*, and *Pseudomonas* sp.), are by far the most common bacterial isolates from patients with haematologic neoplasia and other solid tumours. Gram-negative bacilli account for well over half of all documented infections in these patients, and occur about 10 times more often than fungal organisms. It is common for febrile neutropenic patients to have gram-negative septicaemia with no clinically obvious source. In many of these cases, gastrointestinal ulcerations secondary to chemotherapy may be implicated as the source for opportunistic infection.

II. NEW ORGANISMS AND CHANGING PATTERNS OF INFECTION

The past decade has witnessed the discovery of new bacterial causes of infection and complications in the compromised host. Additionally, the spectrum of infecting organisms is changing. The introduction of the semisynthetic antistaphylococcal penicillins had virtually eliminated *Staph. aureus* as a major cause of infection in leukaemia, but recently it has reappeared as an important pathogen. *Staph. epidermidis* infections, seldom a problem in the past, are now an important cause of morbidity and mortality due to the ubiquitous use of

TABLE 1

Frequent and newly recognized bacteria causing
infection in cancer patients

Staphylococcus aureus	*Pseudomonas aeruginosa*
Staphylococcus epidermidis	*E. coli*
Enterococcus	*Klebsiella* sp.
	Serratia marcescens
JK bacteria	*Legionella pneumophila*
Bacillus cereus	Legionella-like organisms
Listeria monocytogenes	*Clostridium septicum*
Aeromonas hydrophila	*Clostridium perfringes*
Capnocytophaga sp.	*Clostridium difficile*
Eikenella corrodens	*Acinetobacter* sp.

intravenous catheters. Hence, the approach to management of
these patients and the choice of antibiotic regimens needs to
be constantly reevaluated. Table 1 lists many of the bacterial organisms which have been identified as common causes
of infection in cancer patients. Some of the new organisms
recently discovered and the role they play in these patients
will be discussed.

A. *Legionella pneumophila* and Related Organisms

Legionellosis is now recognized as a cause of epidemic and
sporadic pneumonia. The disease was first described in 1976
after an outbreak of pneumonia (Fraser *et al.*, 1977). When
the bacterium was isolated it was found to be closely related
to previous outbreaks (Chandler *et al.*, 1977). In the last
5 years, more than a thousand cases have been confirmed,
many of them affecting compromised patients who have under-
gone renal transplantation or who have cancer and are severely
immunosuppressed, sometimes as nosocomial outbreaks. After
extensive and detailed microbiological and serological work,
more serotypes of Legionella pneumophila as well as other
species have been discovered. Thus far, 5 species (*pneumophila, micdadei, bozemanii, dumoff,* and *gormanii*) are recognized. Also 6 serotypes of pneumophila have been well characterized. Person to person spread has not been documented
and cases of pneumonia due to this organism have been reported from countries. It is a severe multisystemic disease
characterized by fever and pneumonia. Clinically, the pneumonia process is difficult to distinguish from other atypical
pneumonias. It begins as an influenza - like illness with

malaise, myalgias and headache accompanied by fever. Early
in the course a dry cough begins. About 30% of patients
have chest pain and dyspnoea is a common finding later in the
course of the disease. Abdominal pain and vomiting are present in 20% of affected patients, and diarrhoea has been reported to occur in up to 40% of the cases. The spectrum of
the disease ranges from a mild pneumonitis to a multilobar
pneumonia with death from respiratory failure. Some investigators have isolated the organism from transtracheal aspirates and blood (Rodgers, 1979). However, this is unusual and
the diagnosis is still made by the demonstration of organisms
in clinical specimens or by specific seroconversion.

Several epidemics of the disease have been reported and
they usually tend to occur in the summertime. Transmission
appears to be airborne or by aerosol, and in several outbreaks there was some evidence for spread by heat exchange
units such as cooling towers or stream condensers. However,
the organism has also been isolated from streams, lakes, soil
and cooling towers not involved in disease transmission.
Estimates of the proportion of undiagnosed pneumonias that
are caused by this organism have ranged from 0.3% to 4.5%,
some of these figures applying directly to the compromised
patient.

Legionella pneumophila is a fastidious gram-negative bacillus which grows on Mueller-Hinton agar base supplemented
with 1% haemoglobin and 2% Isovitalex or with 0.025% ferric
pyrophosphate and 0.04% L-cysteine HCL, when incubated up to
7 days at 35°C in atmosphere containing 5% CO_2. It is cata-
lase-positive and grows well in the yolk sac of embryonated
eggs and when inoculated intraperitoneally into guinea pigs.
The diagnosis is usually made when lung tissue, bronchial
washings or transtracheal aspirates are stained with direct
fluorescent antibody, or by culturing the organism from lung
tissue, blood or pleural fluid (Watts *et al.*, 1980). Since
the diagnosis usually requires obtaining tissue by invasive
procedures, other techniques are being developed. Kohler *et
al.* have developed a radioimmunoassay to detect *Legionella
pneumophila* serogroup 1 in the urine (Kohler *et al.*, 1981).
The demonstration of a 4-fold rise in antibody titre from
acute phase to convalescent phase also confirms the diagnosis. This titer may rise as early as 7—8 days after onset
of symptoms but it usually reaches its maximum after 3 to 4
weeks. Since the serological tests for antibodies are not
useful for the rapid diagnosis of this disease and the culture of the organism is very difficult because it is so fastidious, the diagnosis usually requires histologic examination. All strains have been sensitive to erythromycin which
is the drug of choice (Edelstein and Meyer, 1980). Rifampin

is also effective but experience with this drug is limited.
However, some authors recommend adding rifampin if the diagnosis has been confirmed and if erythromycin therapy is not
successful.

B. Corynebacteria

JK bacteria and *Corynebacterium* sp. are skin organisms which
if isolated from blood cultures are usually regarded as con-
taminants. However, they can cause septicaemia and pneumonia
in compromised patients. Several species, including *C. haemo-
lyticum, C. xerosis, C. pyogenes* and *C. pseudodiphteriticum*
have been identified as infecting organism. *JK bacteria* are
a group of aerobic, pleomorphic, non-motile, non-sporulating,
gram-positive rods related to Corynebacteria. They can
cause septicaemia and prosthetic valve endocarditis (Gill *et
al.*, 1981; Riley *et al.*, 1979). Infections caused by this
organism have been described in patients with haematological
malignancies who have been hospitalized for prolonged periods,
are receiving antibiotics and remain neutropenic. A break
in the mucocutaneous surface has been identified in the
majority of the infected patients. Characteristically, the
organisms are resistant to the majority of conventional antibiotics. They are susceptible to vancomycin and sometimes
also to erythromycin and rifampin. Young *et al.* have shown
that this organism can be acquired from the hospital environment and they can be cultured from the groin, rectum and
catheters (Young *et al.*, 1981). Thirty of 35 patients in
one series were colonized with JK bacteria while in the hospital and some patients were already colonized on admission
(Gill *et al.*, 1981). About 80% of untreated patients have
died from these infections, as well as 3 of 13 who were
treated appropriately (Lang *et al.*, 1980). In order to
institute appropriate therapy promptly, recognition of these
resistant bacteria is essential.

C. *Clostridium difficile* and *Capnocytophage ochracae*

C. difficile is a gram-positive obligate anaerobic rod, which
has been difficult to isolate in the past. Described in
1935, it was considered non-pathogenic for humans until
recently (Bartlett *et al.*, 1978). Epidemiological studies
on healthy individuals have shown that it can be isolated
from stools in 2-3% of adults (Bartlett, 1979). Pseudomembranous colitis caused by *C. difficile* is a severe form of
diarrhoea usually related to antimicrobial therapy. In 98%
of the cases a toxin can be detected in the faeces. Likewise, 15% of patients with diarrhoea related to antimicrobial

agents will have the toxin present even without signs of colitis. The diagnosis of this entity is made by culturing the organism in special media containing cefoxitin, cyclo-serine and egg yolk and detecting the presence of the toxin by a cytopathic effect on tissue culture cells (Chang *et al.*, 1979) or by counter immunoelectrophoresis. Since necrotizing colitis is a recognized complication of cancer chemotherapy, the role that this organism plays in the disease has recently been evaluated (Fainstein *et al.*, 1980). *C. difficile* can produce colitis in cancer patients when they are treated only with antitumour agents (Fainstein *et al.*, 1981). The organ-ism can colonize patients who are receiving prophylactic antibiotics and predispose them to the development of diarr-hoea and colitis.

Capnocytophaga ochracea formerly called *Bacteroides ochra-ceus*, is a pigment-producing gram-negative, non-sporeforming capnophilic bacillus, which is part of the normal oropharyn-geal flora. However, recently it has been identified as a cause of infection in patients with leukaemia and severe immunosuppression (Florlenza *et al.*, 1980; Gilligan *et al.*, 1981). In the majority of the reported cases, this organism has invaded the bloodstream through disruptions of the epi-thelial surface of the gum and oropharyngeal cavity. The organism is very sensitive to erythromycin, clindamycin, chloramphenicol and cephalosporins. Its ability to cause septicaemia in leukaemia patients and its role in periodontal disease require further investigation.

D. Changing Spectrum of Infection

Several organisms have become important causes of infection in compromised hosts. The gram-positive aerobic cocci which has become infrequent and easily managed cause of infection have again emerged as a major cause of nosocomically acquired infections in leukaemia patients (Kilton *et al.*, 1979). The frequency of *Staph. aureus* infections has increased in many leukaemia units and in some institutions about 10 to 20% of strains are resistant to methicillin and other anti-staphylo-coccal antibiotics. Until recently, the isolation of *Staph. epidermidis* from blood culture specimens indicated skin con-tamination, since it was seldom a cause of infection. How-ever, this organism can no longer be ignored because it has become an increasingly common cause of septicaemia in leukae-mia patients and bone marrow recipients. Infection with this organism is usually related to intravenous catherization or interruption of the integument (Bender and Hughes, 1980).

Many other organisms considered in the past as uncommon or non-pathogenic have become important pathogens in

leukaemia patients. *Aeromonas hydrophila* is an organism
usually found in water which may cause septicaemia and cellulitis. Interestingly, some leukaemia patients have developed
infection with this organism when they have returned to normal activities such as swimming and fishing after achieving
complete remission (Fainstein *et al.*, 1981b; Wolff *et al.*,
1980). The organism is generally resistant to penicillin,
ampicillin, and carbenicillin but susceptible to cephalo-
sporine and aminoglycosides. *Eikenella corrodens* is a gram-
negative aerobic bacterium which is part of the mouth and
throat flora. It has been recognized as a human pathogen
causing endocarditis, meningitis, and septic arthritis (Fainstein *et al.*, 1981c; Ruberstein *et al.*, 1976). *Bacillus* sp.
is a skin contaminant which has been considered essentially
non-pathogenic. Infections caused by these organisms have
been reported in leukaemic patients, usually related to intravenous catheters and hyperalimentation. Infection may be
manifested as septicaemia, cellulitis, conjunctivitis, otitis
or pneumonia (Ihde and Armstrong, 1973).

III. ANTIMICROBIAL THERAPY AND PROPHYLAXIS

Numerous studies of antimicrobial therapy have been conducted
in cancer patients, the majority of them leukaemic, during
recent years. Several important concepts have been examined
such as the role of antibody synergy and the impact of antibiotic schedule on outcome. Several new antibiotics have
been introduced and new combinations have been developed.
Detailed discussion of those studies is beyond the scope of
this review and the reader can find excellent articles on
this subject (Bodey *et al.*, 1972, 1979; Keating *et al.*, 1979;
Klastersky, 1981; Schimpff *et al.*, 1971).

A. New Antibiotics

During the past several years, many new broad-spectrum penicillins and cephalosporins have been synthesized. Mezlocillin, azlocillin and piperacillin are new penicillins of
interest (Schimpff *et al.*, 1971; Fu and Nev, 1978; Issell
and Bodey, 1980; Stewart and Bodey, 1977). Piperacillin and
mezlocillin have a broader spectrum of activity against gram-
negative bacilli than earlier penicillins. Azlocillin is an
antipseudomonal penicillin with substantially greater activity
than carbenicillin or ticarcillin against *P. aeruginosa in
vitro*. A new class of cephalosporins, the "third generation
cephalosporins" are undergoing clinical evaluation at present.
These antibiotics generally are more active than older
cephalosporins against cephalosporin-susceptible gram-nega-

tive bacilli (Bolivar *et al.*, 1981; Hinkle and Bodey, 1980).
Additionally, they are also active against organisms which
are usually resistant to other cephalosporins, such as
Serratia marcescens and *P. aeruginosa*. One of these agents,
cefsulodin, is a cephalosporin with selective activity only
against *P. aeruginosa* (King *et al.*, 1980). Additionally,
several new classes of β-lactam antibiotics have been discovered including thienamycins and monobactams and β-lacta-
mase inhibitions such as clavulanic acid (Neu and Fu, 1978;
Weaver *et al.*, 1979; Fainstein *et al.*, 1981d). All of these
agents are investigational at present and their role in the
treatment of infections in neutropenic patients remains to
be ascertained.

The high frequency of infection mainly in leukaemia patients during periods of neutropenia led to the development of
prophylactic programmes. For many years, physicians have
placed neutropenic patients on reverse isolation in an
attempt to protect them from acquiring infection. Recently,
a group of leukaemia patients were randomly assigned to
reverse isolation or no isolation during periods of neutropenia (Nauseef and Maki, 1981). There were 27 infections
during 23 studies of no isolation (4.4 per 100 days) compared to 28 infections during 20 studies (6.4 per 100 days)
of reverse isolation, hence, this practice is of no value.
Many other techniques have been utilized for prophylaxis of
bacterial infections.

More than a decade ago, techniques utilized with germ-free
animals were modified for clinical application, initially
with the introduction of the plastic tent isolater, and
subsequently with the laminar air flow unit (Bodey, 1969;
Schwartz and Perry, 1966). Most of the early studies attempted to provide maximum reduction in microbial contamination
of the patient and his environment by the use of filtered
air, specially prepared food, oral non-absorbable antibiotics and skin cleansing agents. In the majority of these
studies, the proportion of patients developing severe and
fatal infections was substantially lower among PEPA patients,
and most of these differences were statistically significant.
Patients on the PEPA programme were better able to tolerate
higher doses of chemotherapy that did not increase the complete remission and survival. The PEPA programme definitely
reduces the risk of infection during remission induction
chemotherapy, but its long-term benefits remain to be determined.

The importance of attempting to completely sterilize the
gastrointestinal tract has been questioned by van der Waaij,
who has proposed that the anaerobic flora conveys "colonization resistance" which protects against colonization by

aerobic pathogens. His associates conducted a study in which
patients were randomized to receive an antibiotic regimen
which did not affect the anaerobic flora or to receive no
prophylaxis (Sleijfer *et al.*, 1980). Eighteen gram-negative
infections or fungal infections occurred in 12 of 55 control
patients compared to only 2 infections in 2 of 58 patients on
the prophylactic regimen (p <0.01). Nine control patients
died of infection compared to none on the prophylactic regimen. Another study has shown that the frequency of infection
in patients receiving antibiotic regimens chosen to selectively eliminate the aerobic flora, is related to the efficacy of the regimen in eliminating this flora (Henri *et al.*,
1981). Three infections occurred in 23 patients who became
free of potentially pathogenic organisms whereas 10 infections occurred in 16 patients in whom potential pathogens
persisted in the gastrointestinal tract. In a randomized
trial, cotrimoxazole plus nystatin which has a lesser impact
on the gastrointestinal flora than gentamicin plus nystatin,
was as effective in protecting against acquisition of new
organisms (Wade *et al.*, 1981). Thirty-five infections
occurred in the 27 patients on the former regimen compared to
31 infections in the 26 patients on the latter regimen.
Hence, the data indicate that a variety of regimens, including parenterally administered antibiotics, are effective for
protecting leukaemic patients from infection during periods
of severe neutropenia.

The risk of infection can be reduced by using antibiotic
regimens that do not fully suppress the microbial flora of
the gastrointestinal tract. Children with leukaemia in
remission were prospectively randomized to receive cotrimoxazole or placebo for prevention of pneumocystis pneumonia
(Hughes *et al.*, 1976). Pneumocystis pneumonia occurred in
21% of the control patients but in none of those who received
cotrimoxazole. In addition, bacterial septicaemia, pneumonia,
otitis media, upper respiratory infections, sinusitis and
cellulitis occurred less often in the treated group (p <0.01).
In another study, patients were randomized to receive cotrimoxazole or no prophylaxis during period of neutropenia
(Herzig *et al.*, 1977a). The patients who received prophylaxis spent significantly less time with fever than the control patients. None of these patients developed bacteraemia
compared to 17% of the controls (p = 0.001), and they also
experienced fewer soft tissue and urinary tract infections.

Other techniques like granulocyte transfusions (Gaines and
Remington, 1973; Herzig *et al.*, 1977b), lithium carbonate
(Stein *et al.*, 1979), and immunotherapy have been evaluated
for protecting patients from infectious complications. Since
passive immunotherapy does not depend upon the patients'

ability to respond to antigenic stimuli, hence, this approach might be more effective in cancer patients. Studies have been conducted with Pseudomonas antiserum, hyperimmune globulins and antiserum to core glycolipid of gram-negative bacteria (1-5 antiserum). Since Pseudomonas infections are common in cancer patients, the possibility of preventing this infection has elicited great interest. Until recently, the only available vaccine was a heptavalent vaccine which was tested extensively. When the vaccine was administered to patients with burn wounds, it reduced the frequency of Pseudomonas septicaemia from 18 to 8% (Alexander *et al.*, 1971). A randomized study of this vaccine was conducted in cancer patients (Young *et al.*, 1973). Those with acute leukaemia and solid tumours responded as well as normal individuals, but patients with chronic lymphocytic leukaemia did not, as measured by hemagglutinating antibodies. The frequency of Pseudomonas infection was 16% among the vaccinated patients compared to 24% among the control patients. Mortality due to Pseudomonas infection occurred in 7% and 17% of patients, respectively.

Recently, a new polyvalent Pseudomonas vaccine has been developed (Miller *et al.*, 1977). This vaccine is made from surface components of 16 selected strains of *P. aeruginosa*. A controlled clinical trial of this vaccine was conducted in children and adults with burns (Jones *et al.*, 1979). Thirteen of 35 (43%) blood cultures from unvaccinated patients contained gram-negative bacteria (*P. aeruginosa*, *K. aerogenes*, *E. coli*) and one of 16 blood cultures obtained from vaccinated patients contained gram-negative bacteria.

B. Antisera to Gram-negative Bacteria

Measures to reduce the high frequency of gram-negative bacillary infections especially in leukaemic patients would be beneficial. Braude *et al.* prepared a vaccine from a rough mutant of *E. coli* 0111, known as J-5 (Braude *et al.*, 1981). The antiserum obtained after immunization with this vaccine is designated as J-5 antiserum. Studies in immunosuppressed animals have shown that this antiserum has antitoxic properties which can protect against lethal bacteraemia. The results with this antiserum have been encouraging in patients with gram-negative bacteraemia. The mortality rate in 37 bacteraemic patients given antibiotics plus J-5 antiserum was 14% but it was 26% in 46 patients given antibiotics plus non-immune serum (McGuthan *et al.*, 1979). Among patients with endotoxin shock, the recovery rate was 82% in the former group and 29% in the latter group. The antiserum has been evaluated therapeutically and also prophylactically. Neutro-

penic patients with leukaemia and lymphoma were randomly
assigned to receive either J-5 antiserum or pre-immune control serum as prophylaxis. Patients received serum intravenously every 21 days while neutropenic. The frequency of
febrile days was 44% in controls compared to 18% in those
given the antiserum.

C. Pneumococcal Polysaccharide Vaccine

Although most infections in patients with leukaemia are due
to gram-negative bacilli, infections with pneumococci occur
occasionally. Thirteen (21%) of 60 patients with pneumococcal bacteraemia in one series had leukaemia (Folland *et al.*,
1974). Patients with chronic lymphocytic leukaemia had a
higher mortality from this infection than other patients.
Patients with hairy cell leukaemia frequently undergo splenectomy as part of their therapy. Pneumococcal infections represent a serious threat to splenectomized patients because
infection may be overwhelming and rapidly lead to death,
despite antibiotic therapy. Trials conducted in patients
with multiple myeloma have shown that approximately 30% of
patients demonstrated a significant antibody response to
immunization (Lazarus *et al.*, 1980). IgG and IgM antibody
responses of splenectomized patients to pneumococcal vaccine
was markedly impaired when compared to control patients
(Hosea *et al.*, 1981). Unfortunately, several splenectomized
patients who were vaccinated subsequently died of overwhelming pneumococcal infection, suggesting that penicillin prophylaxis may be necessary in these patients.

D. Passive Immunotherapy

Passive immunotherapy has been evaluated, using immune serum
globulin and hyperimmune gammaglobulin preparations against
viruses and bacteria. Immune serum globulin was not found
to be an effective adjuvant to antibiotic therapy in a randomized trial of infection in patients with acute leukaemia.
The cure rate in patients receiving gammaglobulin was 57%
compared to 56% for patients receiving antibiotic therapy
only (Bodey *et al.*, 1964). The lack of specificity probably
accounted for the lack of efficacy in this study. Hyperimmune globulin directed against specific infectious agents
may prove to be more useful.

IV. REFERENCES

Alexander, J.W., Fisher, M.W. and MacMillan, B.G. (1971). Immunological control of Pseudomonas infection in burn patients: a clinical evaluation. *Arch. Surg.* **102**, 31.

Bartlett, J. (1979). Antibiotic-associated pseudomembranous colitis. *Rev. Infect. Dis.* **1**, 530.

Bartlett, J.G., Chang, T.W., Gurwith, M. and Gorbachsl-Onderdonk, A.E. (1978). Antibiotic associated pseudomembranous colitis due to toxin-producing Clostridia. *N. Engl. J. Med.* **298**, 531.

Bender, J.W. and Hughes, W.T. (1980). Fatal *Staphylococcus epidermis* infection following bone marrow transplantation. *Johns Hopkin Med. J.* **146**, 13.

Bodey, G.P. (1969). Laminar air flow unit for patients undergoing cancer chemotherapy. *In* "Germ-free Biology". (Eds E.A. Mirand and N. Back), p$\frac{1}{2}$ 69, Plenum Press, New York.

Bodey, G.P., Buckley, M., Sathe, Y.S. and Freireich, E.J. (1966). Quantitative relationships between circulating leukocytes and infection in patients with acute leukaemia. *Ann. Intern. Med.* **64**, 328.

Bodey, G.P., Ketchel, S.J. and Rodriguez, V. (1979). A randomized study of carbenicillin plus cefamandole or tobramycin in the treatment of febrile episodes in cancer patients. *Am. J. Med.* **67**, 608.

Bodey, G.P., Middleman, E., Umsawasdi, T. and Rodriguez, V. (1972). Infections in cancer patients — results with gentamicin sulfate therapy. *Cancer* **29**, 1607.

Bodey, G.P., Nies, B.A., Mohberg, N.R. and Freireich, E.J. (1964). Use of gammaglobulin in infection in acute leukaemia patients. *J. A. M. A.* **190**, 73.

Bolivar, R., Fainstein, V. and Bodey, G.P. (1981). Cefoperazone therapy of infections in cancer patients. *Proc. 21st I.C.A.A.C.* (In press).

Braude, A.I., Ziegler, E.J., McCutchan, J.A. and Douglas, H. (1981). Immunization against nosocomial infection. *Am. J. Med.* **70**, 463.

Chandler, F.W., Hicklin, M.D. and Blackman, J.A. (1977). Legionnaires' Disease. Description of an epidemic of pneumonia. *N. Engl. J. Med.* **297**, 1189.

Chang, H.Y., Lavermann, M. and Bartlett, J. (1979). Cytotoxicity assay in antibiotic associated colitis. *J. Infect. Dis.* **140**, 765.

Edelstein, P. and Meyer, R.D. (1980). Susceptibility of *Legionella pneumophila* to 20 antimicrobial agents. *Antimicrob. Ag. Chemother.* **18**, 403.

Fainstein, V., Bodey, G.P. and Fekety, R. (1980). Clostridial enterocolitis in cancer patients. Spectrum of disease. *Proc. 20th I.C.A.A.C.*

Fainstein, V., Bodey, G.P. and Fekety, R. (1981a). Pseudo-
    monas colitis due to cancer chemotherapy. *J. Infect. Dis.*
    143, 865.
Fainstein, V., Hopfer, R.L. and Bodey, G.P. (1981b). *Aero-
    monas hydrophila* septicaemia in cancer patients. *Ann.
    Meeting A.S.M.* Abstr. No. 280.
Fainstein, V., Luna, M. and Bodey, G.P. (1981c). *Eikenella
    corrodens* endocarditis in patient with acute lymphocyte
    leukaemia. *Cancer* **48**, 40.
Fainstein, V., Weaver, S. and Bodey, G.P. (1981d). *In vitro*
    antibacterial activity of monobactum (SQ 26-776) compared
    with that of other β-lactum antibiotics. *Proc. 21st I.C.
    A.A.C.* (In press).
Florlenza, S.W., Newman, M.G., Lipsey, A.I., Siegel, S.E. and
    Blachman, V. (1980). Capnocytophaga sepsis: a newly recog-
    nized clinical entity in granulocytopenic patients.
    Lancet **1**, 567.
Fraser, D.W., Tsai, T.F., Orensrein, W. *et al.* (1977).
    Legionnaires' Disease. Description of an epidemic of
    pneumonia. *N. Engl. J. Med.* **297**, 1189.
Folland, D., Armstrong, S., Seides, S. and Blevins, A. (1974).
    Pneumococcal bacteremia in patients with neoplastic dis-
    ease. *Cancer* **33**, 845.
Fu, K.P. and Neu, H.C. (1978). Piperacillin, a new penicillin
    active against many bacteria resistant to other penicil-
    lins. *Antimicrob. Ag. Chemother.* **13**, 358.
Gaines, D. and Remington, J.S. (1973). Diagnosis of deep
    infection with Candida. A study with *Candida precipitins.*
    Arch. Intern. Med. **132**, 699.
Gill, V.J., Manning, C., Larson, M., Woltering, P. and Pizzo,
    P.A. (1981). Antibiotic resistant group JK bacteria in
    hospitals. *J. Clin. Micro.* **13**, 472.
Gilligan, P.H., McCarthy, L.R. and Bissett, B.K. (1981).
    Capnocytophaga ochracea septicaemia. *J. Clin. Micro.* **13**,
    643.
Henri, F.L., Guit, J., van der Meer, W.M. and van Furth, R.
    (1981). Selective antimicrobial modulation of human
    microbial flora: infection prevention in patients with
    decreased host defence mechanism by selective elimination
    of potentially pathogenic bacteria. *J. Infect. Dis.* **143**,
    No. 5.
Herzig, R.H., Herzig, G.P., Graw, R.G. Jr., Bull, M.I. and
    Ray, K.K. (1977b). *N. Engl. J. Med.* **296**, 701.
Herzig, R.H., Herzig, G.P., Graw, R.G. Jr. and Bull, M.I.
    (1977a). Successful granulocyte transfusion therapy for
    gram-negative septicaemia. *N. Engl. J. Med.* **296**, No. 13.

Hinkle, A.M. and Bodey, G.P. (1980). *In vitro* evaluation of
RO 13-9904. *Antimicrob. Ag. Chemother*. **18**, 574.

Hosea, S.W., Brown, E.J., Burch, C.G., Berg, R.A. and Frank,
M.M. (1981). Impaired immune response of splenectomized
patients to polyvalent pneumococcal vaccine. *Proc. 20th
I.C.A.A.C. Abstract No. 413*.

Hughes, W.T., Kuhn, S., Chaushary, S., Feldman, S., Verzosa,
M., Aur, R.L., Pratt, C. and George, S.L. (1976). Success-
ful chemoprophylaxis for *pneumocytis carinii* pneumonitis.
N. Engl. J. Med. **297**, 1419.

Issell, B.F. and Bodey, G.P. (1980). Mezlocillin for treatment of infections in cancer patients. *Antimicrob. Ag.
Chemother*. **17**, 1008.

Ihde, C.C. and Armstrong, D. (1973). Clinical spectrum of
infection due to Bacillus species. *Am. J. Med*. **55**, 839.

Jones, R.J., Roe, E.A. and Gupta, J.L. (1979). Controlled
trials of a polyvalent Pseudomonas vaccine in burns.
Lancet **2**, 977.

Keating, M.J., Bodey, G.P., Valdivieso, M. and Rodriguez, V.
(1979). A randomized comparative trial of 3 aminoglycosides — comparison of continuous infusions of gentamicin,
amikacin and sisomicin combined with carbenicillin in the
treatment of infections in neutropenic patients with malignancies. *Medicine* **58**, 159.

Kilton, L.J., Fossieck, B.E., Cohen, M.H. and Parker, R.H.
(1979). Bacteremia due to gram-positive cocci in patients
with neoplastic disease. *Am. J. Med*. **66**, 596.

King, A., Shannon, K. and Phillips. I. (1980). *In vitro*
antibacterial activity and susceptibility of cefsulodin,
and antipseudomonal cephalosporin, to beta-lactamases.
Antimicrob. Ag. Chemother. **17**, 165.

Klastersky, J. (1981). Prevention and therapy of infection
in myelosuppressed patients. *In* "Medical Complications in
Cancer Patients". (Eds J. Klastersky and M.J. Staquet),
p. 245, Raven Press, New York.

Kohler, R., Zimmerman, S.E., Wilson, E. *et al*. (1981). Rapid
radioimmunoassay diagnosis of Legionnaires' Disease.
Detection and partial characterization of urinary antigen.
Ann. Int. Med. **94**, 601.

Lange, M., Sobeck, K., Blevins, A., Kiehn, T. and Armstrong,
D. (1980). Sepsis with Corynebacterium species (CDC-JK)
in a cancer hospital. *Proc. Int. Conf. Nosocomial Infect*.
Atlanta.

Lazarus, H.M., Lederman, M., Lubin, A., Herzig, R.H. *et al*.
(1980). Pneumococcal vaccination: the response of patients
with multiple myeloma. *Am. J. Med*. **69**, 419.

McCuthan, J.A., Ziegler, E.J. and Braude, A.I. (1979). Treatment of gram-negative bacteremia with antiserum to core glycolipid. II. A controlled trial of antiserum in patients with bacteremia. *Europ. J. Cancer* **15**, 77.

Miller, J.J., Spillsbury, J.F., Jones, R.J. and Roe, E.A. (1977). A new polyvalent Pseudomonas vaccine. *J. Med. Microbiol.* **10**, 19.

Nauseef, W.M. and Maki, D.G. (1981). A study of the value of simple protective isolation in patients with granulocytopenia. *N. Engl. J. Med.* **304**, 448.

Neu, H.C. and Fu, K.P. (1978). Clavulanic acid, a novel inhibitor of β-lactamases. *Antimicrob. Ag. Chemother.* **14**, 650.

Riley, P.S., Hollis, D.G., Utter, G.B., Weaver, R.E. and Baker, C.N. (1979). Characterization and identification of 95 diphtheroid (group JK) cultures isolated from clinical specimens. *J. Clin. Microbiol.* **9**, 418.

Rodgers, F.G. (1979). Isolation of *Legionella-pneumophila* from blood. *Lancet* **1**, 925.

Ruberstein, J.E., Leiberman, M.F. and Gadoth, N. (1976). Central nervous system infection with *Eikenella corrodens*: Report of 2 cases. *Pediatrics* **57**, 264.

Schimpff, S., Satterlee, W., Young, V.M. and Serpick, A. (1971). Empiric therapy with carbenicillin and gentamicin for febrile patients with cancer and granulocytopenia. *N. Engl. J. Med.* **284**, 1061.

Schwartz, S.A. and Perry, S. (1966). Patient protection in cancer chemotherapy. *J.A.M.A.* **197**, 623.

Sleijfer, D.Th., Mulder, N.H., deVries-Hospers, H.G., Fidler, V., Niewig, H.O., van der Waaij, D. and vanSaebem, H.K.F. (1980). Infection prevention in granulocytopenic patients by selective decontamination of the digestive tract. *Europ. J. Cancer* **16**, 859.

Stein, R., Flexner, J.M. and Graber, S.E. (1979). Lithium and granulocytopenia during induction therapy of acute myelogenous leukaemia. *Blood* **54**, 636.

Stewart, D. and Bodey, G.P. (1977). Azlocillin: *In vitro* studies of a new semisynthetic penicillin. *Antimicrob. Ag. Chemother.* **11**, 865.

Wade, J.C., Schimff, S.C., Hargadon, M.T., Fortner, C.L., Young, V.M. and Wiernik, P.H. (1981). A comparison of trimethoprim-sulfamethoxazole plus nystatin with gentamicin plus nystatin in the prevention of infections in acute leukaemia. *N. Engl. J. Med.* **304**, 1057.

Watts, J.C., Hicklin, M.D., Thomason, B.M. *et al.* (1980). Fatal pneumonia caused by *Legionella pneumophila* serogroup 3: demonstration of the bacilli in extrathoracic organs. *Ann. Intern. Med.* **92**, 186.

Weaver, S.S., Bodey, G.P. and LeBlanc, B.M. (1979). Thiena-
    mycin: New beta-lactam antibiotics with potent broad-spec-
    trum activity. *Antimicrob. Ag. Chemother*. **15**, 518.
Wolff, R.L., Wiseman, S.L. and Kitchens, C.S. (1980).
    Aeromonas hydrophila bacteremia in ambulatory immunocom-
    promised hosts. *Am. J. Med*. **68**, 238.
Young, L.S., Meyer, R.D. and Armstrong, D. (1973). *Pseudo-
    monas aeruginosa* vaccine in cancer patients. *Ann. Intern.
    Med*. **79**, 318.
Young, V.M., Meyers, W.F., Moody, M.R. and Schimff, S.C.
    (1981). The emergence of Coryneform bacteria as a cause
    of nosocomial infections in compromised hosts. *Am. J.
    Med*. **70**, 646.

Dr SEDLACEK: Dr Mitchell, a provocative question: what do you think about the possibility that the induction of suppressor cells by BCG might just be the mechanism of the antitumoural effect of BCG? I think this possibility has also been taken into account if you look at the active role of the cellular and humoral immune response on stimulating tumour growth and metastases.

Dr MITCHELL: It is certainly possible that suppressor macrophages or T-cells assist the host — if you accept the theory that a small amount of immunity is "bad" for the host. However, my bias has been that antitumour immune responses generally aid the host by attacking the tumour. T-suppressor cells are generated by complexes of soluble tumour antigens and host antibody and are, I think, the way in which a tumour specifically escapes destruction by the host. Macrophage suppressors depress skin test reactivity nonspecifically. Whether they play a major role in the rejection of tumours, as you suggest, or in the avoidance of rejection, are both totally unknown. Your question expresses our ignorance of the *in vivo* role of suppressor cells only too well.

Dr PRAGER: When the macrophages from the BCG-treated animals inhibited formation of cytotoxic cells, were these cytotoxic T-cells being generated or induced by some other kind of effector?

Dr MITCHELL: Classical T-cells are generated in the *in vitro* immunization system with alloantigens as immunogens. Non-T-killer cells are also generated but we have not characterized them. It is possible that they may resemble those elicited *in vivo* by BCG and alloantigen, but we really do not have sufficient evidence about that.

Dr BREDE: Allow me an additional remark on the modification of immune response by BCG. A very good example for the modulation of immune responses is the obligatory BCG vaccination of newly borns in endemic leprosy areas as in Africa, where

BCG shifts the cellular immunity to the active side and influences the clinical picture of leprosy in favouring
tuberculoid leprosy.

Dr BAŠIĆ: Dr Kondo, besides plurality of explanations for
the mode of antitumour activity of OK-432 you have shown that
OK-432 causes granulocytosis and that these cells could play
a certain role in antitumour activity of this biomodifier.
Now we come to the question about the role of granulocytes in
defence against cancer: I think this is a very important question because these cells and their role in antitumour actions
have been almost neglected. Would you comment a little bit
more about their eventual place in antitumour defence. The
question is also addressed to Dr Mitchell; he has also found
granulocyte-like cells in the spleen of mice treated with BCG.

Dr KONDO: Antitumour effect of granulocytes of a patient was
observed only against his own tumour cells. Usually granulocytes rapidly appear in the peritoneal cavity and have been
frequently overlooked. I would like to emphasize that we
should focus our attention on the behaviour of neutrophils.

Dr ROSZKOWSKI: Have you tried to show neutrophil cytotoxic
effect in functional tests (^{51}Cr release or inhibition of
^{3}H-thymidine incorporation)?

Dr KONDO: No, we have not done that yet.

Dr SINKOVICS: About granulocytic response, in the 1960s we
experimented with AKR mice which developed lymphatic leukaemia as a rule in 80% of the animals by 1 year of age. When
we treated newborn or suckling AKR mice with killed *B. pertussis* vaccine, we elicited granulocytosis (instead of
lymphocytosis as we expected). The granulocytes showed
rounded (doughnut-shaped) nuclei very much like the cells Dr
Mitchell just showed. These pertussis vaccine-treated AKR
mice with early granulocytosis later had a much reduced incidence of their "spontaneous" leukaemia.

Dr MITCHELL: Granulocytes are very much a part of the immune
system, as Metchnikoff would strongly remind us, but we don't
know what they do against tumours. As I mentioned in my
talk, granulocytes might be the unusual killer cells we elicited with BCG or even the suppressor myeloid cell — although
the latter appears to be a fairly classical macrophage. I
suppose they might simply be responding to BCG, but we would
like to explore the role of granulocytes in rejection through
sorting them by light-scattering microscopy.

Dr PRAGER: Since granulocytes have Fc receptors, I wonder if
an antibody is involved in the tumour target killing by the

neutrophil in Dr Kondo's system or in the killer cell described by Dr Mitchell.

Dr KONDO: It seems reasonable to explain that the antitumour
effect of neutrophils is due to ADCC. But at present we
don't have any data, and it should be further investigated.

Dr MITCHELL: We looked for cytophilic antibodies in the serum
but could not find them. This does not rule out the involvement of another sort of "assuring" principle in the specific
killing noted with BCG and alloantigen-elicited killer cells.

Dr BREDE: Dr Kondo, did you investigate for rheumatic factor
in your cancer patients after treatment with OK-432?

Dr KONDO: We really were worried about producing immune complexes in patients when OK-432 was repeatedly injected. In
some cases rheumatic factor turned to positive, but there was
no evidence of arthralgia nor nephritis.

Helen NAUTS: Did you get good febrile reactions every time
you injected OK-432? Did you use antipyretics?

Dr KONDO: Until now we have used them since patients complained of fever and it often made it difficult for us to
continue the therapy. However, we shall not use antipyretics
later on, following your idea that fever is important as an
antitumour factor.

Dr BREDE: Dr Kondo, would you please characterize your
haemolytic streptococcus a little bit more? To which Lance-
field group does it belong, why did you name the strain OK?

Dr KONDO: The preparation is a mutant of group A Streptococcus haemolyticus, Su. The name OK came from the initials
of Drs Okamoto and Koshimura.

Dr PULVERER: How does the OK preparation inhibit DNA/RNA
synthesis?

Dr KONDO: It has been shown by Dr Okamoto, and I, myself,
don't know precisely. It is said that the inhibition can be
observed when the bacilli are suspended in Bernheimer's basal
medium.

Helen NAUTS: Did you have any 5-year survivals? How often
and how long did you continue injections? Are any of the
streptococcal enzymes present in your product?

Dr KONDO: We do not have any 5-year survival observations
yet. Patients usually received 1 to 5 KE of OK-432 every
day or several times a week as long as possible, preferably
as long as they are alive. Concerning the streptococcal
enzymes, I don't know about it.

Dr SEDLACEK: Do you know anything about the reproduction of
your very impressive data on the effect of Picibanil in lung
and gastric cancer patients for instance in Vienna by Uchida
et al.

Dr KONDO: Survival of patients with lung cancer shown here
was done by Dr Watanabe. I am not familiar with the results
obtained by Uchida *et al.*

SUBJECT INDEX

Acalcinomycin A, 316
Acinetobacter sp., 293
AT-125 acivicin, 316
Actinomycin D, 243,309-312,317,
    318,322
Adjuvants, 2,51,53,54,56-59,69,
    73,84,85,88,119,133,139,185,
    244,249,269,272,309,313,314,
    319,339,356,365,368,373,392,
    417
  activity, 55,78,80,307,388,
    409
  Freund's, 85,185,187,189,372
Adriablastin, 49,50,341,342,345
Adriamycin, 312,314,317-319,321,
    322,343,344,367-371
Aeromonas hydrophilia, 438,442
L-alanosine, 316
L-alanine, 49
Anaerobic bacteria (*see*
    Bacteroides, Clostridium,
    Propionibacterium)
4-Aminopyrazole, 243
Amidopyrine, 110
Antibacterial activity, 133
  (*see* Antibiotics)
Antibiotics, 5,13,17,55,59,75,
    76,310,314,415,436,439,440
  aminoglycosides, 442
  azlocyllin, 442
  ampicillin, 442
  amphotericin, 76,77,442
  carbenicillin, 442
  cefalosporin, 442
  cefazolin, 75,76
  cefoxitin, 441
  chloramphenicol, 441
  clindamycin, 441
  cortimoxazole, 441
  cycloserine, 441

erythromycine, 439-441
gentamycin, 76,77,444
methicillin, 441
mezlocillin, 442
nystatin, 442
penicillin, 76,77,416,442
piperacillin, 442
rifampicin, 440
ticarcillin, 442
vancomycin, 440
Anticancer property, 418,429
Antileukaemic property, 243,291,
    331
Antineoplastic property, 54,129,
    141
Antimetastatic property, 154,157,
    183
Antitumour
  activity, 16,50,53,54,58,60,86,
    88,119,122,129-131,133,134,
    137,139-141,150,152,162,177,
    182,220,231,233-235,241,244,
    259,261,262,291,315,317,319-
    324,331,332,337,340,341,356,
    362,388,402,415,416,422,424,
    430,436,441,453-455
Antiviral activity, 129,133,139,
    219-226,260
Antracycline, 319
Arthus reaction, 170,176
Asparaginase, 291-302,309,310,
    312,313,315,317,318,385
Asparagine, 292,295
Asparagine synthetase, 293,301,
    318
Asperigillus sp., 55
Autoimmune disease, 69
5-Azacytidine, 318,323
Azaserine, 243,313
Azotomycin, 313